2

Topics in Organometallic Chemistry

Topics in Organometallic Chemistry

Forthcoming volumes:

Activation of Unreactive Bonds and Organic Synthesis
Volume Editor: S. Murai

The Metal-Carbon Bond: Theoretical and Fundamental Studies
Volume Editors: J.M. Brown, P. Hofmann

Springer-Verlag
Berlin Heidelberg GmbH

Lanthanides: Chemistry and Use in Organic Synthesis

Volume Editor: S. Kobayashi

With contributions by
R. Anwander, E.C. Dowdy, H. Gröger, Z. Hou, H. Kagan,
S. Kobayashi, G. Molander, J.L. Namy, M. Shibasaki,
Y. Wakatsuki, H. Yasuda

Springer

The series *Topics in Organometallic Chemistry* presents critical overviews of research results in organometallic chemistry, where new developments are having a significant influence on such diverse areas as organic synthesis, pharmaceutical research, biology, polymer research and materials science. Thus the scope of coverage includes a broad range of topics of pure and applied organometallic chemistry. Coverage is designed for a broad academic and industrial scientific readership starting at the graduate level, who want to be informed about new developments of progress and trends in this increasingly interdisciplinary field. Where appropriate, theoretical and mechanistic aspects are included in order to help the reader understand the underlying principles involved.
The individual volumes are thematic and the contributions are invited by the volumes editors.
In references Topics in Organometallic Chemistry is abbreviated
Top. Organomet. Chem. and is cited as a journal

Springer WWW home page: http://www.springer.de

ISSN 1436-6002

Cataloging-in-Publication Data applied for
Die Deutsche Bibliothek - CIP-Einheitsaufnahme
Lanthanides: chemistry and use in organic synthesis / vol. ed.: S. Kobayashi. With contribution by R. Anwander.
(Topics in organometallic chemistry ; 2)
DOI 10.1007/978-3-540-69801-2

Cover: Friedhelm Steinen-Broo, Pau/Spain; MEDIO, Berlin
Typesetting: Data conversion by MEDIO, Berlin

SPIN: 10678986 66/3020 - 5 4 3 2 1 0 – Printed on acid-free paper.
www.springer.com/mycopy

Volume Editor

Prof. Shū Kobayashi
Graduate School of Pharmaceutical Sciences
The University of Tokyo
Hongo, Bunkyo-ku
Tokyo 113-0033, Japan
e-mail: skobayas@mol.f.u-tokyo.ac.jp

Editorial Board

Preface

While the lanthanides (strictly defined as the 14 elements following lanthanum in the periodic table, but as normally used also include lanthanum itself) have several unique characteristics compared to other elements, their appearance in the history of the development of organometallic chemistry is rather recent. Since the f orbitals are filled gradually from lanthanum ($[Xe]4f^0$) to lutetium ($[Xe]4f^{14}$), they are regarded as the f-block elements, which are discriminated from the d-block transition elements.

This book was edited as the second volume of "Topics in Organometallic Chemistry", aiming at an overview of recent advances of chemistry and organic synthesis of lanthanides. Since scandium (Sc) and yttrium (Y) (which lie above the lanthanides and have similar characteristics) are also included, this book covers rare earth chemistry. Recently, especially in this decade, the chemistry and organic synthesis of lanthanides have developed rapidly as one of the most exciting areas. An international team of authors has been brought together in order to provide a timely and concise review of current research efforts such as lanthanide catalysis in small molecule organic synthesis especially focused on carbon-carbon bond-forming reactions, chemistry and organic synthesis using low-valent lanthanides such as diiodosamarium, asymmetric catalysis, lanthanide-catalyzed polymer synthesis, and polymer-supported lanthanide catalysts used in organic synthesis. Principles of organolanthanide chemistry are summarized in the first chapter. I am sincerely grateful to Drs. R. Anwander, E. C. Dowdy, H. Gröger, Z. Hou, H. Kagan, G. Molander, J. L. Namy, M. Shibasaki, Y. Wakatsuki, and H. Yasuda for participating in this volume. J. Richmond, J. Sterritt-Brunner, and B. Benner (Springer) are also acknowledged for encouraging me to organize this work.

Finally, I hope that this volume is helpful to many researchers and students who are or will be involved in or interested in this truly exciting and hot field.

Tokyo, December 1998 Shū Kobayashi

Contents

Principles in Organolanthanide Chemistry
R. Anwander 1

Lanthanide Triflate-Catalyzed Carbon-Carbon Bond-Forming Reactions in Organic Synthesis
S. Kobayashi 63

Lanthanide- and Group 3 Metallocene Catalysis in Small Molecule Synthesis
G. Molander, E. C. Dowdy 119

Influence of Solvents or Additives on the Organic Chemistry Mediated by Diiodosamarium
H. Kagan, J.L. Namy 155

Chiral Heterobimetallic Lanthanoid Complexes: Highly Efficient Multifunctional Catalysts for the Asymmetric Formation of C-C, C-O and C-P Bonds
M. Shibasaki, H. Gröger 199

Reactions of Ketones with Low-Valent Lanthanides: Isolation and Reactivity of Lanthanide Ketyl and Ketone Dianion Complexes
Z. Hou, Y. Wakatsuki 233

Organo Rare Earth Metal Catalysis for the Living Polymerizations of Polar and Nonpolar Monomers
H. Yasuda 255

Polymer-Supported Rare Earth Catalysts Used in Organic Synthesis
S. Kobayashi 285

Author Index 307

Principles in Organolanthanide Chemistry ▪

Reiner Anwander

Anorganisch-chemisches Institut, Technische Universität München, Lichtenbergstraße 4, D-85747 Garching, Germany
e-mail: reiner.anwander@ch.tum.de

During the last decade, the rare earth elements have given enormous stimulus to the field of organic synthesis including stereoselective catalysis. This article outlines both the basic and advanced principles of their organometallic chemistry. The intrinsic electronic features of this 17-element series are reviewed in order to better understand the structural chemistry of their complexes and the resulting structure–activity relationships. Particular emphasis is placed on synthetic aspects, i.e. optimization of established procedures and alternative methods with better access to catalytically relevant species. Accordingly, tailor-made ancillary ligands are reported in detail and the reactivity pattern of lanthanide compounds is examined with representative examples.

Keywords: Lanthanides, Intrinsic properties, Reactivity, Synthesis, Ligands

List of Abbreviations 2

1 **Introduction** 3

2 **Intrinsic Properties of the Lanthanide Elements** 4

2.1 Electronic Features 4
2.2 Steric Features 7

3 **Synthesis of Organolanthanide Compounds** 8

3.1 Thermodynamic and Kinetic Guidelines 9
3.2 Inorganic Reagents 10
3.3 Metalorganic Reagents 15
3.4 Thermal Stability 23

4 **Ligand Concepts** 23

4.1 Steric Bulk and Donor Functionalization 24
4.2 Ancillary Ligands 27
4.3 Immobilization – “Supported Ligands” 31

5 Reactivity Pattern of Organolanthanide Complexes 32

5.1 Donor-Acceptor Interactions 32
5.2 Complex Agglomerization 37
5.3 Ligand Exchange and Redistribution Reactions 39
5.4 Insertion Reactions 41
5.5 Elimination Reactions– Ligand Degradation 42
5.6 Redox Chemistry 44
5.7 Reaction Sequences – Catalytic Cycles 46
5.8 Side Reactions 47

6 Perspectives 50

7 References 50

List of Abbreviations

Ar	aromatic residue
BINOL	binaphthol
CN	coordination number
COT	cyclooctatetraenyl
Cp	η^5-cyclopentadienyl
Cp*	η^5-pentamethylcyclopentadienyl
DME	1,2-dimethoxyethane
HMPA	hexamethylphosphoric triamide
HSAB	hard soft acid base
L	ligand
Ln	lanthanide (Sc, Y, La, Ce-Lu)
MMA	methylmethacrylate
OTf	trifluoromethanesulfonato (“triflate”), CF_3SO_3
Ph	phenyl
PMDETA	N, N, N’,N”,N”-pentamethyldiethylenetriamine
Py	pyridine
R	residue
salen	*N,N’*-bis(3,5-di-*tert*-butylsalicylidene)ethylenediamine
Tp	tris(pyrazolyl)borate
THF	tetrahydrofuran
TMEDA	tetramethylethylenediamine
X	ligand
Z	nuclear charge

1 Introduction

The rare earth elements constitute an integral part of modern organic synthesis [1]. It was about 30 years ago that the peculiar redox behavior of several inorganic reagents was discovered for selective reductive and oxidative conversions [2]. In the interim period fine chemicals and polymer synthesis have increasingly benefited from the application of highly efficient organolanthanide precatalysts [3]. Due to their intrinsic electronic properties expressed in the "lanthanide contraction", the rare earth elements comprising the group 3 metals Sc, Y, La and the inner transition metals Ce-Lu provide new structural and reactivity patterns, emerging in structure-activity relationships unprecedented in main group and *d*-transition metal chemistry. It is also their low toxicity and availability at a moderate price which makes this "17-element series" attractive for organic synthesis. The spectrum of rare earth reagents ranges from inorganic to organometallic compounds as schematically redrawn in Fig. 1 with representative examples.

While highly efficient inorganic reagents such as $SmI_2(thf)_2$ and $Sc(OTf)_3$ are already commercially available, the more sophisticated organometallic reagents are as a rule prepared on a laboratory scale, often under rigorous exclusion of moisture using inert gas techniques [4]. In particular, the latter class of compounds offers access to tailor-made, well-defined molecular species via ligand fine-tuning. The consideration of the intrinsic properties of the lanthanide cati-

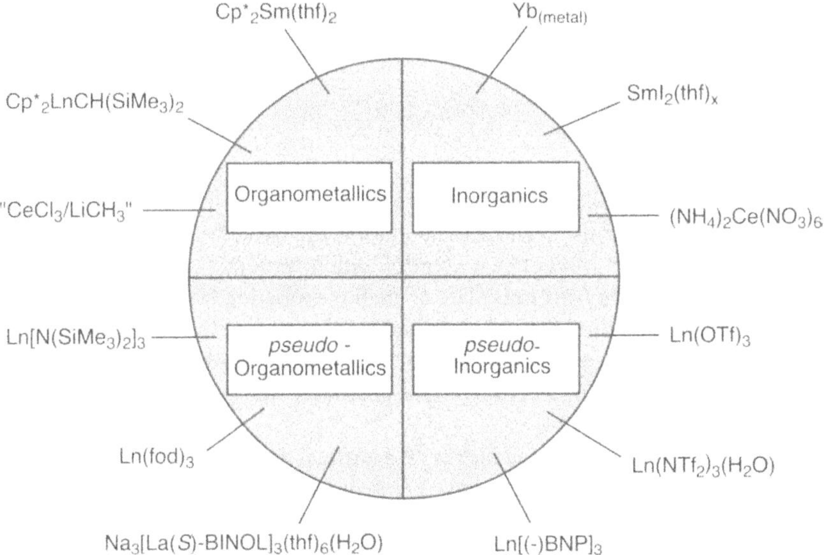

Fig. 1. Rare earth metal reagents in organic synthesis (NTf_2=bis[trifluoromethyl)sulfonyl]amide, (–)BNP=(*R*)-(–)-1,1'-binaphthyl-2,2'-diylphosphato)

ons as well as thermodynamic and kinetic factors are crucial in designing and synthesizing novel molecular compounds. This article also includes reference to highly reactive metalorganic compounds, *pseudo-organometallics*, containing no direct metal carbon linkage; containing, however, otherwise readily hydrolyzable Ln-X bonds. For example, lanthanide compounds such as amide and alkoxide derivatives not only display important synthetic precursors but also exhibit excellent catalytic behavior in organic transformations [5,6]. Macrocyclic ligands exhibiting Ln–N and Ln-O bonds are not considered in this survey [7]. The last 20 years have witnessed a rapid development in organolanthanide chemistry and numerous review articles have been published, emphasizing various aspects including their use in organic transformations. A comprehensive list of relevant articles has been given recently [8]. The purpose of this article is not to give a comprehensive survey of organolanthanide compounds but rather to address the principles of their chemistry.

2 Intrinsic Properties of the Lanthanide Elements

The rare earth elements represent the largest subgroup in the periodic table and offer a unique, gradual variation of those properties which provide the driving force for various catalytic processes. Their peculiar electronic configuration and the concomitant unique physicochemical properties also have to be consulted for the purpose of synthetic considerations. The highly electropositive character of the lanthanide metals, which is comparable to that of the alkali and alkaline earth metals, leads as a rule to the formation of predominantly *ionic* compounds, Ln(III) being the most stable *oxidation state* [9]. This and other intrinsic properties are outlined in Scheme 1 which will serve as a point of reference in this section [10–13].

2.1 Electronic Features

The Ln(III) cations of the series Ce–Lu exhibit the extended Xe-core electronic configuration $[Xe]4f^n$ (n=1–14), a symbol which perfectly pictures the limited radial extension of the *f*-orbitals: The 4*f* shell is embedded in the interior of the ion, well-shielded by the $5s^2$ and $5p^6$ orbitals [14]. A plot of the radial charge densities for the 4*f*, 5*s*, 5*p* and 6*s* electrons for Gd^+ visually explains why Ln(III) cations are commonly thought of as a *"triple-positively charged closed shell inert gas electron cloud"* (Fig. 2) [14].

Ionization energies of the elements [15], optical properties [16], and magnetic moments of numerous complexes [17] prove that the *f*-orbitals are perfectly shielded from ligand effects. Consequently, only minimal perturbation of the *f*-electronic transitions results from the complexation of dipolar molecules. In contrast to the broad $d \rightarrow d$ absorption bands of the outer transition elements, the $f \rightarrow f$ bands of the lanthanides are almost as narrow in solid and in solution as

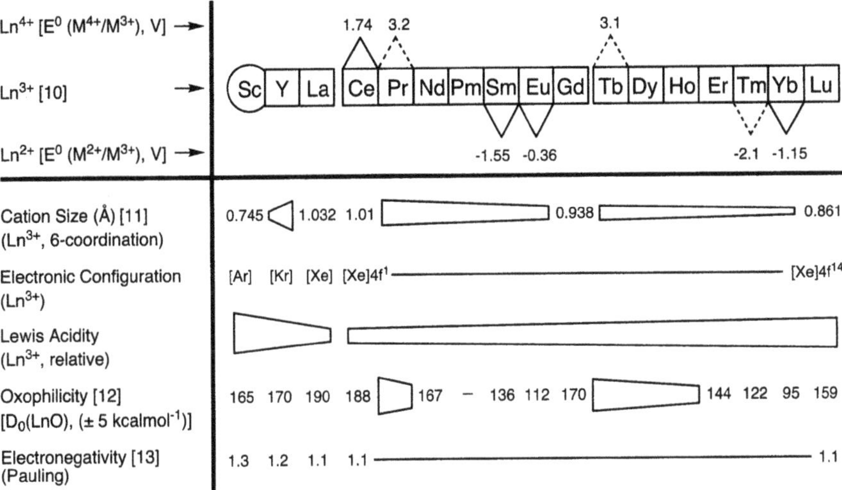

Scheme 1. Trends within intrinsic properties of the lanthanide elements

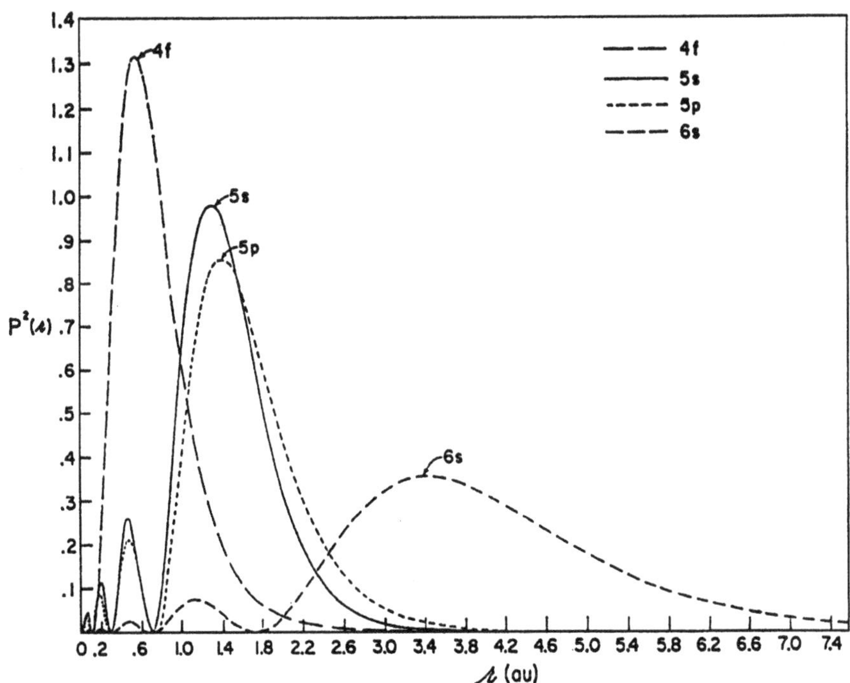

Fig. 2. Plot of the radial charge densities for the 4*f*-, 5*s*-, 5*p*-, and 6*s*-electrons of Gd^+ from [14]

they are for gaseous ions. These transitions are "LaPorte-forbidden" and result in weak intensities which are responsible for the pale color of the trivalent species. General principles of *d*-transiton metal ligand bonding such as σ-donor/π-acceptor interaction, the "18-electron rule", and the formation of classic carbene, carbyne, or carbon monoxide complexes are not observed in lanthanide chemistry, neither do they form Ln=O or Ln≡N multiple bonds. However, the lack of orbital restrictions, e.g. the necessity to maximize orbital overlap as in *d*-transition metal chemistry, allows "orbitally forbidden" reactions. Because of very small crystal-field splitting and very large spin-orbit coupling (high Z) the energy states of the $4f^n$ electronic configurations are usually approximated by the Russel–Saunders coupling scheme [18]. The peculiar electronic properties of the *f*-elements have proved attractive for numerous intriguing opto- and magneto-chemical applications ("probes in life") [15].

The inert gas-core electronic configuration also implies a *conform chemical behavior* of all of the Ln(III) derivatives including Sc(III), Y(III) and La(III). The contracted nature of the 4*f*-orbitals and concomitant poor overlap with the ligand orbitals contribute to the predominantly ionic character of organolanthanide complexes. The existing electrostatic metal ligand interactions are reflected in molecular structures of irregular geometry and varying coordination numbers. According to the HSAB terminology of Pearson [19], lanthanide cations are considered as *hard acids* being located between Sr(II) and Ti(IV). As a consequence, "hard ligands" such as alkoxides and amides, and also cyclopentadienyl ligands show almost constant effective ligand anion radii (alkoxide: 2.21±0.03 Å; amide: 1.46±0.02; cyclopentadienyl: 1.61±0.03) [20] and therefore fit the evaluation criteria of ionic compounds according to Eigenbroth and Raymond [21]. The ionic bonding contributions in combination with the high Lewis acidity cause the strong *oxophilicity* of the lanthanide cations which can be expressed in terms of the dissociation energy of LnO [12]. The interaction of the oxophilic metal center with substrate molecules is often an important factor in governing chemo-, regio- and stereoselectivities in organolanthanide-catalyzed transformations [22]. Complexation of the "softer" phosphorus and sulfur counterions is applied to detect extended covalency in these molecular systems [23,24].

Scheme 1 further indicates the tendency of the Ln(III) cations to form the more unusual oxidation states in solution [25]. Hitherto, organometallic compounds of Ce(IV), Eu(II), Yb(II) and Sm(II) have been described in detail [4]. More sophisticated synthetic approaches involving metal vapor co-condensation give access to lower oxidation states of other lanthanide elements [26]. Charge dependent properties such as cation radii and Lewis acidity significantly differ from those of the trivalent species. Ln(II) and Ce(IV) ions show very intense and ligand-dependent colors attributable to "LaPorte-allowed" 4*f*→5*d* transitions [16b]. Complexes of Ce(IV) and Sm(II) have achieved considerable importance in organic synthesis due to their strongly oxidizing and reducing behavior, respectively [1,27]. Catalytic amounts of compounds containing the "hot oxidation states" also initiate substrate transformations. As a rule this implies a switch to the more stable, catalytically acting Ln(III) species [28].

2.2 Steric Features

Structural changes in homologous rare earth compounds arise from the *lanthanide contraction* [29], i.e. the monotonically decreasing ionic radii with increasing atomic number. The 4*f*-electrons added along the lanthanide series from lanthanum to lutetium do not shield each other efficiently from the growing nuclear charge, resulting in the contraction phenomenon. It is often this varying cationic size which has a tremendous effect on the formation, coordination geometry (coordination numbers) and reactivity of their complexes. Reports have accumulated where organic substrates seem to discriminate not only between ligand environments but also between single lanthanide elements [22]. Successful explanations of these phenomena are based on the systematic theoretical investigation and structural characterization of organolanthanide compounds [4].

Scheme 1 gives the trend of ionic radii of these *"large" cations* which prefer formal coordination numbers in the range of 8–12 [30]. For example, considering the effective Ln(III) radii for 6-coordination, a discrepancy of 0.171 Å between Lu(III) and La(III) allows the steric fine-tuning of the metal center [11]. The structural implications of the lanthanide contraction are illustrated in Fig. 3 with the well-examined homoleptic cyclopentadienyl derivatives [31]. Three structure types are observed depending on the size of the central metal atom: **A**, $[(\eta^5\text{-Cp})_2\text{Ln}(\mu\text{-}\eta^5\text{:}\eta^x\text{-Cp})]_\infty$ $1 \le x \le 2$; **B**, $\text{Ln}(\eta^5\text{-Cp})_3$; **C**, $[(\eta^5\text{-Cp})_2\text{Ln}(\mu\text{-}\eta^1\text{:}\eta^1\text{-Cp})]_\infty$; these exhibit coordination numbers of 11 (10), 9, and 8, respectively. In

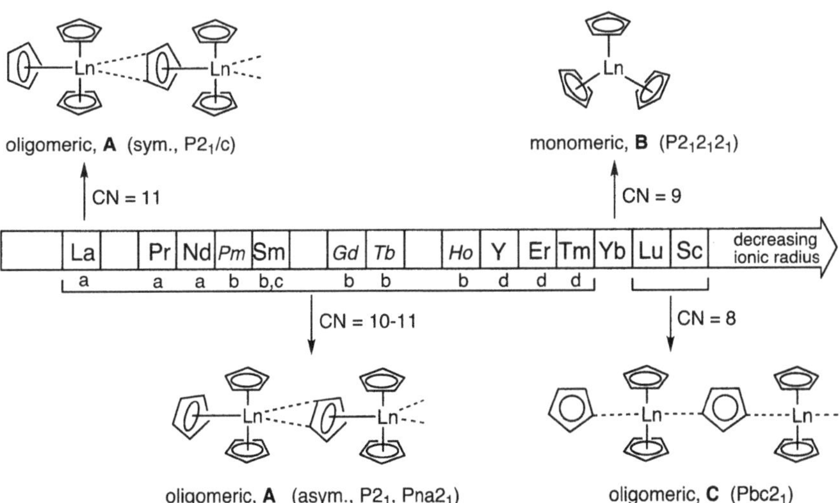

Fig. 3. Coordination modes in homoleptic, ionic LnCp_3 derivatives (*a* belong to space group $P2_1$; *b* indication from powder diffraction pattern; *c* show additional modifications Pbcm and $P2_1/n$ (contact dimer – effect of crystallization conditions [31b]); *d* belong to space group $Pna2_1$ and exhibit lengthened intermolecular Ln-C contacts)

accord with ionic bonding, small changes in ligand substitution lead to changed coordination behavior and number (CN=10), as found in the tetranuclear ring structure of the "MeCp" derivative. Monomeric type **B** is preferentially formed with ligands bearing bulky substituents.

High coordination numbers can usually be accomplished in oligomeric structures or highly solvated complexes. However, both forms are undesirable for synthesizing highly reactive compounds. The reactivity and stability, respectively, of lanthanide complexes is correlated with the steric situation at the metal center. Evaluation criteria as the principle of "steric saturation/unsaturation/oversaturation" have been developed to explain the differences in reactivity [32]. Hence, the main synthetic efforts as in d-metalorganic chemistry are put into the fine-tuning of the ligand sphere to obtain tractable (volatile, catalytically reactive, etc.) compounds. Because of the importance of steric factors, ligand environments have been numerically registrated, e.g. by the "cone-packing model" [33], which represents a 3-D extension of *Tolman's* "cone-angle model" [34]. In this model, solid angles are calculated from structural data employing van der Waals radii [35] and considering effects of second order packing. The introduction of steric coordination numbers for various types of ligands based on solid angle ratios further emphasizes the importance of steric considerations in organo-*f*-element chemistry [36].

The *Lewis acidity* which is affected by the charge density (Z/r) is less distinct in complexes derived from the large Ln(III) cations. Hence, these systems are often reported as mild Lewis acidic catalysts in organic synthesis [1]. However, Sc(III) as by far the smallest Ln(III) cation is located in a "pole position" not only with respect to Lewis acidity. Its "aluminum/lanthanide/early transition metal hybrid character" [37] has revealed its superiority in many catalytic applications [37,38]. Based on their relative preferences for pyridine, Lappert suggested a relative Lewis acidity scale: $Cp_2ScMe > AlMe_3 > Cp_2YMe \approx Cp_2LnMe$ (here: Ln=large lanthanide elements) [39]. Maximum electrostatic metal/ligand interaction and ionic bond strength (enhanced *complex stability*) is also expected for scandium, the smallest element. The Ln(III) charge density and the concomitant complexation tendency also prove useful when studying the nature of Ca^{2+} binding in biological macromolecules exploiting the lanthanide elements as spectroscopic and magnetic probes [15].

3 Synthesis of Organolanthanide Compounds

The availability of pure and well-defined starting materials is crucial for straightforward and high-yield syntheses of organometallic rare earth compounds. The suitability of both synthetic and catalyst precursors can be judged by the consideration of thermodynamic and kinetic factors. For example, the knowledge of metal–ligand bond strengths can assist in a better analysis of the thermodynamics of archetypical ligand exchange reactions and to elaborate the mechanistic scenarios of catalytic transformations [40].

3.1 Thermodynamic and Kinetic Guidelines

Marks and co-workers provided a most valuable examination of absolute bond disruption enthalpies of various relevant metalorganic ligands X in Cp^*_2Sm–X. The data were obtained by anionic titration calorimetry in toluene (Fig. 4) [41]. Although the Ln–X bonds seem to be thermodynamically very stable, they usually display kinetic lability due to high ligand exchange ability, chelating and solubility effects.

Scheme 2 encompasses important synthetic building blocks and preferred synthetic strategies in organolanthanide chemistry. As acid/base-type exchange reactions are fundamental, the ligands are depicted according to their increasing pK_a values (in water). This also correlates with the tendency to hydrolyze (organometallics) or with the competition between solvation and complexation on the basis of the HSAB concept (inorganics).

The central point in this consideration is the Ln–OH moiety, the preferred formation of which is called a "dilemma in organolanthanide chemistry". Organolanthanide and pseudo-organolanthanide compounds readily hydrolyze when exposed to air and moisture, with the formation of hydroxide and oxo-centered ligand cluster intermediates. Lanthanide complexes with Ln–C linkages are considered as "oversensitive" compounds [42]. Even ligands with lower pK_a values than water, as exemplified by substituted phenol ligands, tend to hydrolyze in organic solvents because the insoluble hydroxides formed act as a driving force. However, the presence of hard donor functionalities or multiply charged anions which are capable of chelation, can afford moisture-stable alkoxide and amide complexes, as has been shown for BINOL [43], polypyrazolylborate [44] and porphyrin-like complexes [45]. Nevertheless, all of the organolanthanide complexes should routinely be handled under an inert gas atmosphere by application of high vacuum and glove-box techniques [46].

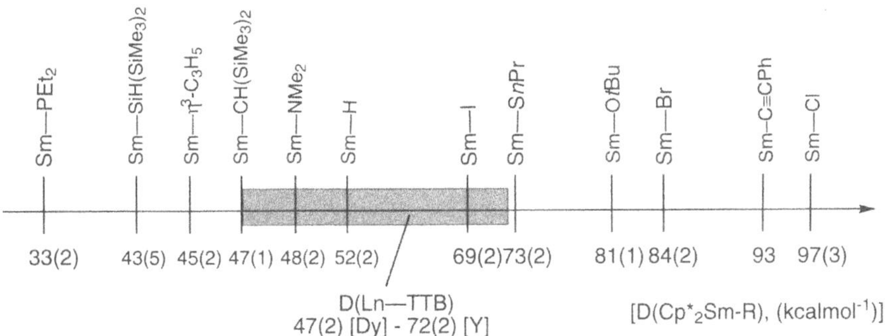

Fig. 4. Bond disruption enthalpies of organolanthanide(III) complexes. The *gray area* indicates the bond disruption enthalpies of organolanthanide(0) arene species (TTB=η^6-$C_6H_3tBu_3$-1,3,5)

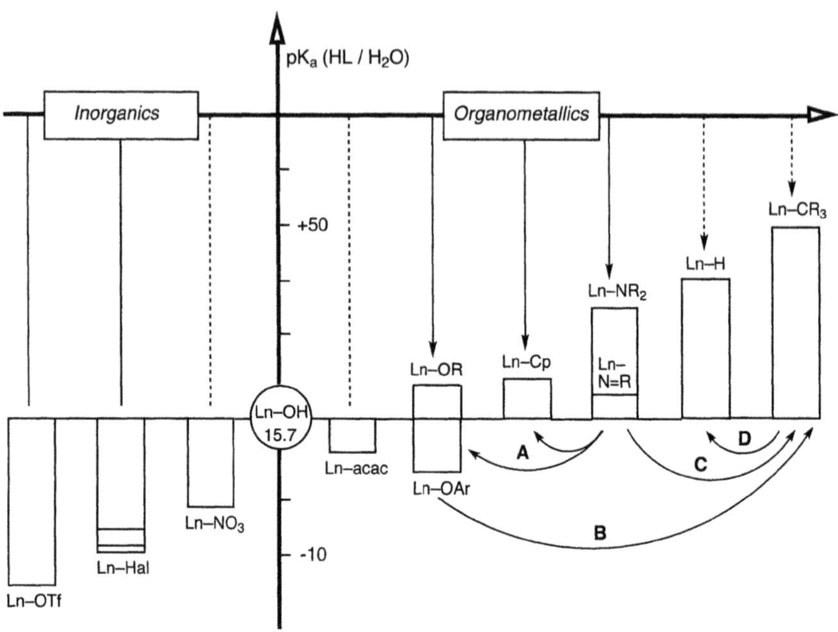

Scheme 2. Synthetic strategies towards organolanthanide compounds [**A:** amine elimination reactions, e.g. *silylamide route*; **B:** alkylation via alkoxide precursor, e.g. *aryloxide route*; **C:** alkylation via amide precursor; **D:** hydrogenolysis of alkyl moieties]

3.2
Inorganic Reagents

Lanthanide halides, nitrates and triflates are not only common reagents in organic synthesis (Fig. 1) but also represent, in dehydrated form, key precursor compounds for the more reactive organometallics (Scheme 2). As a rule, in compounds of strong monobasic acids or even superacids, cation solvation competes with anion complexation, which is revealed by fully or partially separated anions and solvated cations in their solid state structures. The tendency to form outer sphere complexation in coordinating solvents [47] is used as a criterion of the reactivity of inorganic salt precursors in organometallic transformations.

Ln-Halides

Anhydrous lanthanide halides are ionic substances with high melting points which take up water immediately when exposed to air to form hydrates ($I^- > Br^- > Cl^-$) [48]. Straightforward synthetic access and a favorable complexation/solvation behavior make the lanthanide halides the most common precursors in organolanthanide chemistry. Many important Ln–X bonds (X=C, Si, Ge, Sn, N, P, As, Sb, Bi, O, S, Se, Te) can be generated via simple salt metathesis reactions [4,8]. The so-called *ammonium chloride route* either starting from the lanthanide oxides or

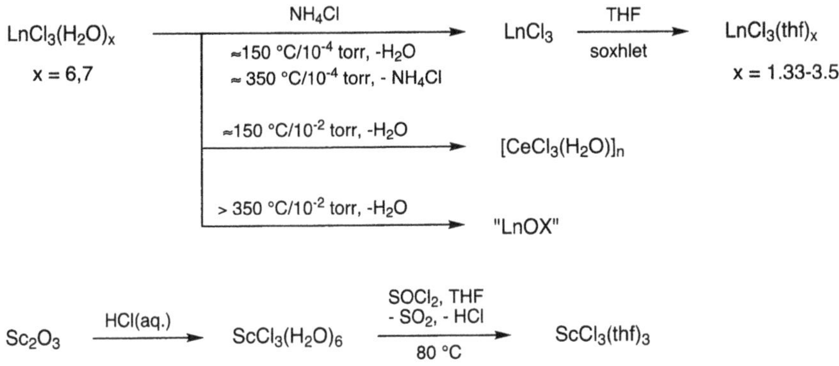

Scheme 3. Laboratory procedures for the preparation of lanthanide chloride precursors

the hydrated halides is the most popular laboratory procedure (upscaling is possible) to anhydrous lanthanide(III) chlorides (Scheme 3) [49]. Simple thermal dehydration which works well for lanthanide triflates leads to the formation of undesired lanthanide oxychlorides. Evans and co-workers have shown that the standard recipe for dehydrating $CeCl_3(H_2O)_7$ to make $CeCl_3$/RLi will produce $[CeCl_3(H_2O)]_n$ [50]. "$CeCl_3$/RLi" is a popular Grignard-type reagent in organic synthesis [51] which, for example, increasingly tolerates functional groups.

A coordinating solvent such as tetrahydrofuran (THF) is often necessary to react the otherwise insoluble lanthanide halides via salt metathesis. These reactions proceed via initial formation of the more soluble compounds $LnX_3(thf)_x$, which are obtained via Soxhlet extraction and are popular, well-defined starting reagents [52]. The extent of THF coordination depends on both the structural type of the anhydrous lanthanide halide and the prevailing crystallization conditions, and affects its solubility and hence its reactivity [53]. "$ScCl_3(thf)_3$" is best synthesized by an alternative procedure utilizing $SOCl_2$ as a dehydrating agent [54]. Neutral donor ligands such as caprolactone [53a], 2,6-dimethyl-4-pyrone [55] or chelating ligands such as DME [56] and crown ethers [57,58] also reveal unforeseen and intriguing coordination chemistry.

Other small-scale laboratory procedures have been developed for the direct synthesis of the more reactive THF adducts, avoiding "inconvenient" high temperature treatment [59–62]. For example, the preparation of "$LnCl_3(thf)_x$" from metal powder and hexachloroethane is facilitated by sonication [Eq. (1)] [59]. Additional metal-based synthetic routes include the redox transmetallation with mercury(II) halides [Eq. (2)] [60] and the reaction with trimethylsilyl chloride and anhydrous methanol [Eq. (3)] [61]. Ammonia has been employed as an alternative donating solvent in the synthesis of lanthanide alkoxides starting from lanthanide chlorides [63].

$$2\,Ln + C_2Cl_6 \xrightarrow[)))]{THF} 2\,LnCl_3(thf)_x + C_2Cl_4 \qquad (1)$$

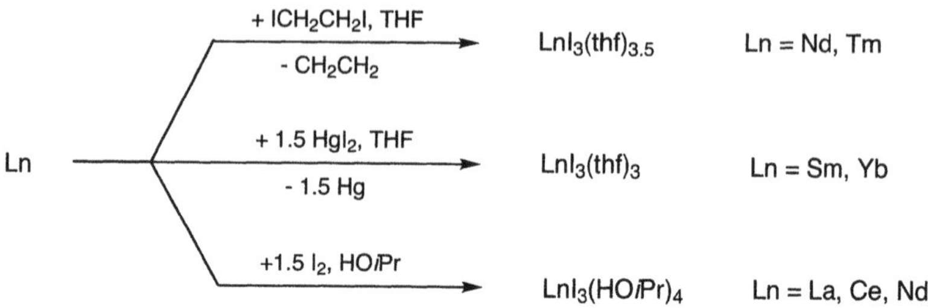

Scheme 4. Small-scale synthesis of solvated Ln(III) iodides

$$2\,Ln + 3\,HgCl_2 \xrightarrow[\Delta]{THF} 2\,LnCl_3(thf)_x + 3\,Hg \tag{2}$$

Ln = Yb (x = 3); Er (x = 3.5); Sm (x = 2); Nd (x = 1.5)

$$2\,Ln + 6\,Me_3SiCl + 6\,MeOH \xrightarrow[\Delta]{THF} 2\,LnCl_3(thf)_x + 6\,Me_3SiOMe + 3\,H_2 \tag{3}$$

Scheme 4 shows small-scale syntheses of solvated iodides [64–66]. Strongly donating solvents such as *N*-methylimidazole (*N*-MeIm) can accomplish complete anion/cation separation as shown for $[Sm(N\text{-}MeIm)_8]I_3$ under anaerobic conditions [67]. The chief factors which affect the often enhanced reactivity of the higher homologous halides are their higher solubility [48a], a thermodynamically more labile Ln–X bond (Fig. 4), the soft Lewis basicity of the iodide anion, and different solubility properties of the eliminated alkali metal salt.

Lanthanide(II) halides, in particular iodides, are prominent synthetic precursors to the corresponding Ln(II) organometallics [32,68,69]. SmI_2 is a well-established reducing reagent in organic synthesis and is commercially available as a THF solution and in solid form [27]. Its THF solvate was synthesized according to Eq. (4) and was structurally characterized as a 7-coordinate $SmI_2(thf)_5$ [70]. The less soluble $YbI_2(thf)_2$ can be obtained analogously [27] and the ammonia complex is readily formed according to Eq. (5) [69]. $TmI_2(dme)_3$ is the only soluble Tm(II) compound synthesized so far [Eq. (6)] [71]. A large-scale synthesis of $SmBr_2$ avoiding the expensive metal precursor has been accomplished according to the reaction sequence shown in Eq. (7) [68].

$$Sm + ICH_2CH_2I \xrightarrow{THF} SmI_2(thf)_x + CH_2{=}CH_2 \tag{4}$$

$$Yb + 2\,NH_4I \xrightarrow{NH_3(liq)} YbI_2(NH_3)_x + H_2 \tag{5}$$

$$TmI_3 + Tm \xrightarrow[\Delta]{DME} 2\,TmI_2(dme)_3 \tag{6}$$

$$Sm_2O_3 \xrightarrow{HBr_{aq.}} SmBr_3x6H_2O \xrightarrow{\text{methyl orthoformate}} SmBr_3$$
$$SmBr_3 \xrightarrow{Li,\ THF} SmBr_2(thf)_x \quad (7)$$

The complex $(pyH)_2(CeCl_6)$ has been discussed as an alternative Ce(IV) precursor [72]. Pseudohalides such as thiocyanates should receive some attention as specific synthetic precursors due to their dual ligation mode [73]. Like the halides [74], their Ln(III) derivatives have been successfully employed as catalysts in organic transformations [75].

Other Inorganic Salts

Alternative inorganic precursors which are referred to in Scheme 2 are also available by treatment of lanthanide oxides Ln_2O_3 with the corresponding acid [76,77]. Nitrate ligands coordinate slightly stronger to the lanthanide centers compared to halides, but are reported to yield coarse precipitates of alkali nitrates in salt metathesis reactions [78]. Nitrates are also preferred as precursors in macrocyclic chemistry where they preferentially occupy the outer ligation sphere [7]. Strong complexation of doubly charged anions ($CO_3^{2-}>SO_4^{2-}$) causes a considerable decrease in solubility of the corresponding Ln_2X_3 and hence precludes their broad use as synthetic precursors [77]. Ln-fluorides [79] and phosphates are totally insoluble in solvents suitable for organometallics [4,15]. Pseudo-inorganic salts derived from superacids, in particular derivatives of triflate, contain weakly coordinating anions and were often found to be superior to lanthanide halides in salt metathesis reactions [80]. Anhydrous $Ln(OTf)_3$ can be easily obtained by thermal dehydration [Eq. (8)] [81]. Lanthanide triflates have attracted considerable attention as reuseable Lewis acidic catalysts in numerous carbon–carbon bond-forming reactions [82].

$$Ln_2O_3 \xrightarrow[100\ ^\circ C,\ 1h]{CF_3SO_3H,\ H_2O} [Ln(H_2O)_9][CF_3SO_3]_3 \xrightarrow[48h]{180\text{-}200\ ^\circ C} Ln(CF_3SO_3)_3 \quad (8)$$

Rare earth borohydrides obtained from the chlorides [Eq. (9)] [83] have been used in salt metathesis reactions and were found to be attractive for the generation of cationic species [84]. The presence of more weakly coordinated BF_4^- anions in $[Eu(MeCN)_3(BF_4)_3]_x$ which can be synthesized according to Eq. (10) promotes several catalytic transformations of non-heteroatom-substituted organic substrates, including the polymerization of styrene [85].

$$NdCl_3 + 3.3\ NaBH_4 \xrightarrow[60\ ^\circ C,\ 48\ h]{THF,\ -3\ NaCl} Nd(BH_4)_3(thf)_2 \quad (9)$$

$$Eu + 3\ NOBF_4 \xrightarrow[rt,\ 1d]{CH_3CN} [Eu(CH_3CN)_3(BF_4)_3]_x + 3\ NO \quad (10)$$

Cerium ammonium nitrate [$(NH_4)_2Ce(NO_3)_6$, CAN], a key oxidizing agent, is the most common Ce(IV) precursor [86]. The use of acetylacetonates of cerium(IV) has been discussed [87] and $Ce(OTf)_4$ should also prove to be a valuable precursor [88].

Metals

Lanthanide metals which are conveniently prepared from the metal halides are commercially available in the form of ingots, chips (filings), foils and powders and are also handled as prominent synthetic precursors. For example, alkoxide complexes derived from cheap and low boiling alcohols are alternatively synthesized from metals under $HgCl_2$ catalysis [89]. Representative examples for transmetallation and transmetallation/ligand exchange reactions are given in Eqs. (11)–(13) [90]. Ammonia solutions of ytterbium and europium react with a variety of Brønsted acidic reagents according to Eq. (14) [91]. Metal oxidation/ligand transfer occurs in THF in the presence of catalytic [Eq. (15)] and stoichiometric amounts of iodine [Eq. (16)] [92]. "Lanthanide Grignard" reagents, formulated as "RLnI" are prepared in situ from the metal and the alkyl(aryl)halide in THF [Eq. (17)] [93]. Utilization of an extremely bulky alkyl ligand allowed the isolation of $\{Yb[C(SiMe_3)_3]I(OEt_2)\}_2$ according to a salt metathesis reaction [94].

$$\mathrm{Ln + 3\,TlCp \xrightarrow[80\,°C,\,20h]{THF} LnCp_3(thf) + 3\,Tl} \tag{11}$$

$$\mathrm{Sm + Hg(C_6F_5)_2 + 2\,HNR_2 \xrightarrow[12h]{THF} Sm(NR_2)_2(thf)_4 + Hg + 2\,HC_6F_5} \tag{12}$$

$$\mathrm{Yb + Sn[N(SiMe_3)_2]_2 \xrightarrow[80\,°C,\,8h]{THF} Yb[N(SiMe_3)_2]_2(thf)_2 + Sn} \tag{13}$$

$$\mathrm{Eu \xrightarrow{NH_3(liq)} Eu^{2+}(NH_3)_6 x 2e^-(NH_3)_n \xrightarrow[-H_2,\,-NH_3]{HR} EuR_2(NH_3)_m} \tag{14}$$

$$\mathrm{Ln + 1.5\,ArSSAr \xrightarrow[50\,°C,\,48\,h]{I_2(!),\,THF,\,Py} Ln(SAr)_3(py)_3} \tag{15}$$

$$\mathrm{2\,Ln + Ph{-}CH{=}CH{-}CH{=}CH{-}Ph + 2\,I_2 \xrightarrow[50\,°C,\,48h]{THF} Ph{-}CH{=}CH{-}CH(LnI_2(thf)_2){-}CH(LnI_2(thf)_2){-}CH{=}CH{-}Ph} \tag{16}$$

$$\mathrm{Yb + CH_3I \xrightarrow[-30\,°C]{THF} [CH_3YbI]} \tag{17}$$

The co-condensation of electron-beam vaporized lanthanide metals with neutral (hetero-)aromatic molecules or 2,2-dimethylpropylidynephosphine (*t*BuCP) affords deeply colored compounds. The isolated, crystalline sandwich

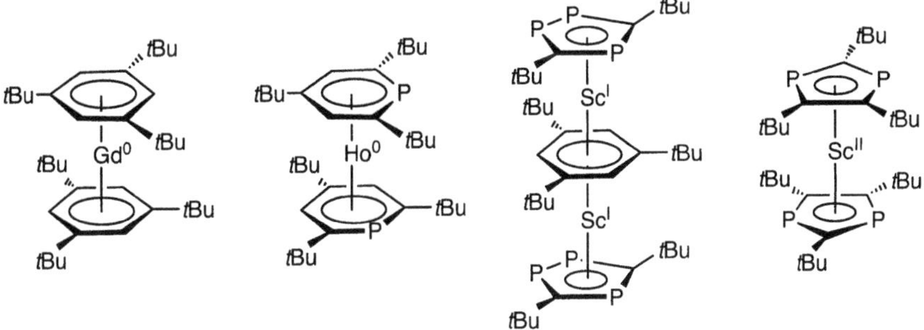

Fig. 5. Arene complexes of low-valent lanthanide elements obtained by co-condensation methods

and triple-decker complexes are thermally stable and exhibit lanthanide centers in the formal oxidation states Ln(0), Sc(I) and Sc(II) (Fig. 5) [95].

3.3 Metalorganic Reagents

According to Scheme 2, inorganics and pseudo-inorganics are suitable precursors for a variety of organometallic compounds. However, incorporation of alkali metal salts and *ate* complex formation according to the traditional metathesis route are often observed. As this is usually an undesired feature and particularly pronounced in rare earth alkyl [96], amide [97] and alkoxide chemistry [98] (Sect. 5.1), new synthetic routes involving well-defined metalorganic precursor compounds have been developed. Considering the (pseudo-)organometallic side of Scheme 2 (right), usually all of the compounds on this side are able to produce the neighboring systems on their left by Brønsted acid/base-type reactions, e.g. alkyls might readily react with amines, cyclopentadienes and alcohols to yield amide, alkoxide and cyclopentadienyl complexes, respectively. Lanthanide silylamide and aryloxide moieties qualify as versatile synthetic precursors due to high-yield and high-purity synthetic procedures. The preparation of their homoleptic derivatives is shown in Eqs. (18) and (19) [99,100].

$$\mathrm{LnCl_3} + 3\,\mathrm{K[N(SiMe_3)_2]} \xrightarrow[\text{2. sublimation}]{\text{1. THF, -3 KCl, 20 °C}} \underset{\textit{pure}}{\mathrm{Ln[N(SiMe_3)_2]_3}} \tag{18}$$

$$\mathrm{LnCl_3} + 3\,\mathrm{KOAr} \xrightarrow[\text{2. sublimation}]{\text{1. THF, -3 KCl, 70 °C}} \underset{\textit{pure}}{\mathrm{Ln(OAr)_3}} \tag{19}$$

$\mathrm{OAr = OC_6H_3\mathit{t}Bu_2\text{-}2,6;\ OC_6H_2\mathit{t}Bu_2\text{-}2,6\text{-}Me\text{-}4}$

The Silylamide Route

Rare earth silylamide complexes have not only attracted enormous attention for the synthesis of precatalyst systems but also for the isolation of well-defined

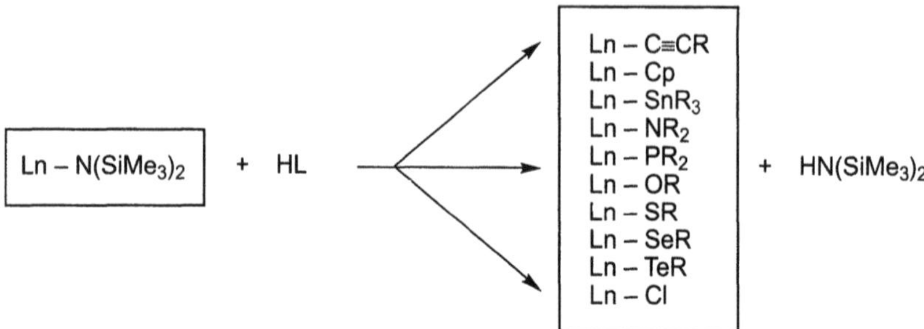

Scheme 5. The *silylamide route*

compounds of relevance in precursor chemistry of ceramic and electronic materials, such as pure alkoxides [101,102]. The general redox stability of the lanthanide cations and the chemical robustness of the silylamide ligand has resulted in numerous ligand exchange reactions with substrate molecules of increased Brønsted acidity such as alcohols, phenols, cyclopentadienyls, acetylenes, phosphanes, and thiols as listed in Scheme 5 [103–113].

Factors which often make the *silylamide route* superior to traditional salt metathesis reactions are (i) the reaction in non-coordinating solvents due to the high solubility of the monomeric metal amides, (ii) mild reaction conditions often at ambient temperature, (iii) avoidance of halide contamination, (iv) ease of product purification [removal of the released amine along with the solvent under vacuum (bp $HN(SiMe_3)_2$: 125 °C)], (v) base-free products (coordination of the sterically demanding, released amine is disfavored), (vi) "quantitative yield", and (vii) the facile availability of mono- and heterobimetallic amide precursors.

A limiting factor of this specific amine elimination route is the steric bulk of the $[N(SiMe_3)_2]$ ligand, obvious in incomplete exchange reactions with similarly bulky ligands such as Cp*H [104], $HOC\mathit{t}Bu_3$ [114] or highly substituted phosphanes [108]. In order to better cope with such sterically suppressed ligand exchange reactions the alternative silylamide precursor $Ln[N(SiHMe_2)_2]_3(thf)_2$, which can be prepared in high yield for all of the lanthanide elements [yttrium: Eq. (20)] [115], has been introduced.

$$YCl_3(thf)_{3.5} + 2.9\ LiN(SiHMe_2)_2 \xrightarrow[\text{rt, 16h}]{n\text{-hexane, - LiCl}} Y[N(SiHMe_2)_2]_3(thf)_2 \quad (20)$$

The bis(dimethylsilylamide) ligand $[N(SiHMe_2)_2]$ not only favors the attack of protic reagents by decreased steric bulk, but amine elimination is also affected by a decreased silylamide basicity, easier workup procedures (bp $HN(SiHMe_2)_2$: 93–99 °C) and the presence of an excellent spectroscopic probe ("Si-H"). According to this *"extended silylamide route"*, catalytically relevant complexes with salen [116], (substituted) linked-indenyl [117], and sulfonamide ligands [118] have been synthesized (Fig. 6). Such controlled ligand associations, which are

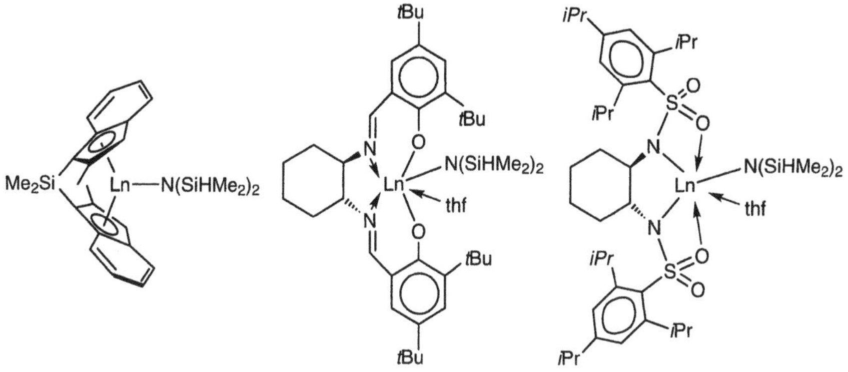

Fig. 6. C_2-symmetric rare earth complexes according to the extended silylamide route

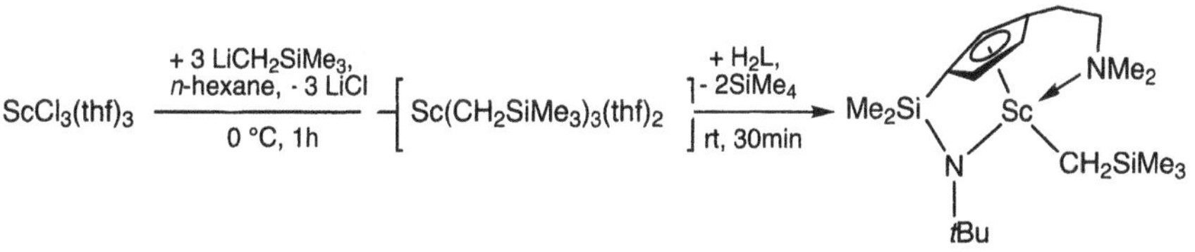

Scheme 6. In situ generation of a reactive alkyl precursor

proposed as proceeding via THF dissociation, are not obtained with the $Ln[N(SiMe_3)_2]_3$ system.

The application of the more basic $Ln(N\mathit{i}Pr_2)_3(thf)$ as a metalorganic precursor compound is controversial [119] because its availability is hampered by ate complexation [Sect. 5.1, $LiLn(N\mathit{i}Pr_2)_4$] and enhanced thermal instability (decomposition at 100 °C/10^{-4} Torr) [120]. An efficient alkane elimination reaction utilizing the in situ formed alkyl species $Ln(CH_2SiMe_3)(thf)_2$ produced complexes with linked amido cyclopentadienyl ligands (Scheme 6) [121]. However, the thermal instability of $Ln(CH_2SiMe_3)(thf)_2$ and ate complex formation seem to be limiting factors [122].

The *silylamide route* can also be applied to lanthanide(II) chemistry (Scheme 7). Although the well-characterized complexes $Ln[N(SiMe_3)_2]_2(thf)_2$ exhibit enhanced steric flexibility [123], the scope of exchange reactions is now limited by the reductive properties of Sm(II). For example, Sm(II) amides tend to get oxidized by enolizable alcohols [124]. However, aryloxides of type $Sm(OAr)_2(thf)_x$ have been isolated and ate complexation as evidenced in $[KSm(OC_6H_3\text{-}2,6\text{-}\mathit{t}Bu_2)_3(thf)]_n$ proves to be a stabilizing factor [125]. According to this latter approach, mixed metallic complexes can be obtained by retention of the original metal ligand composition Partially exchanged heteroleptic complexes such as $\{KSmCp^*_2[N(SiMe_3)_2](thf)_2\}_n$ are available due to steric restrictions [126]. Eu(II) and Yb(II) silylamides are accessible to all of the exchange reactions listed in Scheme 5 [127].

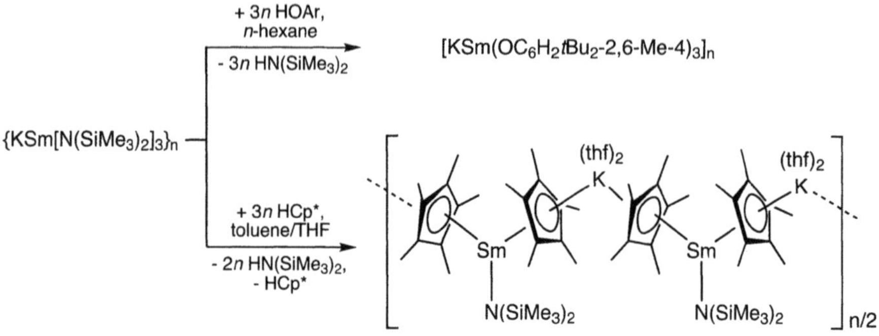

$\{KSm[N(SiMe_3)_2]_3\}_n$ → (+ 3*n* HOAr, *n*-hexane; - 3*n* $HN(SiMe_3)_2$) → $[KSm(OC_6H_2tBu_2\text{-}2,6\text{-}Me\text{-}4)_3]_n$

$\{KSm[N(SiMe_3)_2]_3\}_n$ → (+ 3*n* HCp*, toluene/THF; - 2*n* $HN(SiMe_3)_2$, - HCp*) → $[\ldots Cp^*_2Sm(N(SiMe_3)_2)\,Cp^*K(thf)_2\ldots]_{n/2}$

(thf)$_2$ K Sm N(SiMe$_3$)$_2$ (thf)$_2$ K Sm N(SiMe$_3$)$_2$ n/2

Scheme 7. Heterobimetallic Sm(II) complexes according to the *silylamide route*

Organometallic derivatives of europium and ytterbium are also readily formed via reactions in liquid ammonia. The active species in these reactions are the hexaammine complexes and the only byproducts are hydrogen and ammonia [Eq. (21a-c)]. According to this procedure, the compounds $EuCp_2$ [128], $YbCp^*_2(NH_3)_x$ [91], Ln(COT) (Ln=Eu,Yb) [129, 130], $Eu(C{\equiv}CMe)_2$ [131], $Eu(Ph_2)_2$ [132], $Yb(OC_6H_2tBu_2\text{-}2,6\text{-}Me\text{-}4)_2(thf)_3$ [90c], LnX_2 (Ln=Eu, Yb; X=Cl, Br, I) [133] and decaborates, e.g. $(NH_3)_xYb(B_{10}H_{14})$ [134] have been synthesized.

$$Ln + m\,NH_3(liq) \longrightarrow Ln^{2+}(NH_3)_6 x 2e^-(NH_3)_n \quad (21a)$$

$$2\,HR + 2\,e^-(NH_3)_n \longrightarrow 2\,R^- + H_2 + n\,NH_3 \quad (21b)$$

$$Ln^{2+}(NH_3)_6 + 2\,R^- \longrightarrow LnR_2 + 6\,NH_3 \quad (21c)$$

Generation of Lanthanide Alkyl Bonds

Lanthanide alkyl compounds are important alkyl transfer reagents and initiate a variety of catalytic reactions. The transformation of lanthanide alkoxide bonds to lanthanide alkyl bonds seems to be an attractive alternative to the alkylation of lanthanide halides with alkali metal alkyl compounds. For example, the *aryloxide route* affords homoleptic lanthanide alkyls in good yield [Eq. (22)] [135].

$$Ln(OC_6H_2tBu_2\text{-}2,6)_3 + 3\,LiCH(SiMe_3)_2 \xrightarrow[\text{rt, 30min}]{n\text{-hexane, - 3 LiOAr}} Ln[CH(SiMe_3)_2]_3 \quad (22)$$

Due to the high solubility of the starting and the resulting rare earth complexes, the reaction can be conducted in nonpolar solvents from which the insoluble alkali aryloxides can easily be separated. However, this type of kinetically controllea metathesis reaction is very sensitive towards the reaction conditions including the type of alkoxide (aryloxide) ligand, type of metal, number and type of co-ligands, stability and solubility of the eliminated alkali metal alkoxide (aryloxide), solvent, temperature, etc. As a result, incomplete ligand exchange, exchange of the co-ligand, ate complexation, exchange equilibria and ligand redistribution can occur. Scheme 8 gives an idea of the complexity of these alkoxide-derived alkylation reactions [136–139]. Acetylacetonate complexes have been discussed as alternative alkyl precursors [87].

Aluminum alkyls, in particular trimethylaluminum, produce chelating alkyl alkoxide moieties, $[(\mu\text{-O}t\text{Bu})AlMe_2(\mu\text{-Me})]$, via Lewis acid/base-pair formation [140,141]. In the reaction with $Y_3(\text{O}t\text{Bu})_7Cl_2(thf)_2$, $AlMe_3$ simultaneously acts as a powerful denucleation reagent tolerating ethereal solvents such as THF at the lanthanide center (Scheme 9). Reaction products such as $LnCl_3(dme)_2$ arise from ligand redistributions (Sect. 5.3). The homoleptic complex $Ln[(\mu\text{-O}t\text{Bu})AlMe_2$

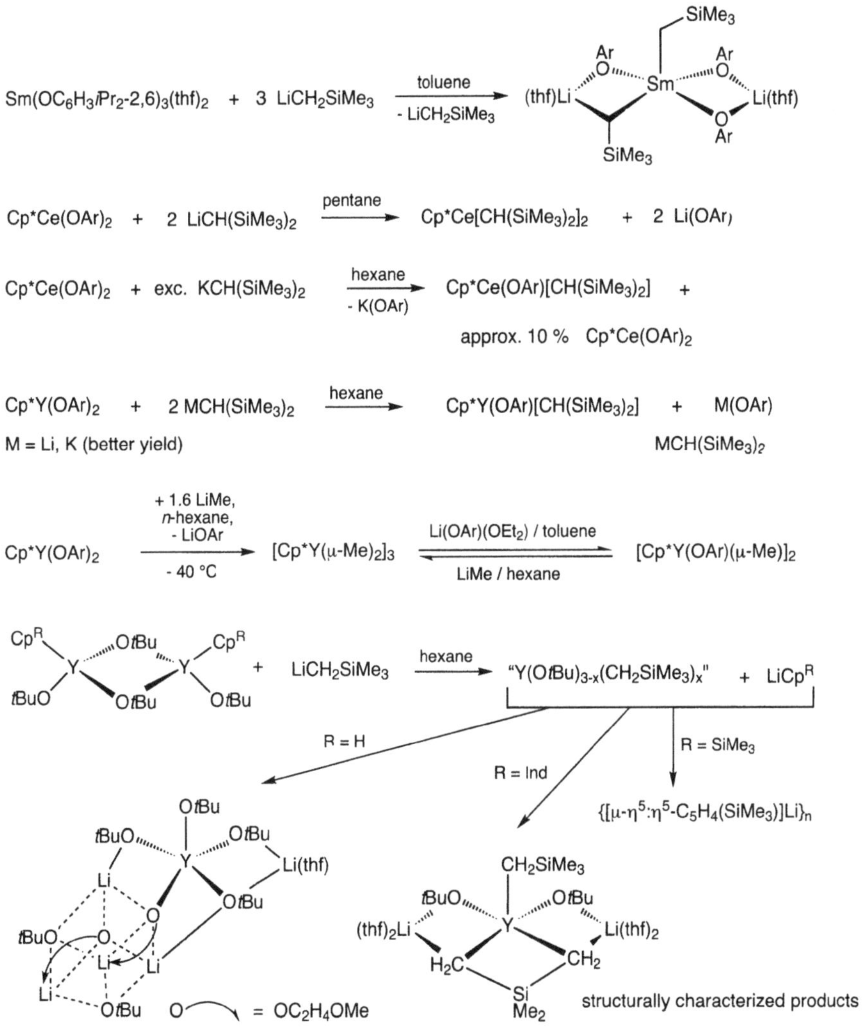

Scheme 8. Reaction behavior of lanthanide aryloxide and alkoxide complexes towards alkali metal alkyl reagents ($OAr=C_6H_3t\text{Bu}_2$-2,6)

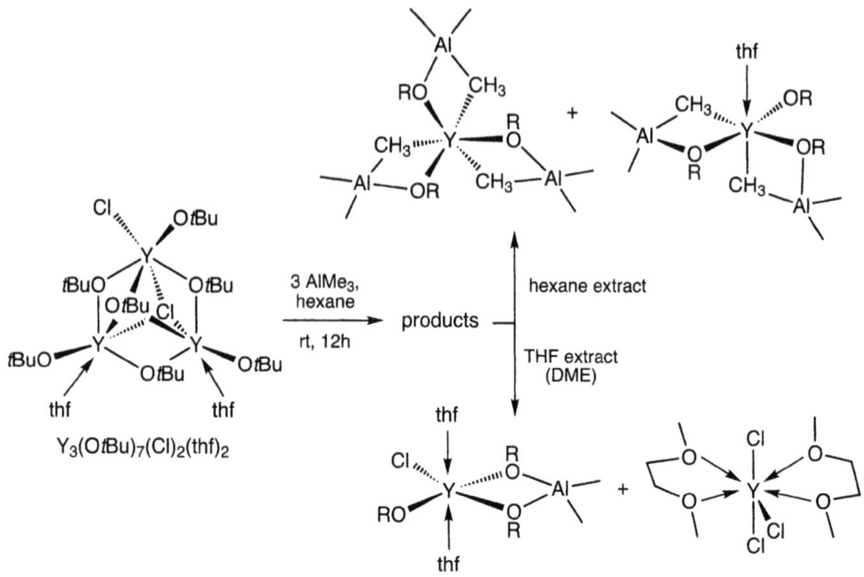

Scheme 9. Reaction of $AlMe_3$ with a mixed alkoxide halide cluster

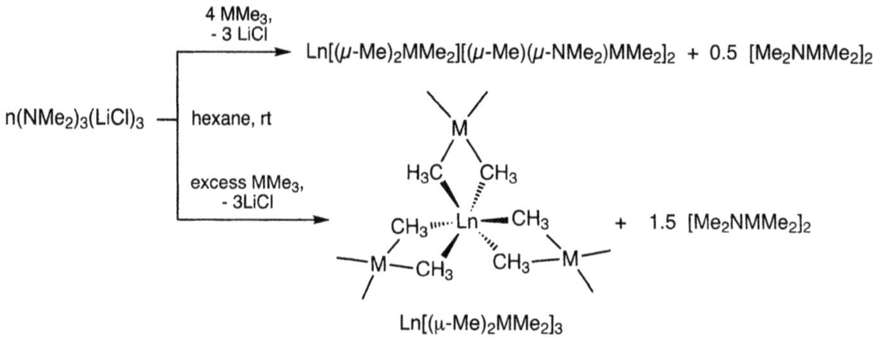

Scheme 10. Alkylation of lanthanide amide complexes with group 13 metal alkyls (M=Al, Ga)

(μ-Me)]$_3$ can be obtained as the sole product from the reaction of $AlMe_3$ with $Ln_3(OtBu)_9$ [142].

Extended alkylation is observed when lanthanide amide complexes are used as synthetic precursors [107,143]. The formation of a strong group 13 metal–N(amide) bond promotes this type of alkylation reaction. Depending on the stoichiometry, the reaction of MMe_3 (M=Al, Ga) with $Ln(NMe_2)_3(LiCl)_3$ yields partially or peralkylated species (Scheme 10). Again, the reactivity is determined by steric factors. For example, the sterically encumbered $La[N(SiMe_3)_2]_3$ does not show any tendency to form a Lewis acid/base adduct with group 13 metal alkyls, a prerequisite for subsequent alkylation under these conditions [115]. However, peralkylated products are obtained from $Ln[N(SiHMe_2)_2]_3(thf)_2$, $Sm[N(SiMe_3)_2]_2(thf)_2$, and $KSm[N(SiMe_3)_2]_3$ [8,115].

Generation of Lanthanide Hydride Bonds

Organolanthanide hydride complexes are also key reagents in rare earth catalysis. The highly reactive lanthanide hydride bonds not only serve as catalytic initiators but are also often assumed key intermediates in catalytic reactions such as the hydrosilylation [144] and olefin polymerization reaction (β-H elimination) [145]. The hydrogenolysis of alkyl complexes is a favorable route for synthesizing lanthanide hydride bonds [Eq. (23)] [146]. However, small changes in the size of the metal, the size of the alkyl group, solvent, degree of association of the complex, or type of co-ligand cause substantial changes in reactivity [137,146].

$$Cp_2LnCH(SiMe_3)_2 + H_2\ (1\ atm) \xrightarrow[0\ °C,\ 3h]{n\text{-hexane}} [Cp_2Ln(\mu\text{-}H)]_2 + H_2C(SiMe_3)_2 \qquad (23)$$

Although solid state structures of these complexes exclusively display bridging "$Ln(\mu\text{-}H)_nLn$" moieties [4], a fluxional behavior in solution with terminal Ln–H bonds as the reactive species has been suggested [147]. $LiAlH_4$ acts as an elegant hydride transfer reagent and depending on the nature of the metal and donor ligand, as well as the cyclopentadienyl substitution, dinuclear species are formed (Scheme 11) [148]. Similar unsolvated dinuclear species were obtained from the reduction of alane by Sm(II) organometallics [Eq. (24)] [149]. Lanthanide hydride species can also be generated by thermal treatment of lanthanide alkyl complexes such as $Cp_2Ln\mathit{t}Bu$ [150] or by salt metathesis reactions employing NaH [151]. The thermal decomposition of the sterically crowded lanthanide alkoxide complex $Ln(OC\mathit{t}Bu_3)_3$ produced a bridged hydride species as a side product (Scheme 12) [109].

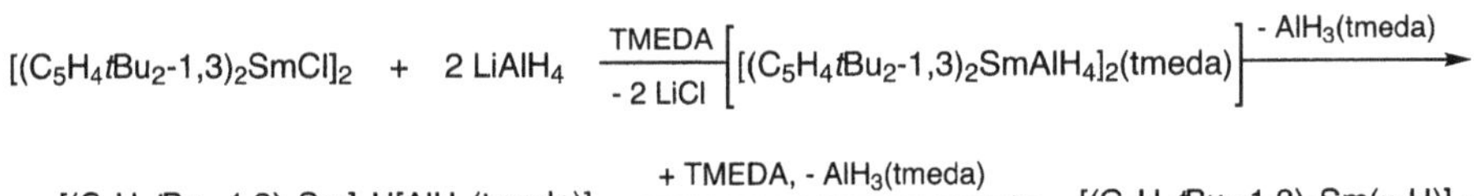

$$[(C_5H_4\mathit{t}Bu_2\text{-}1,3)_2SmCl]_2 + 2\ LiAlH_4 \xrightarrow[-2\ LiCl]{TMEDA} \left[[(C_5H_4\mathit{t}Bu_2\text{-}1,3)_2SmAlH_4]_2(tmeda)\right] \xrightarrow{-AlH_3(tmeda)}$$

$$[(C_5H_4\mathit{t}Bu_2\text{-}1,3)_2Sm]_2H[AlH_4(tmeda)] \xrightarrow{+TMEDA,\ -AlH_3(tmeda)} [(C_5H_4\mathit{t}Bu_2\text{-}1,3)_2Sm(\mu\text{-}H)]_2$$

Scheme 11. Generation of lanthanide hydride bonds via a salt metathesis reaction

$$Ce(OC\mathit{t}Bu_3)_3 \xrightarrow{150\ °C,\ vacuum} \begin{cases} \xrightarrow{90\%} 0.5\ [Ce(OCH\mathit{t}Bu_2)_3]_2 + 3\ \mathit{i}\text{-}C_4H_8 \\ \xrightarrow{10\%} (1/n)\ [Ce(OCH\mathit{t}Bu_2)_2H]_n + 3\ \mathit{i}\text{-}C_4H_8 + \mathit{t}Bu_2CO \end{cases}$$

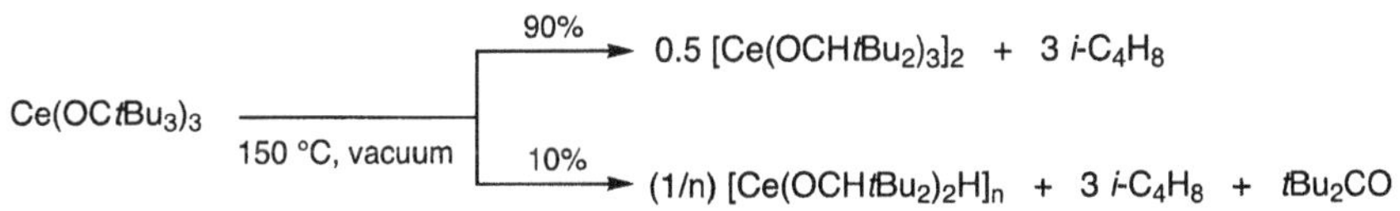

Scheme 12. Generation of lanthanide hydride bonds via thermal degradation

$$2\,(C_5H_4tBu_2\text{-}1,3)_2Sm(thf) + 2\,AlH_3(NEt_3) \xrightarrow[]{Et_2O,\ -2\,Al,\ -2\,H_2} [(C_5H_4tBu_2\text{-}1,3)_2Sm(\mu\text{-}H)]_2 \quad (24)$$

Generation of Cationic Organolanthanide Species

Several routes are currently applied to synthesize cationic organolanthanide species, including the halide abstraction from heteroleptic Ln(III) compounds [Eq. (25)] [152], the oxidation of divalent metallocenes [Eqs. (26) and (27)] [153], the protolysis of lanthanide alkyl and amide moieties [Eqs. (28) and (29)] [154,155], and anion exchange [Eqs. (30) and (31)] [84,156]. In the absence of a coordinating solvent such extremely electrophilic species attain stabilization via arene interactions with the BPh_4^- anion (Sect. 5.1) [153b]. Cationic rare earth species have been considered as promising candidates for Lewis acid catalysis [157].

$$[C_5H_3(SiMe_3)_2\text{-}1,3]_2CeI(NCMe)_2 + AgBF_4 \xrightarrow[rt]{THF,\ -AgI} \{[C_5H_3(SiMe_3)_2\text{-}1,3]_2Ce(NCMe)_2\}[BF_4] \quad (25)$$

$$Cp^*_2Sm(thf)_2 + AgBPh_4 \xrightarrow{THF} [(Cp^*_2Sm(thf)_2][BPh_4] + Ag \quad (26)$$

$$Cp^*_2Sm + AgBPh_4 \xrightarrow{toluene} [Cp^*_2Sm][BPh_4] + \text{“black solids”} \quad (27)$$

$$Cp^*La[CH(SiMe_3)_2]_2 + [NHPhMe_2]BPh_4 \xrightarrow{toluene,\ -CH_2(SiMe_3)_2} [Cp^*LaCH(SiMe_3)_2][BPh_4] \quad (28)$$

$$(salen)Y[N(SiHMe_2)_2](thf) + [NHMe_3]BPh_4 \xrightarrow[rt]{THF,\ -NMe_3,\ -HN(SiHMe_2)_2} [(salen)Y(thf)_3][BPh_4] \quad (29)$$

$$(COT)Nd(BH_4)(thf)_2 + [NHEt_3]BPh_4 \xrightarrow[rt]{THF} [(COT)Nd(thf)_4][BPh_4] + \text{"}[NHEt_3]BH_4\text{"} \quad (30)$$

$$SmCl_3(H_2O)_6 \xrightarrow[2.\ NaBPh_4/H_2O,\ -NaCl]{1.\ KL/MeOH,\ -2\,KCl} [SmL_2][BPh_4] \quad (31)$$

L = tris[3-(2-pyridyl)-pyrazol-1-yl]hydroborate

3.4 Thermal Stability

Despite the kinetic lability of the Ln–X σ-bonds (even the thermodynamically very stable Ln–OR bond undergoes rapid ligand exchange reactions [158]), organolanthanide compounds are thermally robust over a wide range of temperature [99,100,102,104,159–165]. Thermal stability is important for conducting ligand exchange reactions and catalytic transformations at elevated temperatures [1,22]. The sublimation behavior is a criterion of thermal stability, and is frequently consulted to judge the suitability of volatile molecular precursors for chemical vapor deposition techniques (Fig. 7).

Bulky ligands affect the ionic nature of the polarized Ln–X bond by minimizing polar interactions (intra- and intermolecular) and optimizing volatility by the concept of steric shielding. The detection of isolated molecules instead of salt-like arrangements in the solid state confirms this trend. The polarizing effect can also be reduced by introduction of donor-functionlized ligands which can bring about charge transfer to the metal cation. Decomposition pathways can be sterically blocked by filling the coordination sphere of the metal with large ligands. However, sterically overcrowded ligands may degradate at elevated temperature as illustrated for the $Ln(OC\textit{t}Bu_3)_3$ system [109].

4 Ligand Concepts

Ligand design occupies a pivotal role in organolanthanide chemistry. The nature of the ligand, including its size, basicity, and functionalization, promptly affects complex features such as (mono)nuclearity, cation size and electrophilicity. Prolific metal cation/ligand synergisms impart novel reactivity patterns which can

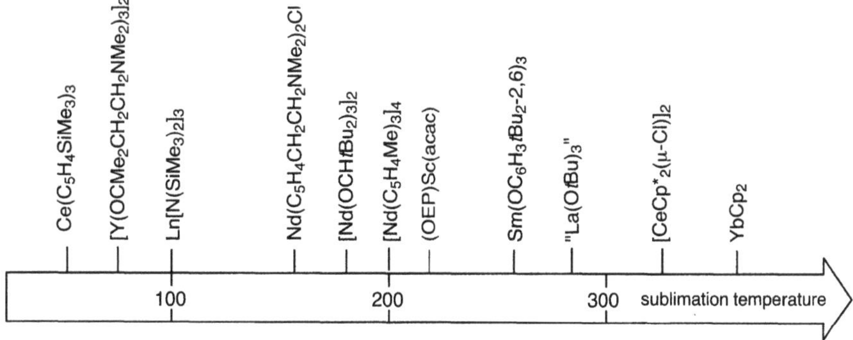

Fig. 7. Sublimation behavior of various lanthanide complexes at 10^{-3} mbar (*OEP* 2,3,7,8,12,13,17,18-octa(ethyl)porphyrin; *acac* acetylacetonate)

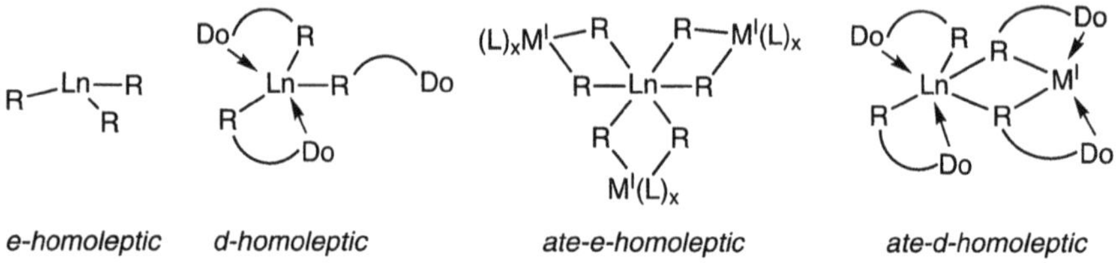

e-homoleptic *d-homoleptic* *ate-e-homoleptic* *ate-d-homoleptic*

Fig. 8. Modes of homoleptic Ln(III) coordination (*Do* donor functionality)

be of interest in, for example, ligand-enhanced stereoselective catalysis [166]. Therefore, ligand classification and adaptation deserve particular attention.

Assuming the ligand interaction to be of electrostatic origin, optimal charge balance of the lanthanide(III) cations should be achieved by three stable anionic ligands. Identical anions accomplish so-called homoleptic systems which can be of neutral type $(MR_n)_x$ or ate type $[MR_n]^{z-}[X_m]^{z+}$ [167]. Homoleptic compounds can be further classified as to whether the ligands are coordinatively *e*qually (*e-homoleptic*) or *d*ifferently (*d-homoleptic*) attached to the metal center (Fig. 8). The *d-homoleptic* mode is found in oligomeric systems where both terminal and bridging ligands are present; however, they are also found in monomeric complexes which contain functionalized ligands. Here, steric oversaturation can prevent the formation of *e-homoleptic* coordination [31,168].

Heteroleptic organolanthanide complexes containing reactive Ln–X bonds and stabilizing ancillary ligands are key precursor compounds in catalytic transformations. Mononuclearity is usually a prerequisite for both good solubility and reactivity. Utilization of bulky ligands, ate complexation, and donor functionalization are applicable procedures for generating mononuclear complexes.

4.1 Steric Bulk and Donor Functionalization

Scheme 13 emphasizes the effect of sterically demanding groups on the generation of homoleptic mononuclear complexes. This modification often gives access to classes of compounds which are not isolable/defined in the presence of correspondingly small ligands. Various examples feature both different Ln–X bonds and different oxidation states [169]. Structurally characterized Ln–C bonded homoleptic systems include alkyl [94,135], cyclopentadienyl [31,170], pentadienyl [171], pentamethylcyclopentadienyl [172], indenyl [173], cyclooctatetraenyl [80e], (aza)allyl [174] and arene derivatives [26]. Representative examples of pseudo-organometallics containing Ln–N bonds comprise silylamide, azabutadiene, benzamidinate and porphyrin complexes [114,175–178]. Aryloxide, alkoxide, β-diketonate and Schiff base ligands can stabilize homoleptic mononuclear Ln–O derived complexes [101, 179–182]. Derivatives featuring phosphorus and sulfur bonds include alkyl phosphides [183] and aryl and alkyl sulfides [90c,184].

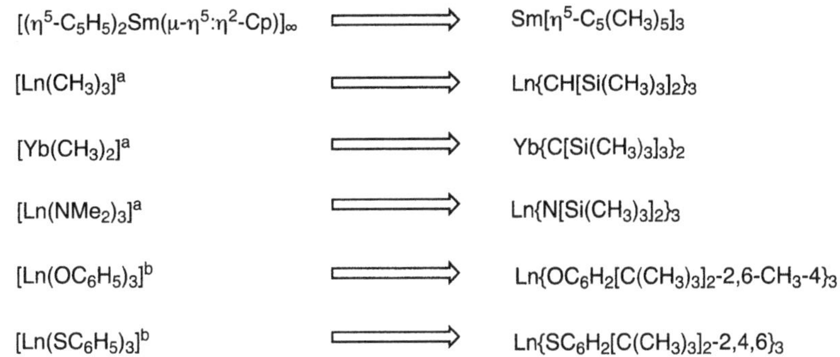

Scheme 13. Tractable, mononuclear species via steric cally bulky ligands ([a]compound not known/ not stable; [b]degree of agglomerization not determined)

The attachment of potentially coordinating groups produces donor-functionalized ligands and is a popular approach to fine-tuned ancillary ligands. This is particularly appealing in organolanthanide chemistry [185], considering the large size of the lanthanide cations and their preference for high coordination numbers. Intramolecular ring formation via "dative bonds" [186] according to the HSAB concept stabilizes mononuclear complexes via the effect of chelation and entropy. Ligand-bonded donor groups successfully compete with donating solvent molecules for coordination sites, implying improved thermal behavior by suppressing oligomerization during heating and sublimation. The presence of intramolecularly active donor functionalities affects the polarity of the lanthanide–counterion bond by more or less pushing electron density into the metal. As a result strong donor groups will significantly decrease the reactivity of an adjacent, kinetically labile Ln–X bond, but on the other hand enhance its stability against hydrolysis. Hence, the balance of donor strength is important for the production of a flexible coordination mode, revealed by "arm on – arm off" processes. Such *hemilabile ligands* are proposed as directing the approach of organic substrates in catalytic transformations [187]. In the absence of ring strain the bond strength of such an intramolecular coordination resembles that of the corresponding intermolecular one [188]. For example, the strength of an intramolecular hard "$\leftarrow OR_2$" coordination will be in the range of 5–7 kcal/mol as determined for intermolecular THF coordination [41].

Figure 9 outlines general strategies of donor functionalization applied in lanthanide alkyl [189–193], amide [194–196] and alkoxide chemistry [180,197,198]. Hard donor functionalities such as NMe_2 and OMe dominate this scenario and it is often the combination of steric bulk and donor ligation which leads to the envisaged mononuclear species. Donor functionalization of alkyl ligands proved to be successful even in stabilizing lanthanide alkyl bonds in a low-valent Sm(II) species [191]. Allylic moieties are also very effective in stabilizing low aggregated organolanthanide species as evidenced in mononuclear homoleptic phosphi-

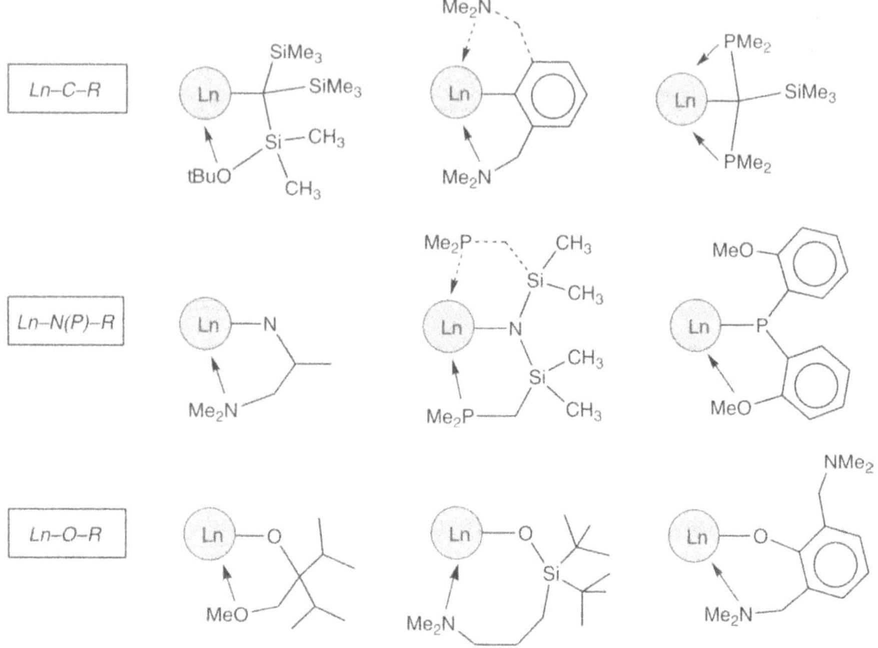

Fig. 9. Donor functionalization of monovalent rare earth complexes

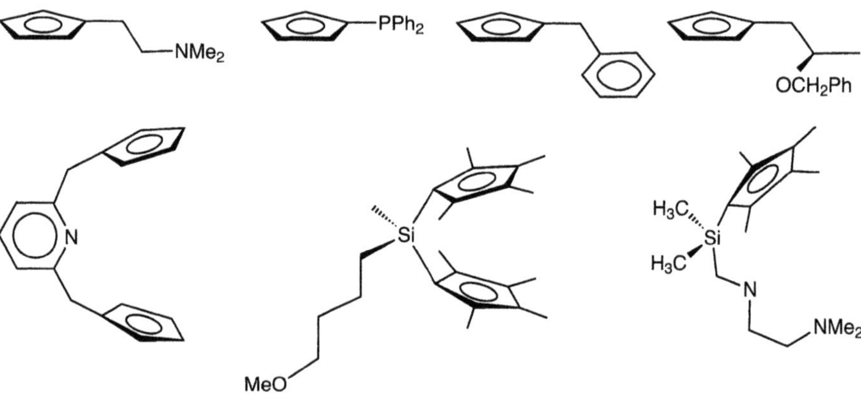

Fig. 10. Functionalization of cyclopentadienyl ligands

nomethanide [199], benzamidinate [200] and azaallyl complexes [177]. Multiple functionalization displays another option. As revealed in aryl [193] and silylamide derivatives [195], donor moieties flank the central bonding unit in a potentially tridentate chelating array.

Not surprisingly, donor functionalization has been extensively applied to tailor cyclopentadienyl ligands (Fig. 10). In addition to the attachment of one pendant donor arm [160, 201], functionalization via donor-linked bis(cyclopentadienyl) ligands [202], including donor-tethered silicon bridges [203], and donor functionalized chiral side chains are found [204]. Only a few examples feature soft donor functionalities such as phosphines [205] and arenes [206].

4.2 Ancillary Ligands

Various terms including "spectator ligand", "auxiliary ligand" or "ancillary ligand" are used to characterize the portion of the ligand sphere which is not directly involved in basic reactivity steps such as insertion or ligand exchange reactions. Primarily, ancillary ligands serve to prevent oligomerization or polymerization of electronically and coordinatively unsaturated derivatives, and to impart kinetic stability for otherwise highly reactive species. The mononuclearity, chemical robustness and rigidity anticipated are common attributes of a well-defined precatalyst system. Furthermore, ancillary ligands can direct catalytic processes if their bulkiness causes metal shielding and steric constraints, if differently polarized Ln–X bonds affect the electrophilicity of the metal center, or if they exhibit a reservoir for chirality and additional flexible coordination sites.

Hence, synthetic organolanthanide chemistry puts the main emphasis on the adaption of prevailing precatalyst types to the requirements of highly enantioselective catalysis. This is impressively demonstrated by tied-back cyclopentadienyl complexes [207], even water-stable BINOL systems [43], and fluorinated β-diketonate complexes (Fig. 11) [208].

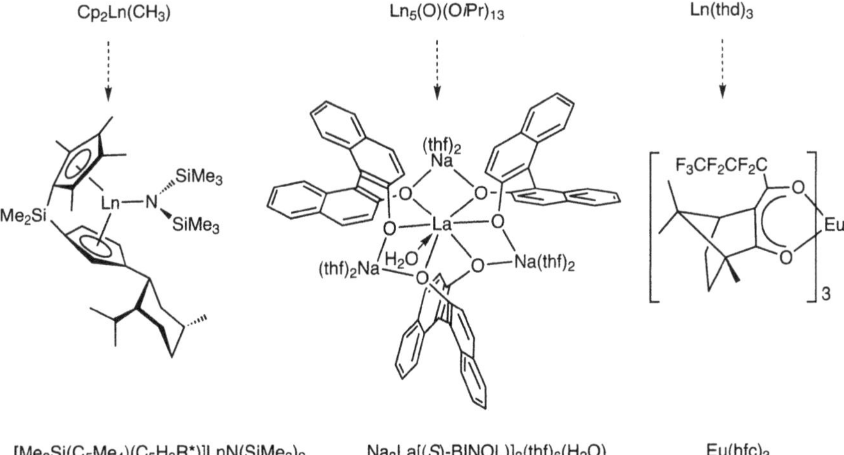

Fig. 11. Development of sophisticated ligand environments for enantioselective catalysis [*thd* 2,2,6,6-tetramethyl-3,5-heptanedione, *hfc* 3-(heptafluoropropylhydroxymethylene)-*d*-camphorato]

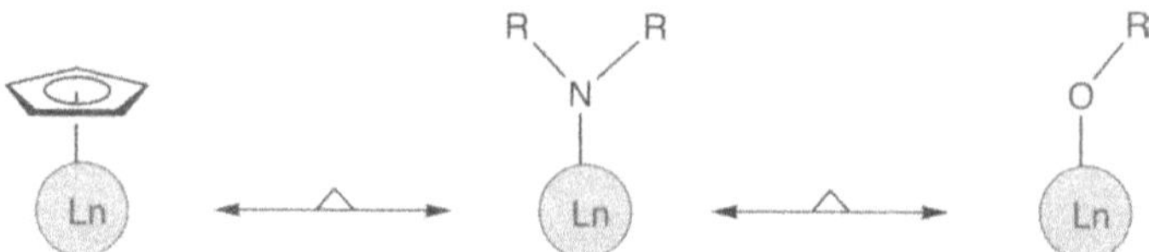

Fig. 12. Amide and alkoxide moieties as cyclopentadienyl-analogous ligands

Cyclopentadienyl derivatives and related systems such as indenyl and fluorenyl still play a dominant role in ligand fine-tuning. However, ligand design has gradually reached a high level of variety and sophistication, e.g. as presented by the new generation of group 4 metal "non-metallocene" catalysts (for olefin polymerization) [209,210]. In particular, chelating nitrogen-containing counter ligands have become the focus of much attention [209]. Similar to the related alkoxide ligands [210], they offer a lower formal electron count, thus rendering a more electrophilic and therefore potentially more active catalyst fragment (Fig. 12) [37]. Ancillary ligands, reported for Ln(III) species, can be classified according to their charge (valency) and coordination mode (Figures 13–15).

Monovalent Ligands

Numerous mono-charged cyclopentadienyl substitutes have been discussed in relevance to the catalysis topic. The attachment of one or two of such ligands is easily perfomed and may render two or one reactive sites, respectively. The resulting ligand sphere offers enhanced steric flexibility, and kinetic inertness of the monovalent ancillary ligand is often achieved via chelation through charge delocalization or donor functionalization. The stability of the resulting complexes, however, can be affected by ligand redistribution (disproportionation) and formation of the homoleptic system (Sect. 5.3). A representative selection of this ligand type is shown (Fig. 13), comprising *N,N'*-bis(*tert*-butyl)glyoxaldiimine $[(dad)Li]^-$ [211], *N*-isopropyl-2-(isopropylamino)troponimine $[(i\mathrm{Pr})_2\mathrm{ati}]$ [212], substituted benzamidinates [213], substituted tris(pyrazolyl)borates $\mathrm{Tp\text{-}R_x}$ [214], (*N,O*-bis(*tert*-butyl) (alkoxydimethylsilyl)amide [215], tri(*tert*-butyl)methoxide (tritox) [102], substituted aryloxides [137], tri(*tert*-butyl)siloxide (silox) [102,216] and functionalized siloxide ligands [217].

Divalent Ligands

Heteroleptic complexes derived from doubly charged, "*linked*" ligands constitute a class of well-defined metallocene-analogous precatalyst species. Various C_1- and C_2-symmetric members of the bis(cyclopentadienyl) fragment have been reported [218], including linked amido-cyclopentadienyl [219] and dihydroanthracene-cyclopentadienyl ligands [220] (Fig. 14). Such divalent ligands when coordinated in a chelating fashion provide a strongly bonded, rigid backbone which not only imparts kinetic stability but is also a prerequisite for asymmetric induction at the metal center. By nature, the synthesis of these ancillary ligands is more costly/lengthy and subsequent complex preparations may re-

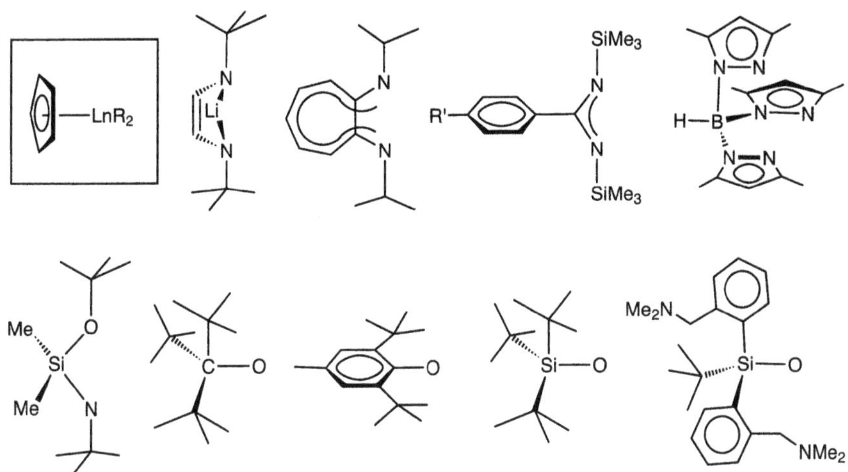

Fig. 13. Monovalent ancillary ligands

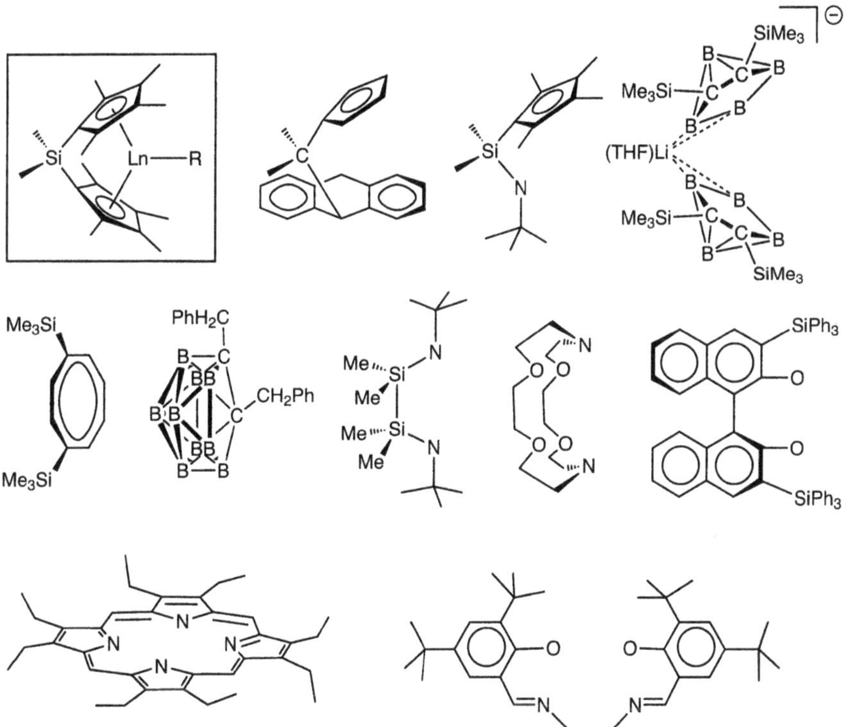

Fig. 14. Divalent ancillary ligands

quire alternative routes to avoid oligomerization by ligand bridging [116–118]. The remaining Ln(III)–R bond, typically an alkyl, hydride or amide moiety, displays a reactive site whose kinetic profile can be stabilized upon variation of the bite angle and chelate ring size of the ancillary ligand, respectively [37]. Together with diamide [221,222], biphenol or BINOL ligands [223,224], salen [116] and sulfonamide ligands [118] have been discussed as alternative spectator environments. The latter can easily be obtained in enantiomerically pure form. *Template*-type ligands such as cyclooctatetraenyl [225] and porphyrin [226] have been shown to be very effective in stabilizing mononuclear complexes. In particular, porphyrin–metal moieties are rather robust and less moisture-sensitive. Carborane derivatives have been employed both in the form of "lithium"-linked [227] and template-type ligands [228].

Trivalent Ligands
Complexes derived from triply charged ligands formally correspond to the homoleptic tris(cyclopentadienyl) complexes $LnCp_3$. These compounds can be of relevance in Lewis acid catalysis where their activity is directed by the formation of Lewis acid (catalyst)/base (substrate) pairs [229]. Hence, metal–ligand bond disruption and formation processes are pushed into the background. Linked cyclopentadienyl-carborane ligands form mononuclear *commo* (sandwiched) metallacarborane complexes [230]. Highly functionalized podate ligands such as triamidoamine ("azatrane") [231] and tribenzyltrifluoroacetoacetate [232] produce formally 4- and 6-coordinated complexes, respectively (Fig. 15). A dinuclear composition has been proven for Ln(III) complexes of the trivalent oligosilsesquioxane ligand, $T_7(OH)_3$ [233]. These incompletely condensed ligands

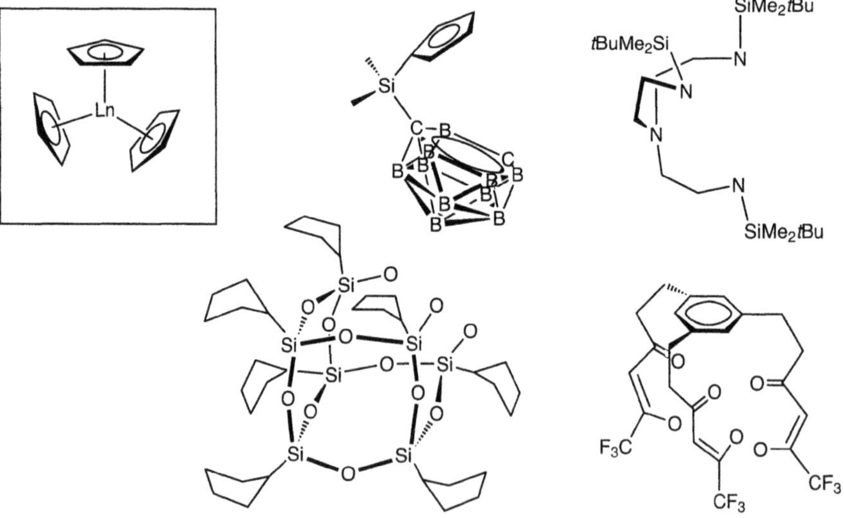

Fig. 15. Trivalent ancillary ligands

appear to be highly electron withdrawing and are currently being studied as models for silica surfaces [234].

4.3
Immobilization – "Supported Ligands"

The ligand sphere can also be manipulated by immobilization (grafting) techniques. The surface of an inorganic oxide or an organic polymer then acts as an alternative ligand environment imposing steric and electronic constraints. The so-called heterogenization of homogeneous catalysts is a popular method for the generation of hybrid catalysts, featuring both the advantages of homogeneous and heterogeneous catalysis [235]. Figure 16 shows various types of such supramolecular entities which have been successfully employed in catalytic transformations. Neodymium centers bonded to a carboxylated divinylbenzene crosslinked polystyrene matrix mediate the polymerization of butadiene to *cis*-1,4-polybutadiene in the presence of organoaluminum reagents [236]. Polymer-supported scandium triflate has been employed as a reusable catalyst in several fundamental Lewis acid catalyzed carbon–carbon bond-forming reactions such as aldol, Michael, and Friedel–Crafts acylation reactions [237] and the preparation of diverse quinoline derivatives according to a three-component coupling reaction [238]. Immobilization techniques such as chemisorbed polyacrylonitrile derivatives [PA-Sc-TAD, polyallyl scandium trifylamide ditriflate] and microencapsulated scandium triflate [MC-$Sc(OTf)_3$] have been used. In the latter hybrid system the Sc(III) is probably stabilized by π-interactions with the poly-

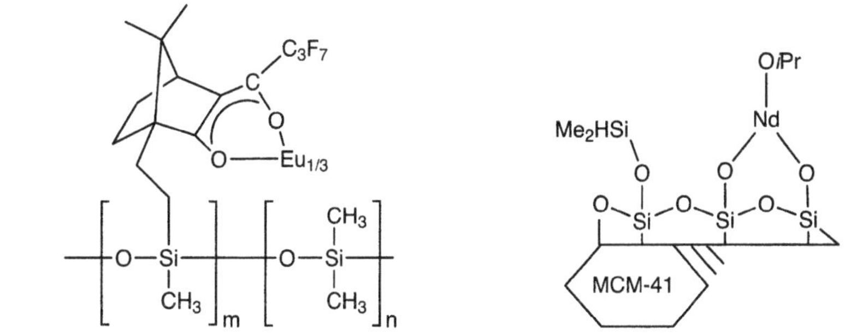

Fig. 16. Heterogenized homogeneous catalysts

styrene backbone. Chiral polysiloxane-fixed europium β-diketonates were found to be reusable catalysts in the Danishefsky hetero-Diels–Alder reaction [239].

Mesoporous silicates of type MCM-41 display versatile, rigid and thermally and chemically robust host materials for the grafting of pseudo-organometallic compounds such as silylamides and alkoxides (Fig. 16) [115,240]. These materials effectively catalyze carbon–carbon bond-forming reactions and functional group transformations, as shown for the Meerwein–Ponndorf–Verley reduction [241]. Heterogeneous reactions of lanthanide metals dissolved in liquid ammonia with free hydroxy groups of silica and alumina [242] or zeolites [243] yielded supported lanthanide species which exhibit catalytic activity in (de-)hydrogenation, isomerization and Michael reactions. Lanthanide amide and imide moieties have been discussed as catalytically active surface species.

5 Reactivity Pattern of Organolanthanide Complexes

The intrinsic properties of the rare earth cations as revealed by their oxophilicity, "hardness", and large size govern the reactivity of organolanthanide complexes. Hence, parallels to the chemistry of aluminum, the group 2 and group 4 elements, and the actinides are often detected. In this section the most important reaction pathways of these highly reactive complexes are surveyed with representative examples. The stoichiometric reactions outlined also represent the elementary steps in catalytic reaction sequences.

5.1 Donor–Acceptor Interactions

The interaction of the hard, Lewis acidic Ln(III) centers with neutral electron-donating moieties is a ubiquitous feature of organolanthanide complexes. In addition to the electron deficiency of the lanthanide center, steric unsaturation directs this stabilization of the complexes via adduct formation.

Simple Lewis Acid/Base-Type Interactions

Solvent complexation usually results from salt metathesis reactions which have been carried out in ethereal solvents such as THF or DME. Solvent coordination as a rule decreases the reactivity of Ln–R bonds by depolarization, steric saturation, and competition reactions. On the other hand, donor coordination leads to an enhanced reducing ability of Ln(II) compounds (e.g. HMPA coordination [244]). Lewis base coordination often forces crystallization (e.g. $OCPh_2$, $OPPh_3$, TMEDA) [Eq. (32)] [245,246] and strongly donating ligands such as acetonitrile or pyridine were shown to act as denucleating agents in cyclopentadienyl and alkoxide chemistry [Eq. (33)] [20,247]. Substituted imidazol-2-ylidenes, carbene-type ligands, form strong adduct complexes with Ln(II) and Ln(III) metal centers as indicated by THF displacement [Eq. (34)] [248]. However, coordination of soft Lewis bases was observed in the case of terminally bonded PMe_3 in a highly electrophilic scandium complex [Eq. (35)] [249]. Even water complexes could be trapped as the initial step of the hydrolysis reaction [250]. The vast majority of organic transformations mediated by lanthanide centers depend on pre-coordination of a neutral, functionalized substrate and subsequent formation of an activated species.

$$Ln[N(SiMe_3)_2]_3 + O{=}CPh_2 \xrightarrow{\text{hexane}} Ln[N(SiMe_3)_2]_3(O{=}CPh_2) \quad (32)$$

$$[Dy(OCH\textit{t}Bu)_3]_2 + 4\,CH_3CN \xrightarrow{\text{hexane}} 2\,Dy(OCH\textit{t}Bu)_3(CH_3CN)_2 \quad (33)$$

$$(C_5Me_4Et)_2Yb(thf) + \text{"carbene"} \xrightarrow{\text{THF}} (C_5Me_4Et)_2Yb(carbene) + THF \quad (34)$$

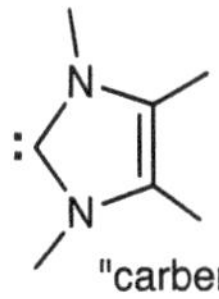

$$2\,[(\eta^5\text{-}C_5Me_4)SiMe_2(\eta^1\text{-}N\textit{t}Bu)]ScCH(SiMe_3)_2 \xrightarrow[-CH_2(SiMe_3)_2]{H_2,\ PMe_3,\ n\text{-hexane}} \quad (35)$$

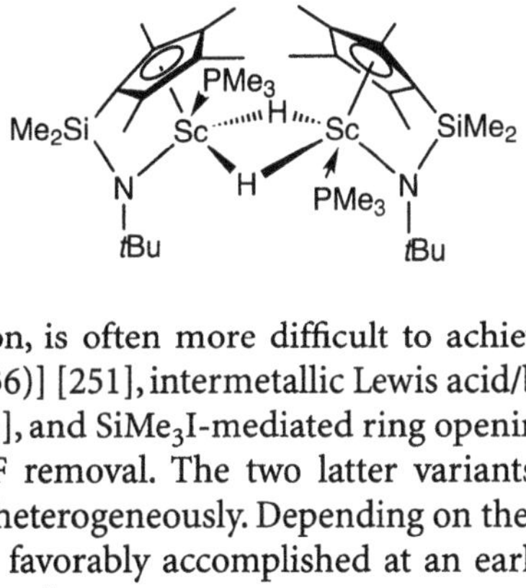

Me2Si SiMe2 PMe3 Sc H Sc H PMe3 N N tBu tBu

The reverse process, desolvation, is often more difficult to achieve. Thermal treatment (*toluene method*) [Eq. (36)] [251], intermetallic Lewis acid/base competition reactions [Eq. (37)] [252,253], and $SiMe_3I$-mediated ring opening [Eq. (38)] [254] have been applied for THF removal. The two latter variants have been probed both homogeneously and heterogeneously. Depending on the desolvating reagent, donor solvent removal is favorably accomplished at an early stage of a multistep synthesis (Scheme 14) [254].

$$Ce[C_5H_3(SiMe_3)_2\text{-}1,3]_3(thf) \xrightarrow[\Delta,\ \text{vacuum}]{\text{toluene, - THF}} Ce[C_5H_3(SiMe_3)_2\text{-}1,3]_3 \quad (36)$$

$$Ln[N(SiHMe_2)_2]_3(thf)_2 + AlMe_3 \xrightarrow{n\text{-hexane}} Ln[N(SiHMe_2)_2]_3(thf) + AlMe_3(thf) \quad (37)$$

$$Cp^*La[CH(SiMe_3)_2]_2(thf) + [\text{Merrifield polymer}]\text{-}CH_2SiMe_2I \xrightarrow{\text{toluene}} Cp^*La[CH(SiMe_3)_2]_2 + [\text{Merrifield polymer}]\text{-}CH_2SiMe_2O(CH_2)_4I \quad (38)$$

route A

$$Cp^*LaI_2(thf)_3 + \text{excess } Me_3SiI \xrightarrow{\text{toluene}} [Cp^*LaI_2]_n + 3\ Me_3SiO(CH_2)_4I$$

$$\xrightarrow{+ 2\ K[CH(SiMe_3)_2],\ OEt_2} Cp^*La[CH(SiMe_3)_2]_2$$

route B

$$Cp^*LaI_2(thf)_3 + 2\ K[CH(SiMe_3)_2] \xrightarrow{\text{hexane/OEt}_2} Cp^*La[CH(SiMe_3)_2]_2(thf) + 2\ KI$$

$$\xrightarrow[0\ °C]{\text{excess } Me_3SiI,\ \text{toluene},\ - Me_3SiO(CH_2)_4I} \left[Cp^*La[CH(SiMe_3)_2]_2 \right] \longrightarrow [Cp^*LaI_2]_n + 2\ Me_3SiCH(SiMe_3)_2$$

Scheme 14. Solvent-free rare earth alkyl species – THF removal at an early stage

Ate Complexation

The formation of anionic rare earth metal ligand moieties or ate complexation are commonly observed features of salt metathesis reactions when alkali metal cyclopentadienyl [147], alkyl [96], amide [97,255] and alkoxide derivatives are employed [Eqs. (39)-(42)] [256].

$$NdCl_3 + 2\ LiCp^* \xrightarrow[2.\ OEt_2]{1.\ THF} Cp^*_2\,Nd(\mu\text{-}Cl)_2Li(OEt_2)_2 + LiCl \tag{39}$$

$$LaCl_3 + 3\ LiCH(SiMe_3)_2 \xrightarrow[rt]{THF,\ PMDETA,\ -2\ LiCl} La[CH(SiMe_3)_2]_3(\mu\text{-}Cl)Li(pmdeta) \tag{40}$$

$$SmCl_3(thf)_3 + 2\ LiN\textit{i}Pr_2 \xrightarrow[rt,\ 3h]{THF/TMEDA} (N\textit{i}Pr_2)_2Sm(\mu\text{-}Cl)_3Li_2(tmeda) \tag{41}$$

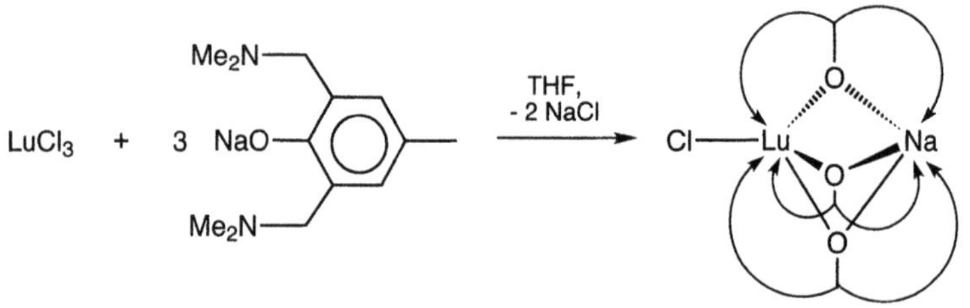

$$LuCl_3 + 3\ NaO\text{-}C_6H_2(CH_2NMe_2)_2\text{-}CH_3 \xrightarrow{THF,\ -2\ NaCl} Cl\text{—}Lu(\mu\text{-}O)_3Na \tag{42}$$

Ate complexation, as main product or as contamination, occurs via coordination of additional counter ligands or alkali metal halide incorporation. The removal of alkali metal halides can often only be achieved by tedious extraction experiments [257] or in the case of thermally stable compounds via sublimation [258]. Due to the additional electronic and steric saturation of the metal environment, the reactivity of ate complexes is significantly decreased. Ate complexation can be

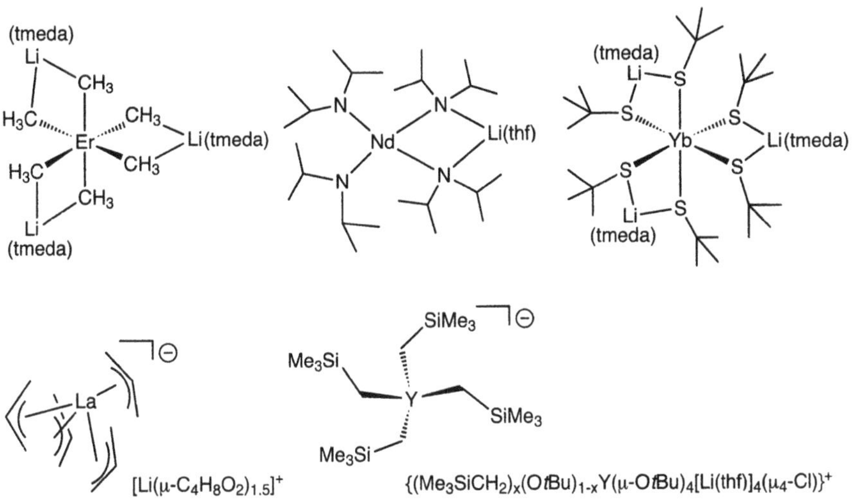

$[Li(\mu\text{-}C_4H_8O_2)_{1.5}]^+$ $\{(Me_3SiCH_2)_x(OtBu)_{1-x}Y(\mu\text{-}OtBu)_4[Li(thf)]_4(\mu_4\text{-}Cl)\}^+$

Fig. 17. Stabilization of reactive homoleptic complexes via ate complexation

avoided by the application of alternative synthetic procedures such as the *silylamide* route (Sect. 3.3). Although ate complexation is considered a nuisance for the preparation of highly reactive lanthanide species, promising synthetic and catalytic behavior has been ascribed. For example, ligand exchange reactions involving heterobimetallic ate systems can proceed via retention of the metal composition [125]. Their potential for cation exchange reactions still has to be examined. Rare earth binolate complexes of the type $Na_3La[(S)\text{-BINOL}]_3(thf)_6(H_2O)$ effectively catalyze enantioselective carbon–carbon bond-forming reactions [43] and heterobimetallic complexes containing linked amido-cyclopentadienyl ligands, $Li[Ln(\eta^5{:}\eta^1\text{-}C_5Me_4SiMe_2NCH_2CH_2NMe_2)_2]$, have been reported to be active in the ring-opening polymerization of lactones and lactides [219].

The stabilization of otherwise labile homoleptic complexes is another useful option of ate complexation. This can occur via ion pair formation [174,259] or entrapment of the cationic moieties in the ligand sphere (Fig. 17) [120,184,260].

Neutral π-Donor Ligation – Arene and Olefin Coordination

The coordination of an alkene to an electron-deficient metal center is the proposed preceding step to the insertion into Ln–C, Ln–H and Ln–N bonds [22,145]. η^2-Olefin and alkyne complexes of the lanthanide elements are particularly unstable due to the lack of $d\pi$-$p\pi$ back-bonding to the coordinated alkene [261–268]. However, tailoring of the olefin environment by maximization of the Lewis basicity of the alkene [262] or utilizing chelation effects [263] has led to complexes indentified by structure analysis and NMR spectroscopy, respectively (Fig. 18). η^6-Arene coordination is considerably more stable than η^2-olefin coordination and can occur in a mono(arene) [265], bis(arene) (Sect. 3.2) [26] and intermolecular fashion [247]. Lanthanide π-donor ligation is observed in the ab-

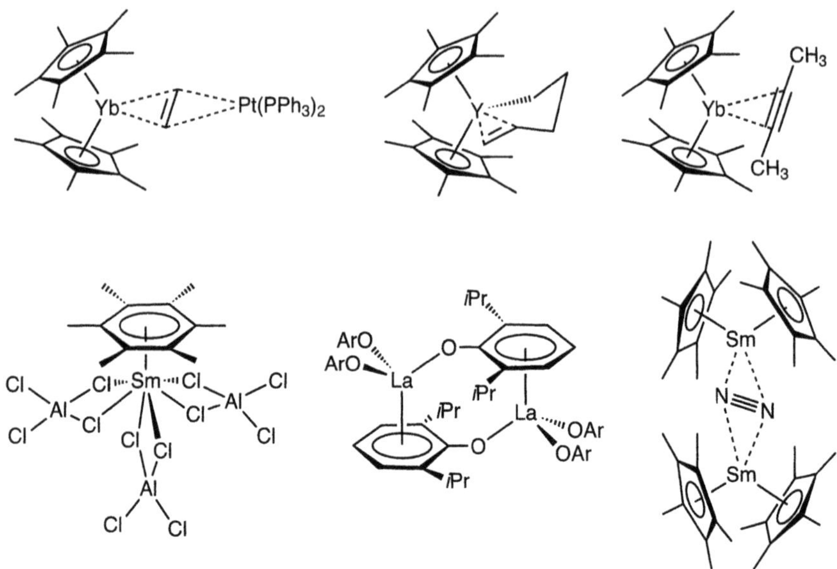

Fig. 18. π-Donor interactions in organolanthanide chemistry

sence of coordinating solvent molecules only. Pseudo-η^2-π-bonded complexes involving hydrogen [266] and nitrogen [267] also show reversible coordination with Ln(II) fragments. In contrast, the reaction of $SmCp^*_2$ with hydrazine produces a dinuclear Sm(III) complex featuring a bridging $(NHNH)^{2-}$ moiety [268].

Agostic Interactions

The term "agostic" bonding, originally proposed for the formation of two-electron three-center bonds of the type C–H→M [269], is now often used in lanthanide chemistry to describe the interaction of a highly electron-deficient, sterically unsaturated metal center with "CH", "SiMe", and "SiH" ligand fragments. These intramolecular, chelate-type interactions are of predominantly electrostatic nature (Fig. 19) [270]. Although the agostic bonding is weak and usually not observed in solution [271], it can have significant implications for the molecular and electronic structure and hence reactivity of the molecule. The solid state structures often reveal quite remarkable angle distortions within the agostically interacting fragment. Detailed studies performed on complexes which are active in olefin polymerization propose that α-C-H agostic interactions assist chain propagation (transition state) [272], while β-C-H interactions retard ethylene insertion (ground state) [273]. Strong β-Si-H diagostic interactions can be formed even in the presence of a coordinating solvent. Such potentially tridentate chelating arrays direct the *rac*/*meso* ratio in *ansa*-lanthanidocene complexes [117]. The intramolecular agostic approach is routinely observed in electron-deficient complexes of the bulky ligand $CH(SiMe_3)_2$ [274]. The presence of a

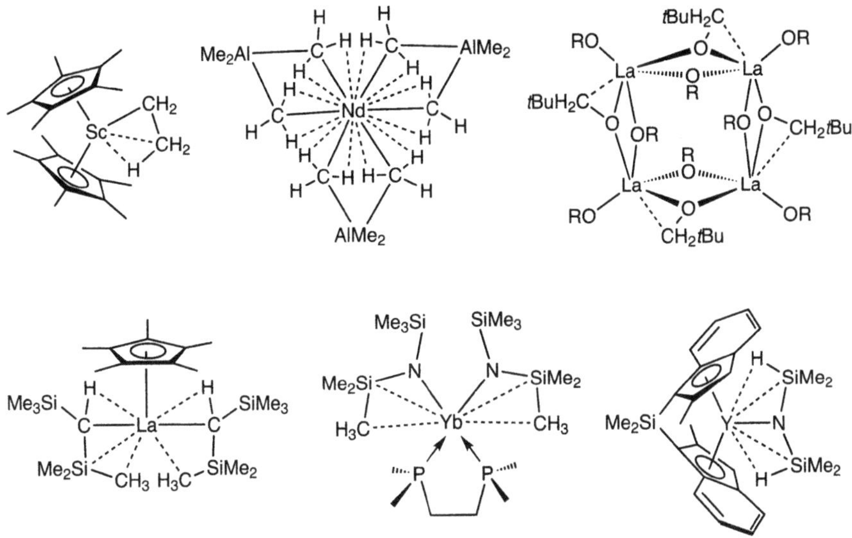

Fig. 19. Agostic interactions in organolanthanide chemistry

multi-agostic interaction was proposed in permethylated complexes $Ln(AlMe_4)_3$, yielding formal coordination numbers as high as 18 [275]. The formation of agostic interactions is not restricted to compounds containing metal–carbon bonds, as evidenced by the ring structure of $[La(OCH_2tBu)_3]_4$ [276]. Bulky amide and aryloxide (sulfide) ligands such as $N(SiMe_3)_2$ [277], $N(SiHMe_2)_2$ [117], and $NiPr_2$ [120] offer another agostic reservoir as emphasized in the important synthetic precursors $Ln[N(SiMe_3)_2]_3$ [114] and $Ln(OC_6H_3tBu_2\text{-}2,6)_3$ [101].

5.2 Complex Agglomerization

Steric and electronic factors often force the stabilization of monometallic species via agglomerization. As a rule, the formation of di- and multinuclear species is achieved by intermolecular bridging of the smallest, most reactive and labile Ln–X bond and, hence, leads to decreased reactivity.

Dinuclear Complexes

The formation of dinuclear complexes is routinely observed along with the important class of lanthanidocene complexes of type Cp_2LnR when R is a sterically less demanding ligand such as H, Me or a small alkoxide group [4]. Depending on the metal and ligand size, bridging can also occur in an asymmetric fashion as evidenced in $[Cp^*_2Lu(CH_3)]_2$ [Eq. (43)] [145]. The formation of dinuclear ligand bridged species is also observed in organo-Ln(II) chemistry. Two representative examples are given in Eqs. (44) und (45) [127,278]. Rare earth amide and alkoxide complexes show an even more pronounced tendency of agglomeriza-tion [Eq. (46)] [159].

$$Cp^*_2Lu(\mu\text{-}CH_3)_2AlMe_2 \xrightarrow[\text{2. NEt}_3\text{, vacuum}]{\text{1. OEt}_2\text{, - AlMe}_3\text{(OEt}_2\text{), -40°C}} Cp^*_2Lu(\mu\text{-}CH_3)LuCp^*_2(CH_3) \quad (43)$$

$$2\,SmI_2(thf)_2 + 2\,KCp^* \xrightarrow{\text{THF}} (thf)_2Cp^*Sm(\mu\text{-}I)_2SmCp^*(thf)_2 \quad (44)$$

$$\{Yb[N(SiMe_3)_2]_2\}_2 \xrightarrow[\text{rt, 16h}]{\text{+ 2 HOC}t\text{Bu}_3\text{, hexane, - 2 HN(SiMe}_3)_2} (Me_3Si)_2N\text{–}Yb(\mu\text{-}OCtBu_3)_2Yb\text{–}N(SiMe_3)_2 \quad (45)$$

$$2\,Lu[N(SiMe_3)_2]_3 + 6\,HOCMe_2CH_2OMe \xrightarrow[\text{rt, 16h}]{n\text{-hexane, - 6 HN(SiMe}_3)_2} (\eta^2\text{-}OR)Lu(\mu,\eta^2\text{-}OR)_3Lu(\eta^1\text{-}OR)_2 \quad (46)$$

Self Assembly – Rings and Clusters

In the absence of strongly coordinating donor molecules, specific ligand arrangements can direct the formation of higher agglomerated systems (Fig. 20). Depending on the Ln/Cp ratio, the formation of rings (Cp/Ln=2) and clusters (Cp/Ln<2) can be observed in cyclopentadienyl (pseudo-)halide complexes [279]. $[Cp^*_2Sm(\mu\text{-}CN)]_6$ obtained by a thermal hetero-ligand degradation reaction involving $N(SiMe_3)CH(Ph)(N{=}CHPh)$ is composed of a 18-membered ring, exhibiting a S_6-symmetric chair conformation [280]. In the presence of coordinating isonitrile ligands $[Cp^*_2Sm(\mu\text{-}CN)(CN\textit{c}\text{-}C_6H_{11})]$ is formed featuring a 9-membered ring [281]. Lower Cp/Ln ratios inevitably create further coordination sites at the metal center and lead to cluster formation. $[Cp^*_6Yb(\mu\text{-}F)_4]$ (Cp/Ln=1.2) is a rare example of a ring-cluster hybrid molecule [282]. A highly symmetric icosaeder arrangement of the metal centers was observed in $[Cp_{12}Sm_{12}(\mu\text{-}Cl)_{24}]$ (Cp/Ln=1) containing a centered Cl_4 tetrahedron [283]. Ring formation is less predictable in alkoxide chemistry, as revealed by the decameric constitution of $[Y(OCH_2CH_2OMe)_3]_{10}$ [284]. In the presence of small charge densities such as Cl, OH, and O alkoxide, cluster formation is the preferred agglomerization pathway [285]. A few cluster compounds featuring only amide ligands have been reported [97].

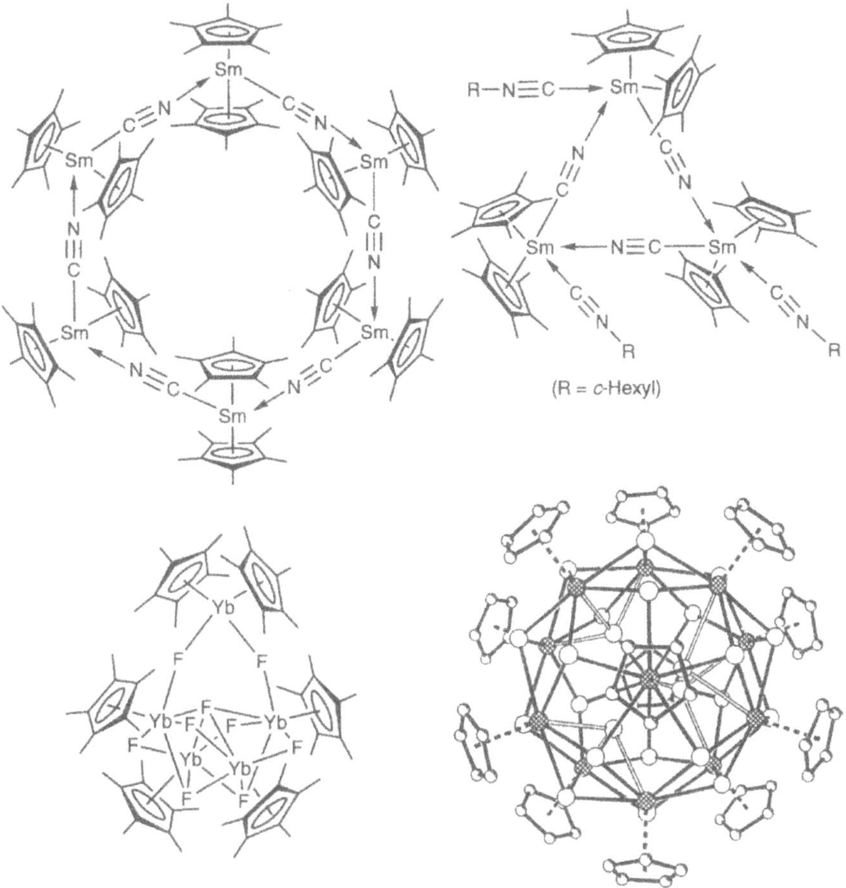

Fig. 20. Self assembly of cyclopentadienyl derivatives – rings and clusters

5.3 Ligand Exchange and Redistribution Reactions

The vast majority of ligand exchange reactions such as salt metathesis, amine elimination, and hydrogenolysis have already been addressed to in the previous sections. The high ligand exchange ability is a peculiar feature of lanthanide complexes and of fundamental importance for their catalytic application. Even the thermodynamically very strong lanthanide alkoxide and amide bonds are kinetically labile as found in transamination and transalcoholysis reactions [158]. Donor ligand exchange as a rule occurs via dissociation processes and the exchange rate of water has been studied in detail in the context of biologically and medically relevant processes [286]. Counter ligand exchange proceeds via

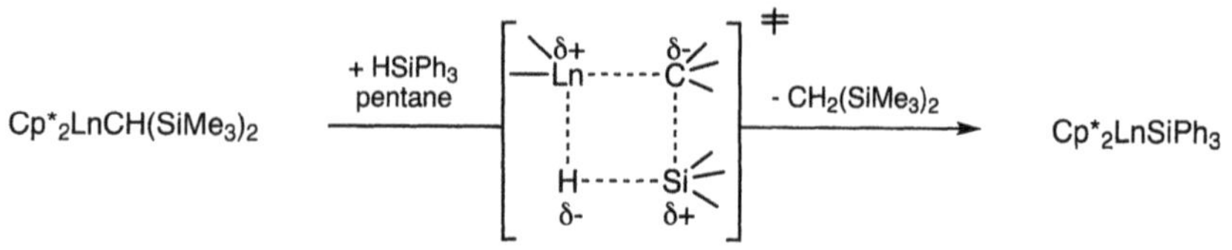

$$Cp^*_2LnCH(SiMe_3)_2 \xrightarrow[\text{pentane}]{+\,HSiPh_3} \left[\begin{array}{ccc} \overset{\delta+}{Ln} & \cdots & \overset{\delta-}{C} \\ \vdots & & \vdots \\ \underset{\delta-}{H} & \cdots & \underset{\delta+}{Si} \end{array}\right]^{\ddagger} \xrightarrow{-\,CH_2(SiMe_3)_2} Cp^*_2LnSiPh_3$$

Scheme 15. Ln-C/Si-H transposition in the σ-bond metathesis reaction

$$[Me_2Si(NtBu)(OtBu)]_2YCH(SiMe_3)_2 \xrightarrow[50\,°C,\,3d]{H_2,\,c\text{-hexane}} \left[\{[Me_2Si(NtBu)(OtBu)]_2Y(\mu\text{-}H)\}_2\right] \longrightarrow Y[Me_2Si(NtBu)(OtBu)]_3$$

Scheme 16. Ligand redistribution (disproportionation) in a heteroleptic alkyl complex

$$\{[C_5H_4(SiMe_3)]Y(OtBu)_2\}_2 + 2\ LiN(SiMe_3)_2 \xrightarrow{THF} \xrightarrow{\text{hexane extract}} [C_5H_4(SiMe_3)]_2Y(OtBu)_2Li(thf)_2 + Y[N(SiMe_3)_2]_3$$

Scheme 17. Ligand redistribution in a heteroleptic cyclopentadienyl complex

σ-bond metathesis involving four-centered transition states. Such converted processes are often the proposed key intermediates (turnover-limiting steps) in hydrogenolysis reactions, and are shown in Scheme 15 for the Si-H/Ln-C transposition of the catalytic hydrosilylation reaction [144]. The pronounced electrophilicity of the lanthanide centers mediates the activation of C–H bonds of a range of hydrocarbons including methane [Eq. (47)] [287]. C-H activation of propene by organolanthanide hydride systems leads to allyl complexes, an important deactivation step in organolanthanide-catalyzed propene polymerization [Eq. (48)] [147,288]. Thermal treatment of $Cp^*_2LaCH(SiMe_3)_2$ led to agglomerization via intramolecular C-H activation [Eq. (49)] (cf. Sect. 5.5) [289].

$$Cp^*_2LuCH_3 \;+\; {}^{13}CH_4 \xrightleftharpoons{\text{benzene}} Cp^*_2Lu^{13}CH_3 \;+\; CH_4 \tag{47}$$

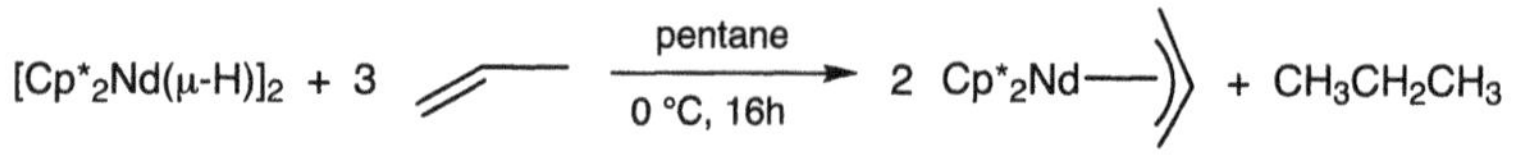

$$[Cp^*_2Nd(\mu\text{-}H)]_2 \;+\; 3\ \text{propene} \xrightarrow[0\,°C,\,16h]{\text{pentane}} 2\ Cp^*_2Nd(\eta^3\text{-allyl}) \;+\; CH_3CH_2CH_3 \tag{48}$$

$$4\ Cp^*_2CeCH(SiMe_3)_2 \xrightarrow[125\,°C,\,24h]{c\text{-hexane},\ -4\,CH_2(SiMe_3)_2} \{Cp^*_3Ce_2[\mu_3\text{-}\eta^5,\eta^1,\eta^1\text{-}C_5Me_3(CH_2)_2]\}_2 \tag{49}$$

Ligand redistribution (scrambling, disproportionation) can occur in hetereoleptic compounds, the formation of the most stable homoleptic derivatives being the driving force [Schemes 16 and 17, Eq. (50)] [139,215,290].

$$1.5\,\{Eu[N(SiMe_3)_2]_2(\mu\text{-}Cl)(thf)\}_2 \xrightarrow{\text{toluene}} 2\,Eu[N(SiMe_3)_2]_3 + EuCl_3(thf)_x \quad (50)$$

5.4 Insertion Reactions

The formal insertion of double and triple bonds into reactive Ln–H, Ln–C, and Ln–N bonds is of fundamental significance in a large number of catalytic reactions. Direct insertion of olefins such as propene into a Lu–Me bond has been detected as part of the "lanthanide model for Ziegler–Natta polymerization" [Eq. (51)] [291]. Alkene and alkyne insertion is also a key step in the hydroamination/cyclization reaction of *N*-unprotected aminoolefins [292]. Insertion of polarized double bonds such as carbon monoxide and acetonitrile often leads to highly functionalized reaction products. Carbon monoxide insertion into lanthanide alkyl bonds can occur in a single and multiple fashion with the formation of η^2-acyl and enedione diolate moieties, respectively (Scheme 18) [293]. The insertion of the metal-bonded carbonyl of $CpCo(CO)_2$ into a Sc–Me bond led to a heterobimetallic system [Eq. (52)] [294]. A non-classical carbocation species was isolated from the insertion of CO into one cyclopentadienyl moiety of the sterically crowded complex $SmCp^*_3$ [Eq. (53)] [295]. This is a remarkable reaction promoted by steric constraints. A "Sc-Si(SiMe$_3$)$_3$" moiety mediates the insertion and coupling of the isocyanide CN(Xyl) (Scheme 19) [296]. The subsequent rapid intramolecular rearrangement reactions probably proceed via a reactive silene intermediate.

$$Cp^*_2Lu{-}CH_3 + H_2C{=}C(H)CH_3 \xrightarrow{c\text{-hexane}} Cp^*_2Lu{-}CH_2{-}CH(CH_3){-}CH_3 \quad (51)$$

$$Cp^*_2Sc{-}CH_3 + CpCo(CO)_2 \xrightarrow[\text{rt, 12h}]{\text{toluene}} CpCo(C{\equiv}O)(C(CH_3){-}O)ScCp^*_2 \quad (52)$$

$$Cp^*_3Sm + 2\,CO \xrightarrow[\text{rt, 5min}]{\text{toluene}} Cp^*_2Sm^{\ominus}(O)_2[\text{carbocation}]^{\oplus} \quad (53)$$

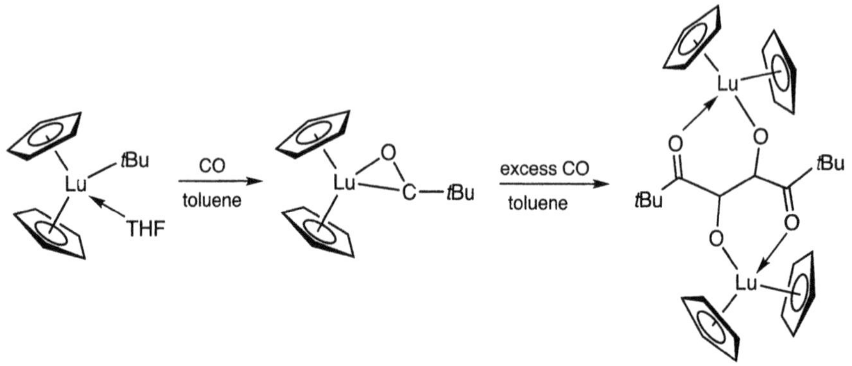

Scheme 18. Single and multiple carbon monoxide insertion into a lutetium alkyl bond

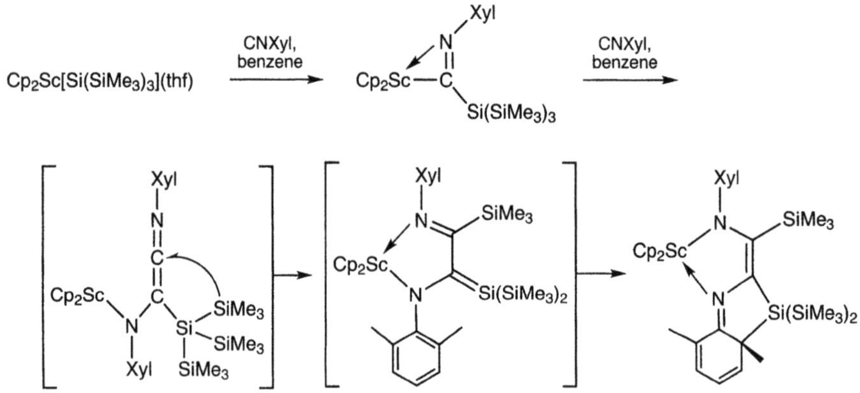

Scheme 19. Double insertion of CN(Xyl) into a scandium silicon bond (Xyl=$C_6H_3Me_2$-2,6)

5.5 Elimination Reactions – Ligand Degradation

β-H and β-alkyl elimination have been reported as major decomposition routes of lanthanide alkyl bonds. These extrusion reactions were initially observed in Cp^*_2Ln-R catalysts and can be considered as models for chain termination occuring during propene polymerization [Eqs. (54) and (55)] [297]. The intramolecular ligand metallation and concomitant hydrocarbon extrusion found in amide templated complexes depend on the metal size and alkyl ligand (Scheme 20) [298]. Silylamine fragmentation has been observed along with the synthesis of the alkoxide complexe $Tm(OCtBu_3)_3$ according to the silylamide route [Eq. (56)] [114]. In corresponding exchange reactions with excess of fluorinated alcohols HOR^F, ammonia is the final degradation product of the si-

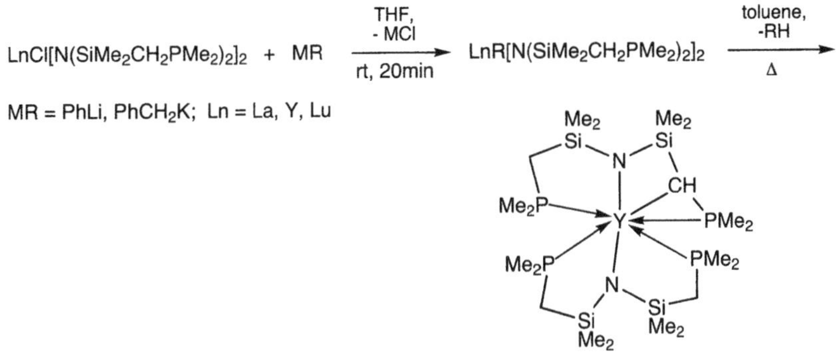

$$LnCl[N(SiMe_2CH_2PMe_2)_2]_2 + MR \xrightarrow[\text{rt, 20min}]{\text{THF, - MCl}} LnR[N(SiMe_2CH_2PMe_2)_2]_2 \xrightarrow[\Delta]{\text{toluene, -RH}}$$

MR = PhLi, $PhCH_2K$; Ln = La, Y, Lu

Scheme 20. Thermally induced intramolecular alkane (benzene) elimination

lylamine, trapped as an ammine complex. The bulkiness and the basicity of the alkoxide ligand, and the type of solvent, effect the ammonia formation [299]. Due to its relevance to the conversion of oxofunctionalized substrates the unexpected cleavage of acetylacetone in the presence of an yttrium alkoxide is shown in Eq. (57) [300].

$$Cp^*_2LuCH_2CH(CH_3)_2 \xrightleftharpoons{c\text{-hexane}} Cp^*_2LuH + CH_2{=}C(CH_3)_2 \quad (54)$$

$$Cp^*_2LuCH_2CH(CH_3)_2 \xrightleftharpoons{c\text{-hexane}} Cp^*_2LuCH_3 + CH_2{=}CH(CH_3) \quad (55)$$

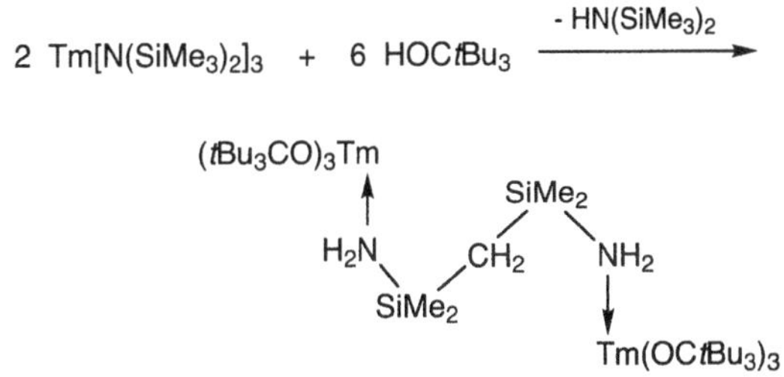

$$2\ Tm[N(SiMe_3)_2]_3 + 6\ HOCtBu_3 \xrightarrow{-\ HN(SiMe_3)_2} \quad (56)$$

$$Y_5O(OiPr)_{13} + \text{excess Hacac} \xrightarrow[20\ °C,\ 24\ h]{\text{toluene}} \quad (57)$$

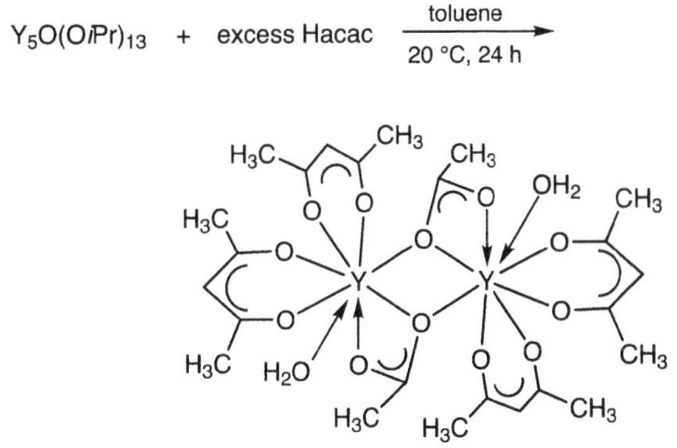

5.6 Redox Chemistry

Kagan and co-workers pioneered the work on the reductive behavior of the low oxidation states of the lanthanide elements in organic synthesis [2b]. Ln metals and Ln(II) derivatives were subsequently found to promote a number of important individual reactions [301]. "The combination of one- and two-electron chemistry sets SmI_2 apart from virtually every other reductive coupling agent currently available" and exhibits exceptional properties for sequential conversions tolerating unprotected functional groups [1b].

Evans and co-workers have worked out the peculiar reducing ability of Sm(II) with the corresponding organometallic reagents, in particular $Cp^*{}_2Sm(thf)_x$, and characterized many metal-bonded products by X-ray crystallography [32]. Their standard reagent is readily oxidized by oxygen to form an oxo-bridged dimer [Eq. (58)] [302]. In contrast, the tris(3,5-dimethylpyrazole)hydroborate ligand produces a superoxo complex under an atmosphere of oxygen [Eq. (59)] [303]. Again, steric and electronic constraints at the metal center, induced by the ligand environment, seem to force this different reaction behavior. Examples of the unique reductive potential of $Cp^*{}_2Sm$ include the functionalization of unsaturated hydrocarbon substrates with carbon monoxide (Scheme 21) [304, 305].

$$Cp^*{}_2Sm(thf)_2 + O_2(g) \xrightarrow[rt]{toluene} Cp^*{}_2Sm{-}O{-}SmCp^*{}_2 \quad (58)$$

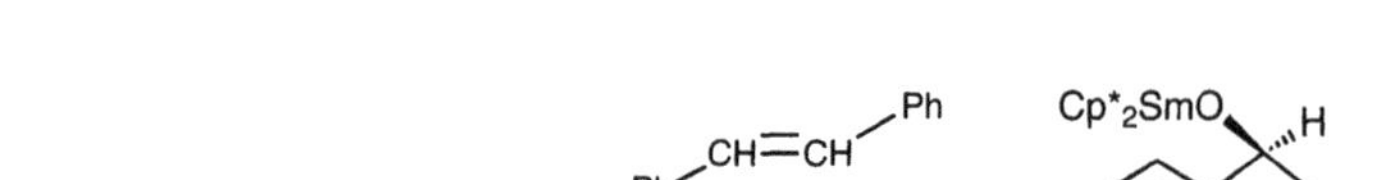

$$Sm(TpMe_2\text{-}3,5)_2 + O_2(g) \xrightarrow[-78\,°C \rightarrow rt]{toluene} (TpMe_2\text{-}3,5)Sm(\eta^2\text{-}O_2) \quad (59)$$

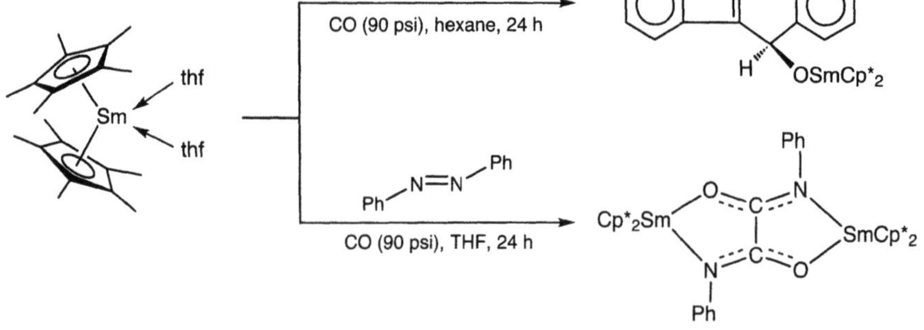

Scheme 21. The reductive potential of $Cp^*{}_2Sm(thf)_2$

The reaction of one-electron reducing agents with ketones yields radical anions (ketyls) which are key intermediates in a variety of carbonyl group transformations. Reduction of aromatic ketones by low-valent samarium and ytterbium compounds [Ln metal and Ln(II)] allowed the first trapping of these radicals in the coordination sphere of a metal [306]. The isolation of an Yb(II) benzophenone dianion [Eq. (60)] [307], a lanthanoid-imine azametallacyclopropane complex [Eq. (61)] [308], and a heteroleptic fluorenone ketyl organosamarium(III) complex [Eq. (62)] [309] by Hou and co-workers emphasize the usefulness of hexamethylphosphoramide (HMPA) and sterically demanding groups as stabilizing ligands. The reaction of La[η^5-$C_5H_3(SiMe_3)_2$-1,3] with a potassium mirror in dimethoxyethane produced a lanthanocene(III) methoxide complex via persistent paramagnetic La(II) intermediates (Scheme 22) [310].

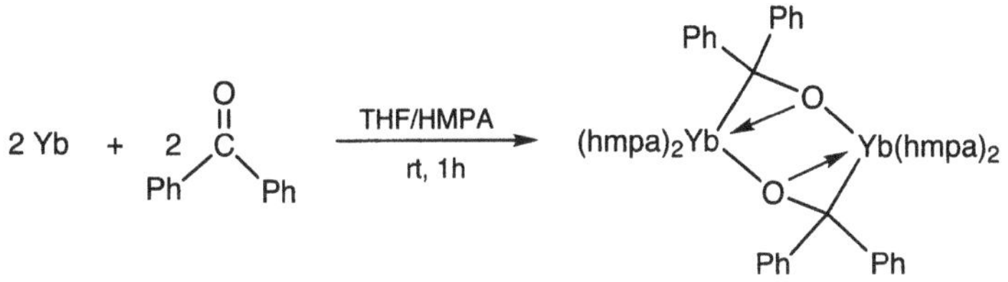

2 Yb + 2 $Ph_2C{=}O$ $\xrightarrow[\text{rt, 1h}]{\text{THF/HMPA}}$ $(hmpa)_2Yb(\mu\text{-}OCPh_2)_2Yb(hmpa)_2$ (60)

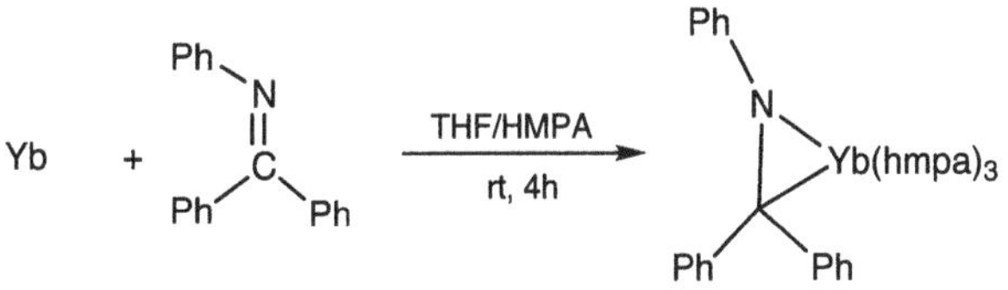

Yb + $PhN{=}CPh_2$ $\xrightarrow[\text{rt, 4h}]{\text{THF/HMPA}}$ $Ph_2C(NPh)Yb(hmpa)_3$ (61)

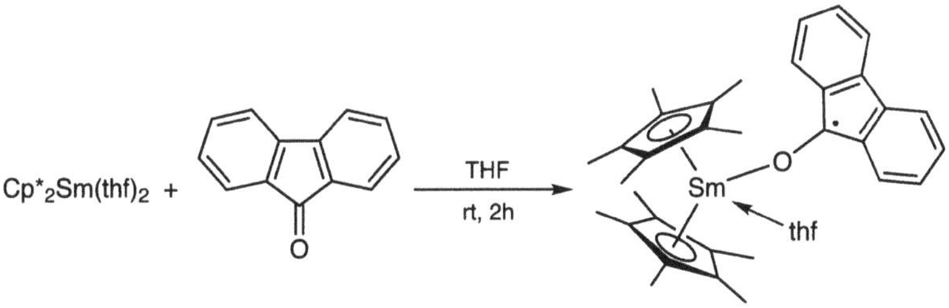

$Cp^*_2Sm(thf)_2$ + fluorenone $\xrightarrow[\text{rt, 2h}]{\text{THF}}$ $Cp^*_2Sm(O\text{-fluorenyl})(thf)$ (62)

$LaCp''_3$ $\xrightarrow[\text{1. -40 °C, 3h; 2. rt, 45h}]{\text{K(mirror), DME}}$ $[K(dme)_x][LaCp''_3]$ $\rightleftharpoons$ $[LaCp''_2(dme)_y]$ + $[KCp''(dme)_z]$

$\xrightarrow{\text{- KCp''}}$ $[LaCp''_2(\mu\text{-}OMe)]_2$ + ?

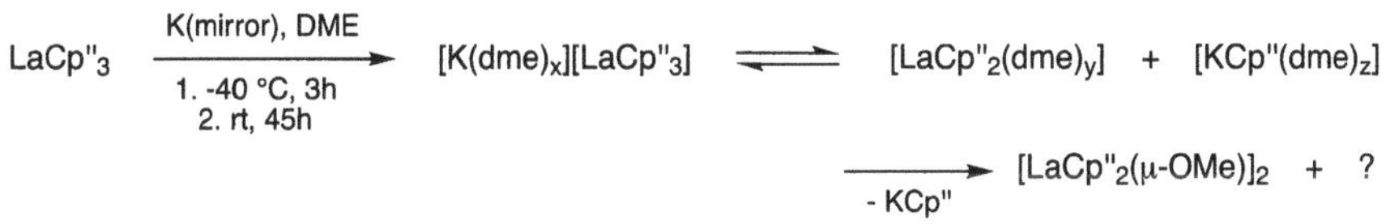

Scheme 22. Generation of low-valent lanthanum reaction intermediates [Cp''=$C_5H_3(SiMe_3)_2$-1,3]

Only a few complexes containing Ce(IV)–carbon bonds have been structurally characterized so far. Salt metathesis reactions utilizing mixed alkoxide nitrate Ce(IV) precursors [Eq. (63)] [311], as well as the oxidative salt elimination of $K[Ce(COT)]_2$ ate complexes with silver iodide, led to isolable organocerium(IV) species [Eq. (64)] [80e]. The oxidation of the homoleptic Ce(III) alkoxide $Ce(OCtBu_3)_3$ with benzoquinone yielded a dinuclear heteroleptic Ce(IV) complex [Eq. (65)] [312].

$$2\ Ce(OtBu)(NO_3)_3(thf)_2 + 6\ NaCp \xrightarrow{THF} Cp_2Ce(OtBu)_2 + 6\ NaNO_3 + \text{other products} \tag{63}$$

+ Ce OtBu

$$K\{Ce[C_8H_5(SiMe_3)_3\text{-}1,3,6]_2\} + \text{excess AgI} \xrightarrow[\text{rt, 1h}]{\text{THF, - Ag, - KI}} \tag{64}$$

Me3Si SiMe3 SiMe3 Ce Me3Si SiMe3 Me3Si

$$2\ Ce(OCtBu_3)_3 + C_6H_4O_2 \xrightarrow[\text{rt, 30min}]{\text{benzene}} (Bu_3CtO)_3Ce\text{–O–}C_6H_4\text{–O–}Ce(OCtBu_3)_3 \tag{65}$$

5.7 Reaction Sequences – Catalytic Cycles

The vast majority of the reaction pathways outlined in the preceding sections can be rediscovered as basic steps in many organolanthanide-mediated organic transformations. The well-examined mechanistic scenarios shown as follows

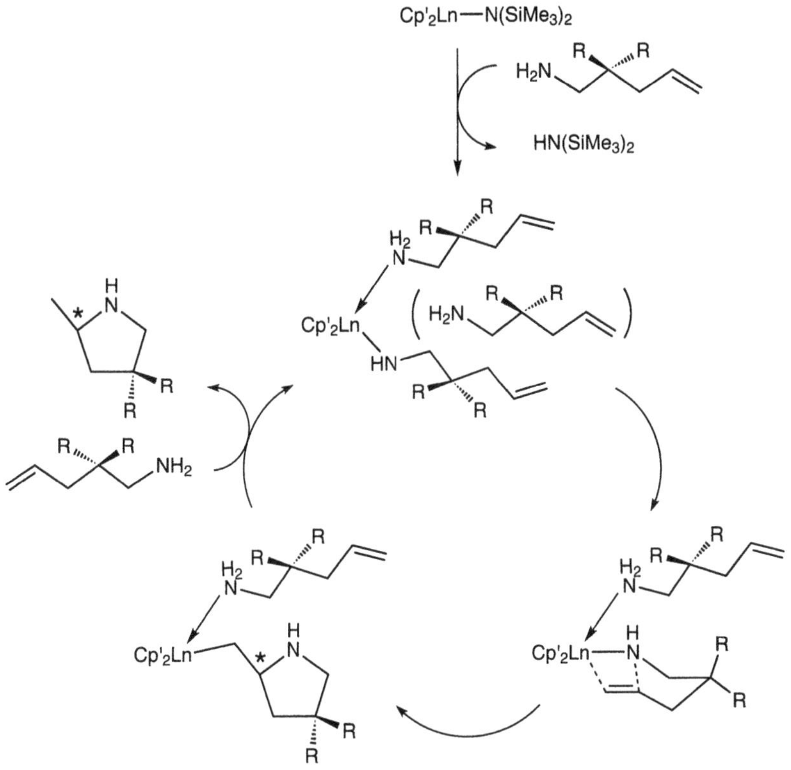

Scheme 23. Organolanthanide-catalyzed hydroamination

with the hydroamination (Scheme 23) [218a,292], Michael addition (Scheme 24) [313], and MMA polymerization reactions (Schemes 25 and 26) [314,315] summarize this section.

5.8 Side Reactions

The intrinsic properties and concomitant high reactivity of organolanthanide moieties often lead to unforeseen and undesired products under the prevailing reaction and storage conditions and, finally, some of these cause accidental reactions such as ether cleavage [Eqs. (66)–(68)] [94, 177,316–318], carbon dioxide insertion [Eq. (69)] [319], or partial hydrolysis [Eq. (70)] [320]. Cluster formation [285] of complexes containing small alkoxide ligands is often observed in the presence of traces of water and proceeds via hydrolysis/dehydration/dealkoxylation processes [Eq. (71)] [321]. In these systems the oxo and hydroxo moieties occupy central cluster positions where they can accomplish high metal coordination (CN=6).

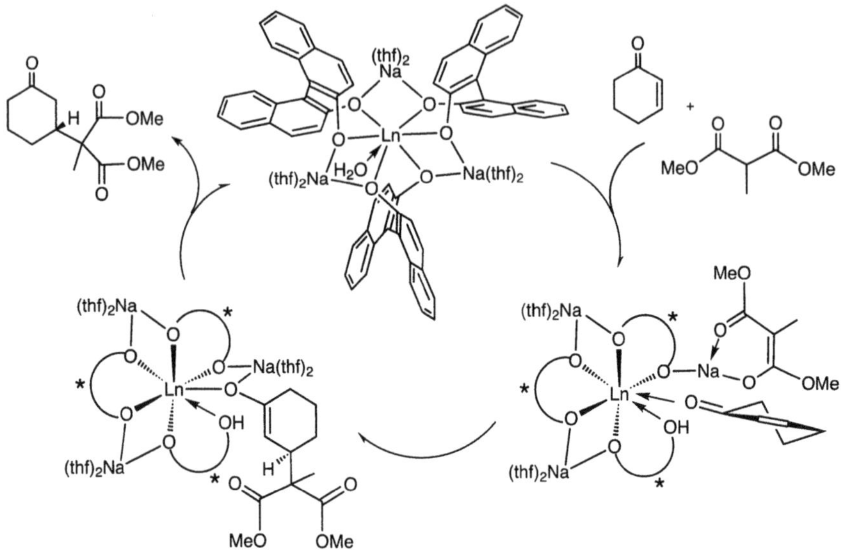

Scheme 24. Heterobimetallic multifunctional catalysis of a Michael reaction by $Na_3[La(S)\text{-}BINOL]_3(thf)_6(H_2O)$

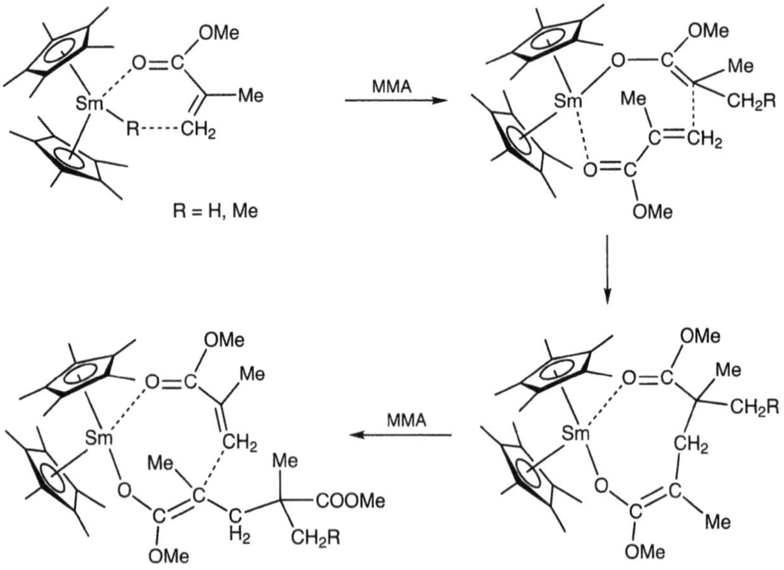

Scheme 25. Organolanthanide(III)-catalyzed methylmethacrylate polymerization

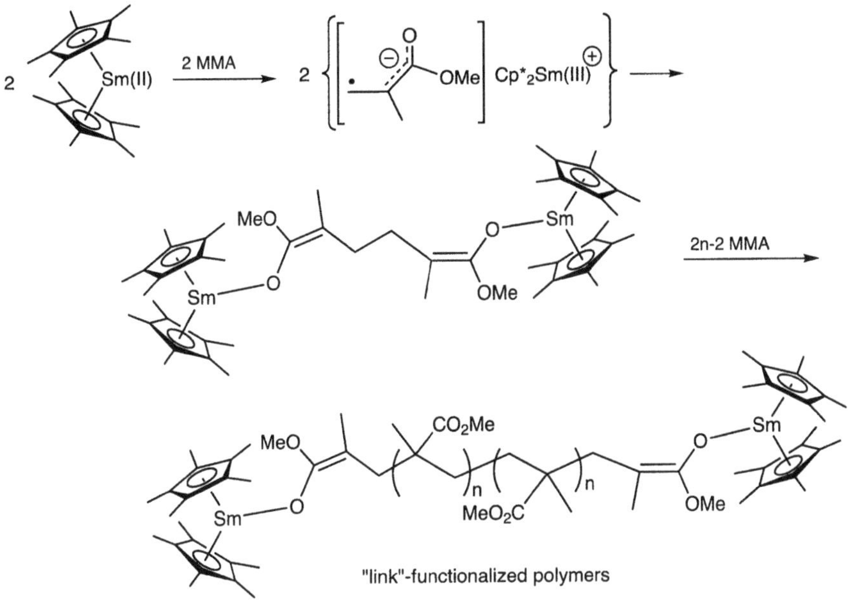

2 MMA

"link"-functionalized polymers

Scheme 26. Organosamarium(II)-initiated methylmethacrylate polymerization

$$Yb[C(SiMe_3)_3]_2 + 2\ Et_2O \xrightarrow[20\ °C]{\text{benzene, } -HC(SiMe_3)_3,\ -C_2H_4} 0.5\ [C(SiMe_3)_3Yb(\mu\text{-}OEt)(OEt_2)]_2 \quad (66)$$

$$Cp_2Yb(CH_3)(thf)\ /\ LiCl(thf)_x \xrightarrow[\text{reflux, 72 h}]{\text{toluene, } -THF} [Cp_2Yb(\mu\text{-}OCH{=}CH_2)]_2 \quad (67)$$

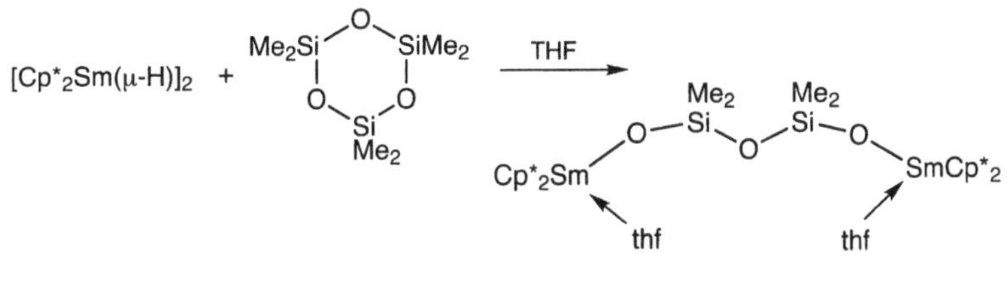

$$[Cp^*_2Sm(\mu\text{-}H)]_2 + (Me_2SiO)_3 \xrightarrow{THF} Cp^*_2Sm(thf){-}O{-}SiMe_2{-}O{-}SiMe_2{-}O{-}SmCp^*_2(thf) \quad (68)$$

$$YCl_3 + 2\ NaCp + Li(CH_2)_3NMe_2 \xrightarrow{\text{THF / } CO_2 \text{ (dry ice)}, -2\ NaCl, -2\ LiCl} Cp_2Y(O_2C(CH_2)_3NMe_2) \quad (69)$$

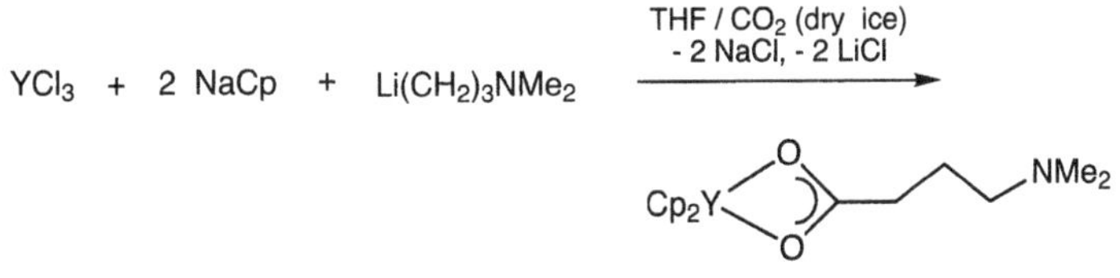

$$2\ (C_5H_4tBu)_3Ln \xrightarrow[\text{-35 °C}]{\text{crystallization, hexane / }H_2O\text{ (traces)}} [(C_5H_4tBu)_2Ln(\mu\text{-}OH)]_2 + 2\ HCptBu \quad (70)$$

$$3\ Ce(OtBu)_4(thf)_2 \xrightarrow[\text{rt, 2-3d}]{\text{toluene, -THF}} Ce_3(OtBu)_{10}O \quad (71)$$

6 Perspectives

Examples of the exceptional and intriguing potential of rare earth reagents in organic synthesis will be treated in comprehensive form in the following chapters of this volume. The text of this chapter on "Principles in Organolanthanide Chemistry" has been directed towards a basic understanding of the chemistry of the most reactive members of this family. Improved and alternative synthetic procedures ensure the availability of both inorganic and organometallic reagents in pure and well-defined form. Each of the important areas of reactivity of organolanthanide compounds which have been addressed to in this survey should prove fertile for further development. Examination of such highly reactive species will provide important details to explain the reaction pathways of the inorganic reagents in organic synthesis by means of spectroscopy and structure determination. So far, the vast majority of active components and reaction intermediates is under-determined due to the application of in situ reaction sequences. The operating system is often a "black-box" and process optimization is achieved by empirical methods. This dearth of data of the active components should further stimulate the interaction between organic synthesis and organometallic chemistry. On the other hand, chiral organometallic and pseudo-organometallic reagents challenge the field of enantioselective catalysis. For this relatively young branch of lanthanide chemistry, ligand design has become indispensable. Since rare earth ligand interactions are ruled by simple principles such as ionic bonding and the HSAB theory, combinatorial chemistry could prove a valuable tool for ligand fine-tuning. In addition to the evaluation of novel ancillary ligand sets, detailed studies on supramolecular rare earth chemistry are to be expected, tackling host-guest interactions [322], the topic of immobilization [235], and dendrimer chemistry [323]. Clearly, organolanthanide chemistry has much to offer to the field of organic synthesis.

7 References

1. For recent reviews, see: (a) Imamoto T (1994) Lanthanides in organic synthesis. Academic Press, London; (b) Molander GA, Harris CR (1996) Chem Rev 96:307; (c) Nair V, Mathew J, Prabhakaran J (1997) Chem Soc Rev 127

2. (a) Evans DF, Fazakerley GV, Phillips RF (1970) Chem Commun 244; (b) Kagan HB, Namy JL (1984) In: Gschneidner KA Jr, Eyring L (eds) Handbook on the physics and chemistry of the rare earths. North-Holland Publishing Company, Amsterdam, chap 50
3. For recent examples, see (a) Li Y, Marks TJ (1998) J Am Chem Soc 120:1757; (b) Gröger H, Saida Y, Sasai H, Yamaguchi K, Martens J, Shibasaki M (1998) J Am Chem Soc 120:3089
4. (a) Schumann H, Meese-Marktscheffel JA, Esser L (1995) Chem Rev 95:865; (b) Edelmann FT (1995) In: Abel EW, Stone FGA, Wilkinson G (eds) Comprehensive organometallic chemistry II, vol 4. Pergamon, Oxford, chap 2
5. Anwander R (1996) Top Curr Chem 179:33
6. Anwander R (1996) Top Curr Chem 179:149
7. For recent reviews, see: (a) Guerriero P, Tamburini S, Vigato PA (1995) Coord Chem Rev 139:17; (b) Alexander V (1995) Chem Rev 95:273
8. Anwander R, Herrmann WA (1996) Top Curr Chem 179:1
9. Klemm (1929) Z Anorg Chem 184:352
10. (a) Morss LR (1976) Chem Rev 76:827; (b) Morss LR (1994) In: Gschneidner KA Jr, Eyring L, Choppin GR, Lander GH (eds) Handbook on the physics and chemistry of rare earths. Elsevier, Amsterdam, chap 122
11. Shannon RD (1976) Acta Cryst A32:751
12. (a) Murad E, Hildenbrand DL (1980) J Chem Phys 73:4005; (b) Liu MB, Wahlbeck PG (1974) High Temp Sci 6:179; (c) Samsonov GV (1982) The oxide handbook, 2nd ed. IFI/Plenum, New York, pp 86–105
13. Husain M, Batra A, Srivastava KS (1989) Polyhedron 8:1233
14. Freeman AJ, Watson RE (1962) Phys Rev 127:2058
15. Bünzli J-CG, Choppin CR (eds) (1989) Lanthanide probes in life, chemical and earth sciences, theory and practice. Elsevier, Amsterdam
16. (a) Peacock RD (1975) Struct Bonding (Berl.) 22:83; (b) Carnall WT (1979) In: Gschneidner KA Jr, Eyring L (eds) Handbook on the physics and chemistry of the rare earths. North-Holland Publishing Company, Amsterdam, chap 24
17. Moeller T (1980) In: Moeller T, Schleitzer-Rust E (eds) Gmelin handbook of inorganic chemistry, Sc, Y, La-Lu rare earth elements, part D1, 8th edn, Springer, Berlin Heidelberg New York, chap 1
18. Moeller T (1973) In: Bailar JC, Emeleus HJ, Nyholm RS, Trotman-Dickenson AF (eds) Comprehensive inorganic chemistry, vol 4. Pergamon, Oxford, chap 44
19. (a) Pearson RG (1963) J Am Chem Soc 85:3533; (b) Pearson RG (ed) (1973) Hard and soft acids and bases. Dowden, Hutchinson, and Ross, Stroudsburg, PA
20. Herrmann WA, Anwander R, Scherer W (1993) Chem Ber 126:1533
21. Raymond KN, Eigenbroth CW Jr (1980) Acc Chem Res 13:276
22. (a) Anwander R (1996) In: Cornils B, Herrmann WA (eds) Applied homogeneous catalysis by organometallic complexes. VCH, Weinheim, p 866; (b) Edelmann FT (1996) Top Curr Chem 179:247
23. Bradley DC, Hursthouse MB, Aspinall HC, Sales KD, Walker NPC (1985) J Chem Soc Chem Commun 1585
24. For further literature dealing with the ionic/covalent bonding behavior in lanthanide complexes, see: (a) Jørgensen CK, Pappalardo R, Schmidtke H-H (1963) J Chem Phys 39:1422; (b) Katzin LI, Barnett ML (1964) J Phys Chem 68:3779; (c) Burns CJ, Bursten BE (1989) Comments Inorg Chem 9:61
25. Johnson DA (1977) Adv Inorg Chem Radiochem 20:1
26. Cloke FGN (1993) Chem Soc Rev 17
27. (a) Kagan HB (1987) Inorg Chim Acta 140:3; (b) Evans WJ (1987) Inorg Chim Acta 139:169; (c) Soderquist JA (1991) Aldrichim Acta 24:15; (d) Sasaki M, Collin J, Kagan HB (1992) New J Chem 16:89

28. For examples, see: (a) Evans DA, Hoveyda AH (1990) J Am Chem Soc 112:6447; (b) Yokoo K, Mine N, Taniguchi H, Fujiwara Y (1985) J Organomet Chem 279:C19
29. For a recent theoretical treatment, see: Seth M, Dolg M, Fulde P, Schwerdtfeger P (1995) J Am Chem Soc 117:6597
30. Thompson LC (1979) In: Gschneidner KA Jr, Eyring L (eds) Handbook on the physics and chemistry of the rare earths. North-Holland Publishing Company, Amsterdam, chap 25
31. (a) Herrmann WA, Anwander R, Scherer W, Munck FC (1993) J Organomet Chem 462:163 and references therein; (b) Bel'skii VK, Gun'ko YK, Soloveichik GD, Bulychev BM (1991) Organomet Chem USSR 4:281
32. Evans WJ (1987) Polyhedron 6:803
33. (a) Bagnell K, Xing-Fu L (1982) J Chem Soc Dalton Trans 1365; (b) Fischer RD, Li X-F (1985) J Less-Common Met 112:303; (c) Li X-F, Guo A-L (1986) Inorg Chim Acta 134:143
34. Tolman CA (1977) Chem Rev 77:313
35. Bondi A (1964) J Phys Chem 68:441
36. Marcalo J, DeMatos AP (1989) Polyhedron 8:2431
37. Piers WE, Shapiro PJ, Bunel EE, Bercaw JE (1990) Synlett 74
38. Kobayashi S, Moriwaki M, Hachiya I (1995) J Chem Soc Chem Commun 1527 and references therein
39. Holton J, Lappert MF, Ballard DGH, Pearce R, Atwood JL, Hunter WE (1979) J Chem Soc Dalton Trans 54
40. Marks TJ, Gagne MR, Nolan SP, Schock LE, Seyam AM, Stern D (1989) Pure Appl Chem 61:1665
41. (a) Nolan SP, Stern D, Marks TJ (1989) J Am Chem Soc 111:7844; (b) King WA, Marks TJ, Anderson DM, Duncalf DJ, Cloke FGN (1992) J Am Chem Soc 114:9221; (c) King WA, Marks TJ (1995) Inorg Chim Acta 229:343
42. Schumann H (1979) Nachr Chem Tech Lab 27:393
43. Sasai H, Suzuki T, Itoh N, Tanaka K, Date T, Okamura K, Shibasaki M (1993) J Am Chem Soc 115:10372
44. Reger DL, Lindeman JA, Lebioda L (1987) Inorg Chim Acta 139:71
45. Sessler JL, Hemmi G, Mody TD, Murai T, Burrell A, Young SW (1994) Acc Chem Res 27:43
46. Shriver DF, Drezdzon MA (1986) In: The manipulation of air-sensitive compounds, 2nd edn. Wiley, New York
47. Choppin GR, Rizkalla EN (1994) In: Gschneidner KA Jr, Eyring L, Choppin GR, Lander GH (eds) Handbook on the physics and chemistry of the rare earths. Elsevier, Amsterdam, chap 128
48. (a) Burgess J, Kijowski J (1981) Adv Inorg Chem Radiochem 24:57; (b) Eick HA (1994) In: Gschneidner KA Jr, Eyring L, Choppin GR, Lander GH (eds) Handbook on the physics and chemistry of the rare earths. Elsevier, Amsterdam, chap 124 and references therein
49. (a) Reed JB, Hopkins BS, Audrieth LF (1939) Inorg Synth 1:28; (b) Taylor MD (1962) Chem Rev 62:503; (c) Meyer, G (1989) Inorg Synth 25:146
50. Evans WJ, Feldman JD, Ziller JW (1996) J Am Chem Soc 118:4581
51. Imamoto T (1991) In: Trost BM, Fleming I, Schreiber SL (eds) Comprehensive organic synthesis. Pergamon, London, chap 1.8
52. Rossmanith K, Auer-Welsbach C (1965) Mh Chem 96:602
53. (a) Evans WJ, Shreeve JL, Ziller JW, Doedens RJ (1995) Inorg Chem 34:576; (b) Willey GR, Woodman TJ, Drew MGB (1997) Polyhedron 16:3385; (c) Deacon GB, Feng T, Junk PC, Skelton BW, Sobolev AN, White AH (1998) Aust J Chem 51:75 and references therein
54. (a) Freeman JH, Smith ML (1958) J Inorg Nucl Chem 7:224; (b) Manzer LE (1982) Inorg Synth 21:135

55. Castellani Bisi C, Coda A (1985) Acta Cryst C41:186
56. Evans WJ, Boyle TJ, Ziller JW (1993) J Am Chem Soc 115:5084
57. Bünzli J-CG, Wessner D (1984) Coord Chem Rev 60:191
58. For examples, see: (a) Rogers RD, Voss EJ, Etzenhouser RD (1988) Inorg Chem 27:533; (b) Willey GR, Lakin MT, Alcock NW (1992) J Chem Soc Chem Commun 1619
59. Deacon GB, Feng T, Nickel S, Skelton BW, White AH (1993) J Chem Soc Chem Commun 1328
60. Deacon GB, Tuong TD, Wilkinson DL (1990) Inorg Synth 27:136
61. Wu S-H, Ding Z-B, Li X-J (1994) Polyhedron 13:2679
62. Belli Dell'Amico D, Calderazzo F, della Porta C, Merigo A, Biagini P, Lugli G, Wagner T (1995) Inorg Chim Acta 240:1
63. Butcher RJ, Clark DL, Grumbine SK, Vincent-Hollis RL, Scott BL, Watkin JG (1995) Inorg Chem 34:5468
64. Rabe GW, Riede J, Schier A (1995) J Chem Soc Chem Commun 577
65. Deacon GB, Koplick AJ (1979) Inorg Nucl Chem Lett 15:263
66. Barnhart DM, Frankcom TM, Gordon PL, Sauer NN, Thompson JA, Watkin JG (1995) Inorg Chem 34:4862
67. Evans WJ, Rabe GW, Ziller JW (1994) Inorg Chem 33:3072
68. Lebrun A, Rantze E, Namy J-L, Kagan HB (1995) New J Chem 19:699
69. Tilley TD, Boncella JM, Berg DJ, Burns CJ, Andersen RA (1990) Inorg Synth 27:146
70. Evans WJ, Gummersheimer TS, Ziller JW (1995) J Am Chem Soc 117:8999
71. Bochkarev MN, Fedushkin IL, Fagin AA, Petrovskaya TV, Ziller JW, Broomhall-Dillard RNR, Evans WJ (1997) Angew Chem 109:123; Angew Chem Int Ed Eng 36:133
72. Jacob K, Thiele K-H (1986) Z Anorg Allg Chem 536:147
73. For an example, see: Depaoli G, Ganis P, Zanonato PL, Valli G (1993) Polyhedron 12:1933
74. For examples, see: (a) Yamanaka I, Nakagaki K, Otsuka K (1995) J Chem Soc Chem Commun 1185; (b) Gong L, Steitwieser A (1990) J Org Chem 55:6235
75. Matsubara S, Onishi H, Utimoto K (1990) Tetrahedron Lett 31:6209
76. Bergmann H (chief ed) (1974): Gmelin handbook of inorganic chemistry, Sc, Y, La-Lu rare earth elements, part C2, 8th edn. Springer, Berlin Heidelberg New York, chap 5.6
77. Bergmann H (chief ed) (1981): Gmelin handbook of inorganic chemistry, Sc, Y, La-Lu rare earth elements, part C8, 8th edn. Springer, Berlin Heidelberg New York, chap 28.7
78. Gradeff PS, Yunlu K, Deming TJ, Olofson JM, Doedens RJ, Evans WJ (1990) Inorg Chem 29:420
79. Greis O, Haschke JM (1982) In: Gschneidner KA Jr, Eyring L (eds) Handbook on the physics and chemistry of the rare earths. North-Holland Publishing Company, Amsterdam, chap 45 and references therein
80. For examples, see: (a) Schumann H, Meese-Marktscheffel JA, Dietrich A (1989) J Organomet Chem 377:C5; (b) Stehr J, Fischer RD (1992) J Organomet Chem 430:C1; (c) Frankland AD, Hitchcock PB, Lappert MF, Lawless GA (1994) J Chem Soc Chem Commun 2435; (d) Sofield CD, Andersen RA (1995) J Org Chem 501:271; (e) Kilimann U, Herbst-Irmer R, Stalke D, Edelmann FT (1994) Angew Chem 106:1684; Angew Chem Int Ed Engl 33:1618; (f) Liu S-Y, Maunder GH, Sella A, Stevenson M, Tocher DA (1996) Inorg Chem 35:76
81. Hamidi MEM, Pascal J-L (1994) Tetrahedron 13:1787
82. Kobayashi S, Hachiya I, Takahori T (1993) Synthesis 371
83. (a) Morris JH, Smith WE (1970) J Chem Soc Chem Commun 245; (b) Marks TJ, Grynkewich GW (1976) Inorg Chem 15:1302; (c) Sagal BG, Lippard SJ (1978) Inorg Chem 17:844; (d) White III JP, Deng H, Shore SG (1991) Inorg Chem 30:2337
84. (a) Arliguie T, Lance M, Nierlich M, Ephritikhine M (1997) J Chem Soc Dalton Trans 2501; (b) Cendrowski-Guillaume S, Nierlich M, Lance M, Ephritikhine M (1998) Organometallics 17:786
85. Thomas RR, Chebolu V, Sen A (1986) J Am Chem Soc 108:4096

86. Evans WJ, Deming TJ, Olofson JM, Ziller JW (1989) Inorg Chem 28:4027
87. Jacob K, Thiele K-H (1986) Z Anorg Allg Chem 543:192
88. Imamoto T, Koide Y, Hiyama S (1990) Chem Lett 1445
89. Brown LM, Mazdiyasni KS (1970) Inorg Chem 9:2783
90. (a) Deacon GB, Pain GN, Tuong TD (1989) Inorg Synth 26:17; (b) Deacon GB, Forsyth CM (1988) Inorg Chim Acta 154:121; (c) Cetinkaya B, Hitchcock PB, Lappert MF, Smith RG (1992) J Chem Soc Chem Commun 932
91. For an example, see: Wayda AL, Dye JL, Rogers RD (1984) Organometallics 3:1605
92. (a) Mashima K, Sugiyama H, Nakamura A (1994) J Chem Soc Chem Commun 1581; (b) Mashima K, Nakayama Y, Fukumoto H, Kanehisa N, Kai Y, Nakamura A (1994) J Chem Soc Chem Commun 2523
93. (a) Fukagawa T, Fujiwara Y, Yokoo K, Taniguchi H (1981) Chem Lett 1771; (b) Hou Z, Fujiwara Y, Jintoku T, Mine N, Yokoo K, Taniguchi H (1987) J Org Chem 3524
94. Eaborn C, Hitchcock PB, Izod K, Smith JD (1994) J Am Chem Soc 116:12071
95. (a) Brennan JG, Cloke FGN, Sameh AA, Zalkin A (1987) J Chem Soc Chem Commun 1668; (b) Arnold PL, Cloke FGN, Hitchcock PB, Nixon JF (1996) J Am Chem Soc 118:7630; (c) Arnold PL, Cloke FGN, Hitchcock PB (1997) Chem Commun 481; (d) Arnold PL, Cloke FGN, Nixon JF (1998) Chem Commun 797
96. For an example, see: Atwood JL, Lappert MF, Smith RG, Zhang H (1988) J Chem Soc Chem Commun 1308
97. Minhas RK, Ma Y, Song J-I, Gambarotta S (1996) Inorg Chem 35:1866
98. Edelmann FT, Steiner A, Stalke D, Gilje JW, Jagner S, Håkansson M (1994) Polyhedron 13:539
99. Bradley DC, Ghotra JS, Hart FA (1973) J Chem Soc Dalton Trans 1021
100. Lappert MF, Singh A, Smith RG (1990) Inorg Synth 27:164
101. Hitchcook PB, Lappert MF, Singh A (1983) J Chem Soc Chem Commun 1499
102. Herrmann WA, Anwander R, Kleine M, Scherer W (1992) Chem Ber 125:1971
103. Bochkarev LN, Shustov SB, Guseva TV, Zhil'tsov SF (1988) J Gen Chem USSR 58:819
104. (a) Booij M, Kiers NH, Heeres HJ, Teuben JH (1989) J Organomet Chem 364:79; (b) Stults SD, Andersen RA, Zalkin A (1990) Organometallics 9:115
105. Mu Y, Piers WE, MacDonald M-A, Zaworotko MJ (1995) Can J Chem 73:2233
106. Razuvaev GA, Kalinina GS, Fedorova EA (1980) J Organomet Chem 190:157
107. Evans WJ, Ansari MA, Ziller JW, Khan SI (1996) Inorg Chem 35:5435
108. Allen M, Aspinall HC, Moore SR, Hursthouse MB, Karvalov AI (1992) Polyhedron 11:409
109. Stecher HA, Sen A, Rheingold AL (1989) Inorg Chem 28:3280
110. McGeary MJ, Coan PS, Folting K, Streib WE, Caulton KG (1989) Inorg Chem 28:3283
111. Barash EH, Coan PS, Lobkovsky EB, Streib WE, Caulton KG (1993) Inorg Chem 32:497
112. (a) Rad'kov YF, Fedorova EA, Khorshev SY, Kalinina GS, Bochkarev MN, Razuvaev GA (1986) J Gen Chem USSR 55:1911; (b) Cary DR, Arnold J (1993) J Am Chem Soc 115:2520; (c) Strzelecki AR, Likar CL, Helsel BA, Utz T, Lin MC, Bianconi PA (1994) Inorg Chem 33:5188
113. Fedorova EA, Kalinina GS, Bochkarev MN, Razuvaev GA (1982) J Gen Chem USSR 52:1041
114. Herrmann WA, Anwander R, Munck FC, Scherer W, Dufaud V, Huber NW, Artus GRJ (1994) Z Naturforsch B49:1789
115. Anwander R, Runte O, Eppinger J, Gerstberger G, Herdtweck E, Spiegler M (1998) J Chem Soc Dalton Trans 847
116. Runte O, Priermeier T, Anwander R (1996) Chem Commun 1385
117. Herrmann WA, Eppinger J, Spiegler M, Runte O, Anwander R (1997) Organometallics 16:1813
118. Görlitzer HW, Spiegler M, Anwander R (1998) Eur J Inorg Chem 1009
119. Aspinall H, Moore SR, Smith AK (1993) J Chem Soc, Dalton Trans 993

120. Evans WJ, Anwander R, Ziller JW, Khan SI (1995) Inorg Chem 34:5927
121. Mu Y, Piers WE, MacQuarrie DC, Zaworotko MJ, Young VG Jr (1996) Organometallics 15:2720
122. Schumann H, Müller J (1979) J Organomet Chem 169:C1
123. Evans WJ, Drummond DK, Zhang H, Atwood JL (1988) Inorg Chem 27:575
124. Evans WJ, Anwander R, Berlekamp UH, Ziller JW (1995) Inorg Chem 34:3583
125. Evans WJ, Anwander R, Ansari MA, Ziller JW (1995) Inorg Chem 34:5
126. Evans WJ, Fries G, Anwander R, Ziller JW (unpublished results)
127. For an example, see: van den Hende JR, Hitchcock PB, Lappert MF (1994) J Chem Soc Chem Commun 1413
128. (a) Fischer EO, Fischer H (1964) Angew Chem 76:52; (b) Fischer EO, Fischer H (1965) J Organomet Chem 3:181
129. Hayes RG, Thomas JL (1969) J Am Chem Soc 91:6876
130. Wayda AL, Mukerji I, Dye JL, Rogers RD (1987) Organometallics 6:1328
131. Murphy E, Toogood GE (1971) Inorg Nucl Chem Lett 7:755
132. Pytlewski LL, Howell JK (1967) Chem Commun 1280
133. Howell JK, Pytlewski LL (1969) J Less-Common Met 18:437
134. (a) White JP III, Deng H-B, Shore SG (1989) J Am Chem Soc 111:8946; (b) White JP III, Shore SG (1992) Inorg Chem 31:2756
135. Hitchcock PB, Lappert MF, Smith RG, Bartlett RA, Power PP (1988) J Chem Soc Chem Commun 1007
136. (a) Heeres HJ, Meetsma A, Teuben JH (1988) J Chem Soc Chem Commun 962; (b) Teuben JH, Rogers RD, Heeres HJ, Meetsma A (1989) Organometallics 8:2637; (c) Schaverien CJ, Frijns JHG, Herres HJ, Van den Hende JR, Teuben JH, Spek AL (1991) J Chem Soc Chem Commun 642
137. (a) Schaverien CJ (1992) J Chem Soc Chem Commun 11; (b) Schaverien CJ (1994) Organometallics 13:69
138. Clark DL, Gordon JC, Huffmann JC, Watkin JG, Zwick BD (1994) Organometallics 13:4266
139. (a) Evans WJ, Boyle TJ, Ziller JW (1993) Organometallics 12:3998; (b) Evans WJ, Boyle TJ, Ziller JW (1993) J Organomet Chem 462:141
140. Evans WJ, Boyle TJ, Ziller JW (1993) J Am Chem Soc 115:5084
141. Evans WJ, Ansari MA, Ziller JW (1995) Inorg Chem 34:3079
142. Biagini P, Lugli G, Abis L, Millini R (1994) J Organomet Chem 474:C16
143. (a) Evans WJ, Anwander R, Doedens RJ, Ziller JW (1994) Angew Chem 106:1725; Angew Chem Int Ed Engl 36:1641; (b) Evans WJ, Anwander R, Ziller JW (1995) Organometallics 14:1107
144. Fu P-F, Brard L, Li Y, Marks TJ (1995) J Am Chem Soc 117:7157
145. Watson PL, Parshall GW (1985) Acc Chem Res 18:51
146. (a) Evans WJ (1983) J Organomet Chem 250:217; (b) Evans WJ, Dominguez R, Hanusa TP (1986) Organometallics 5:263; (c) Evans WJ, Drummond DK, Hanusa TP, Doedens RJ (1987) Organometallics 6:2279
147. Jeske G, Schock LE, Swepston PN, Schumann H, Marks T J (1985) J Am Chem Soc 107:8103
148. Soloveichik GL, Knyazhanskii SY, Bulychev BM, Bel'skii VK (1992) Organomet Chem USSR 5:73
149. Soloveichik GL (1995) New J Chem 19:597
150. Evans WJ, Meadows JH, Hunter WE, Atwood JL (1984) J Am Chem Soc 106:1291
151. Schumann H, Genthe W, Hahn Ekkehardt, Hossain MB, Van der Helm D (1986) J Organomet Chem 299:67
152. Hazin PN, Bruno JW, Schulte GK (1990) Organometallics 9:416
153. (a) Evans WJ, Ulibarri TA, Chamberlain LR, Ziller JW, Alvarez Jr D (1990) Organometallics 9:2124; (b) Evans WJ, Seibel CA, Ziller JW (1998) J Am Chem Soc 120:6745

154. (a) Heeres HJ, Meetsma A, Teuben JH (1991) J Organomet Chem 414:351; (b) Schaverien CJ (1992) Organometallics 11:3476
155. Anwander R, Görlitzer HW, Runte O, Priermeier T (unpublished results)
156. Amoroso AJ, Jeffery JC, Jones PL, McCleverty JA, Rees L, Rheingold AL, Sun Y, Takats J, Trofimenko S, Ward MD, Yap GPA (1995) J Chem Soc Chem Commun 1881
157. For reviews, see: (a) Wipf P, Xu W (1995) Tetrahedron 51:4551; (b) Hoveyda AH, Morken JP (1996) Angew Chem 108:1378; Angew Chem Int Ed Eng 35:1263
158. (a) Hubert-Pfalzgraf LG (1987) New J Chem 11:663; (b) Hubert-Pfalzgraf LG (1995) New J Chem 19:727
159. Anwander R, Munck FC, Priermeier T, Scherer W, Runte O, Herrmann WA (1997) Inorg Chem 36:3545
160. Herrmann WA, Anwander R, Munck FC, Scherer W (1993) Chem Ber 126:331
161. Reynolds LT, Wilkinson G (1959) J Inorg Nucl Chem 9:86
162. Buchler JW, Eikelmann G, Puppe L, Rohbock K, Schneehage HH, Weck D (1971) Liebigs Ann Chem 745:135
163. Bradley DC, Faktor MM (1958) Chem Ind (London) 1332
164. Rausch MD, Morlarty KJ, Atwood JL, Weeks JA, Hunter WE, Brittain HG (1986) Organometallics 5:1281
165. Calderazzo F, Pappalardo R, Losi S (1966) J Inorg Nucl Chem 28:987
166. For reviews, see: (a) Reetz MT (1984) Angew Chem 96:542; Angew Chem Int Ed Engl 23:556; (b) Hoveyda AH, Evans DA, Fu GG (1993) Chem Rev 93:1307; (c) Berrisford DJ, Bolm C, Sharpless KB (1995) Angew Chem 107:1159; Angew Chem Int Ed Engl 34:1059
167. (a) Davidson PJ, Lappert MF, Pearce R (1974) Acc Chem Res 7:209; (b) Schumann H (1985) J Less-Commun Met 112:327
168. For examples, see : (a) Hogerheide MP, Jastrzebski JTBH, Boersma J, Smeets WJJ, Spek AL, van Koten G (1994) Inorg Chem 33:4431; (b) Kessler VG, Hubert-Pfalzgraf LG, Halut S, Daran J-C (1994) J Chem Soc Chem Commun 705
169. For average bond lengths, see: Orpen AG, Brammer L, Allen FH, Kennard O, Watson DG, Taylor R (1989) J Chem Soc Dalton Trans S1
170. Birmingham JM, Wilkinson G (1956) J Am Chem Soc 78:42
171. (a) Ernst RD, Cymbaluk TH (1982) Organometallics 1:708; (b) Schumann H, Dietrich A (1991) J Organomet Chem 401:C33; (c) Baudry D, Nief F, Ricard L (1994) J Organomet Chem 482:125
172. (a) Evans WJ, Hughes LA, Hanusa TP (1986) Organometallics 5:1285; (b) Evans WJ, Forrestal KJ, Leman JT, Ziller JW (1996) Organometallics 15:527
173. Atwood JL, Burns JH, Laubereau PG (1973) J Am Chem Soc 95:1830
174. (a) Huang Z, Chen M, Qiu W, Wu W (1987) Inorg Chim Acta 139:203; (b) Taube R, Windisch H, Görlitz FH, Schumann H (1993) J Organomet Chem 445:85
175. Ghotra JS, Hursthouse MB, Welch AJ (1973) J Chem Soc Chem Commun 669
176. Wedler M, Knösel F, Pieper U, Stalke D, Edelmann FT, Amberger H-D (1992) Chem Ber 125:2171
177. Hitchcock PB, Holmes SA, Lappert MF, Tian S (1994) J Chem Soc Chem Commun 2691
178. Buchler JW, De Cian A, Fischer J, Kihn-Botulinski M, Paulus H, Weiss R (1986) J Am Chem Soc 108:3652
179. Stecher H, Sen A, Rheingold AL (1988) Inorg Chem 27:1130
180. Hitchcook PB, Lappert MF, Mackinnon IA (1988) J Chem Soc Chem Commun 1557
181. De Villiers JPR, Boeyens JCA (1972) Acta Cryst B28:2335
182. Terzis A, Mentzafos D, Tajmir-Riahi H-A (1984) Inorg Chim Acta 84:187
183. (a) Rabe GW, Riede J, Schier A (1996) Inorg Chem 35:40; (b) Rabe GW, Riede J, Schier A (1996) Inorg Chem 35:2680
184. Tatsumi K, Amemiya T, Kawaguchi H, Tani K (1993) J Chem Soc Chem Commun 773
185. For a review, see: Hogerheide MP, Boersma J, van Koten G (1996) Coord Chem Rev 155:87

186. Haaland A (1989) Angew Chem 101:1017; Angew Chem Int Ed Engl 28:992
187. For a review, see: Bader A, Lindner E (1991) coord Chem Rev 108:27
188. Gruter G-JM, van Klink GPM, Akkerman OS, Bickelhaupt F (1995) Chem Rev 95:2405
189. Schumann H, Meese-Marktscheffel JA, Dietrich A, Pickardt J (1992) J Organomet Chem 433:241
190. Herbrich T, Thiele K-H, Thewalt U (1996) Z Anorg Allg Chem 622:1609
191. Clegg W, Eaborn C, Izod K, O'Shaughnessy P, Smith JD (1997) Angew Chem 109:2925; Angew Chem Int Ed Eng 36:2815
192. Wayda AL, Rogers RD (1985) Organometallics 4:1440
193. (a) Hogerheide MP, Grove DM, Boersma J, Jastrzebski JTBH, Kooijman H, Spek AL, van Koten G (1995) Chem Eur J 1:343; (b) Hogerheide MP, Boersma J, Spek AL, van Koten G (1996) Organometallics 15:1505
194. Schumann H, Lee PR, Loebel J (1989) Chem Ber 122:1897
195. Fryzuk MD (1992) Can J Chem 70:2839
196. Deacon GB, Forsyth CM, Junk PC, Skelton BW, White AH (1998) J Chem Soc Dalton Trans 1381
197. Herrmann WA, Anwander R, Denk M (1992) Chem Ber 125:2399
198. Shao P, Berg DJ, Bushnell GW (1994) Inorg Chem 33:3452
199. Karsch HH, Ferazin G, Steigelmann O, Kooijman H, Hiller W (1993) Angew Chem 105:1814; Angew Chem Int Ed Engl 32:1739
200. (a) Edelmann FT (1994) J Alloys Compounds 207/208:182; (b) Edelmann FT (1996) Top Curr Chem 179:113
201. (a) Wang B, Deng D, Qian C (1995) New J Chem 19:515 and references therein; (b) Zhang S, Zhung X, Wei G, Chen W, Liu J (1994) Polyhedron 13:2867; (c) Van den Hende JR, Hitchcock PB, Lappert MF, Nile TA (1994) J Organomet Chem 472:79; (d) Jutzi P, Dahlhaus J, Kristen MO (1993) J Organomet Chem 450:C1; (e) Schumann H, Erbstein F, Weimann R, Demtschuk J (1997) J Organomet Chem 536:541
202. (a) Schumann H, Görlitz FH, Hahn FE, Pickardt J, Qian C, Xie Z (1992) Z Anorg Allg Chem 609:131; (b) Qian C, Zhu D (1994) J Chem Soc Dalton Trans 1599; (c) Gräper J, Fischer RD, Paolucci G (1994) J Organomet Chem 471:87; (d) Qian C, Zhu D (1993) J Organomet Chem 445:79; (e) Paolucci G, D'Ippolito R, Ye C, Qian C, Gräper J, Fischer RD (1994) J Organomet Chem 471:97
203. Roesky PW, Stern CL, Marks TJ (1997) Organometallics 16:4705
204. (a) Van de Weghe P, Bied C, Collin J, Marcalo J, Santos I (1994) J Organomet Chem 475:121; (b) Molander GA, Schumann H, Rosenthal ECE, Demtschuk J (1996) Organometallics 15:3817; (c) Trifonov AA, Van de Weghe P, Collin J, Domingos A, Santos I (1997) J Organomet Chem 527:225
205. (a) Schumann H, Meese-Markstscheffel JA, Gorella B, Görlitz FH (1992) J Organomet Chem 428:C27; (b) Deacon GB, Fallon GD, Forsyth CM (1993) J Organomet Chem 462:183; (c) Broussier R, Delmas G, Perron P, Gautheron B, Petersen JL (1996) J Organomet Chem 511:185; (d) Lin G, Wong W-T (1996) J Organomet Chem 523:93
206. Xia J, Zhuang X, Jin Z, Chen W (1996) Polyhedron 15:3399
207. For an example, see: Giardello MA, Conticello VP, Brard L, Sabat M, Rheingold AL, Stern CL, Marks TJ (1994) J Am Chem Soc 116:10212
208. For an example, see: Tsukube H, Shiba H, Uenishi J-I (1995) J Chem Soc Dalton Trans 181
209. For recent examples, see: (a) Tsuie B, Swenson DC, Jordan RF, Petersen JL (1997) Organometallics 16:1392; (b) Scollard JD, McConville DH, Vittal JJ (1997) Organometallics 16:4415; (c) Gibson VC, Kimberley BS, White AJP, Williams DJ, Howard P (1998) Chem Commun 313; (d) Schrock RR, Schattenmann F, Aizenberg M, Davis WM (1998) Chem Commun 199
210. For recent examples, see: (a) Fokken S, Spaniol TP, Okuda J, Sernetz FG, Mülhaupt R (1997) Organometallics 16:4240; (b) Repo T, Klinga M, Pietikäinen P, Leskelä M, Uusitalo A-M, Pakkanen T, Hakala K, Aaltonen P, Löfgren B (1997) Macromolecules 30:171

211. Görls H, Neumüller B, Scholz A, Scholz J (1995) Angew Chem 107:732; Angew Chem Int Ed Eng 34:673
212. Bürgstein MR, Berberich H, Roesky PW (1998) Organometallics 17:1452
213. Duchateau R, van Wee CT, Meetsma A, Teuben JH (1993) J Am Chem Soc 115:4931
214. (a) Long DP, Bianconi PA (1996) J Am Chem Soc 118:12453; (b) Santos I, Marques N (1995) New J Chem 19:551
215. (a) Duchateau R, van Wee CT, Teuben JH (1996) Organometallics 15:2291; (b) Duchateau R, Tuinstra T, Brussee EAC, Meetsma A, van Duijnen PT, Teuben JH (1996) Organometallics 16:3511
216. Covert KJ, Neithamer DR, Zonnevylle MC, LaPointe RE, Schaller CP, Wolczanski PT (1991) Inorg Chem 30:2494
217. Shao P, Berg DJ, Bushnell GW (1994) Inorg Chem 33:6334
218. (a) Giardello MA, Conticelli VP, Brard L, Gagne MR, Marks TJ (1994) J Am Chem Soc 116:10241; (b) Mitchell JP, Hajela S, Brookhart SK, Hardcastle KI, Henling LM, Bercaw JE (1996) J Am Chem Soc 118:1045
219. Hultzsch KC, Spaniol TP, Okuda J (1997) Organometallics 16:4845
220. Chauvin Y, Heyworth S, Olivier H, Robert F, Saussine L (1993) J Organomet Chem 455:89
221. Shah SAA, Dorn H, Roesky HW, Lubini P, Schmidt H-G (1997) Inorg Chem 36:1102
222. Lee L, Berg DJ, Einstein FW, Batchelor RJ (1997) Organometallics 16:1819
223. Schaverien CJ, Meijboom N, Orpen AG (1992) J Chem Soc Chem Commun 124
224. Sasai H, Arai T, Satow Y, Houk KN, Shibasaki M (1995) J Am Chem Soc 117:6194
225. Edelmann FT (1995) New J Chem 19:535
226. Arnold J, Hoffman CG, Dawson DY, Hollander FJ (1993) Organometallics 12:3645
227. (a) Oki AR, Zhang H, Hosmane NS (1991) Organometallics 10:3964; (b) Hosmane NS, Wang Y, Zhang H, Maguire JA, McInnis M, Gray TG, Collins JD, Kremer RK, Bider H, Waldhör E, Kaim W (1996) Organometallics 15:1006
228. (a) Manning MJ, Knobler CB, Hawthorne MF (1988) J Am Chem Soc 110:4458; (b) Bazan GC, Schaefer WP, Bercaw JE (1993) Organometallics 12:2126; (c) Xie Z, Liu Z, Chiu K, Xue F, Mak TCW (1997) Organometallics 16:2460
229. For an example, see: Bednarski M, Danishefsky (1986) J Am Chem Soc 108:7060
230. Xie Z, Wang S, Zhou Z-Y, Mak TCW (1998) Organometallics 17:1907
231. (a) Aspinall HC, Tillotson MR (1996) Inorg Chem 35:2163; (b) Roussel P, Alcock NW, Scott P (1998) Chem Commun 801
232. Spino C, Clouston L, Berg DJ (1996) Can J Chem 74:1762
233. Herrmann WA, Anwander R, Dufaud V, Scherer W (1994) Angew Chem 106:1338; Angew Chem Int Ed Engl 33:1285
234. Feher FJ, Budzichowski TA (1995) Polyhedron 14:3239
235. (a) Robinson AL (1976) Science 194:1261; (b) Sachtler WMH, Zhang Z (1993) Adv Catal 39:129; (c) Marks TJ (1992) Acc Chem Res 25:57
236. (a) Bergbreiter DE, Chen L-B, Chandran R (1985) Macromolecules 18:1055; (b) Yu G, Li Y, Qu Y, Li X (1993) Macromolecules 26:6702
237. (a) Kobayashi S, Nagayama S (1996) J Org Chem 61:2256; (b) Kobayashi S, Nagayama S (1998) J Am Chem Soc 120:2985
238. Kobayashi S, Nagayama S (1996) J Am Chem Soc 118:8977
239. Keller F, Weinmann H, Schurig V (1997) Chem Ber/Recueil 130:879
240. Anwander R, Roesky R (1997) J Chem Soc Dalton Trans 137
241. Anwander R, Palm C (1998) Stud Surf Sci Catal 117:413
242. (a) Imamura H, Konishi T, Sakata Y, Tsuchiya S (1991) J Chem Soc Chem Commun 1527; (b) Imamura H, Konishi T, Sakata Y, Tsuchiya S (1992) J Chem Soc Faraday Trans 88:2251; (c) Imamura H, Konishi T, Sakata Y, Tsuchiya S (1993) J Chem Soc Chem Commun 1852; (d) Imamura H, Suda E, Konishi T, Sakata Y, Tsuchiya S (1995) Chem Lett 215 and references therein

243. (a) Baba T, Koide R, Ono Y (1991) J Chem Soc Chem Commun 691; (b) Baba T, Kim GJ, Ono Y (1992) J Chem Soc Faraday Trans 88:891; (c) Baba T, Hikita S, Koide R, Ono Y, Hanada T, Tanaka T, Yoshida S (1993) I Chem Soc Faraday Trans 89:3177
244. Takaki K, Fujiwara Y (1990) Appl Organomet Chem 4:311
245. Bradley DC, Ghotra JS, Hart FA, Hursthouse MB, Raithby PR (1977) J Chem Soc Dalton Trans 1166
246. Aspinall HC, Moore SR, Smith AK (1992) J Chem Soc Dalton Trans 153
247. Barnhart DM, Clark DL, Gordon JC, Huffman JC, Vincent RL, Watkin JG, Zwick BD (1994) Inorg Chem 33:3487
248. (a) Schumann H, Glanz M, Winterfeld J, Hemling H, Kuhn N, Kratz T (1994) Angew Chem 106:1829; Angew Chem Int Ed Engl 33:1733; (b) Arduengo AJ III, Tamm M, McLAin SJ, Calabrese LC, Davidson F, Marshall WJ (1994) J Am Chem Soc 116:7927; (c) Herrmann WA, Munck FC, Artus GRJ, Runte O, Anwander R (1997) Organometallics 16:682
249. (a) Shapiro PJ, Bunel E, Schaefer WP, Bercaw JE (1990) Organometallics 9:867; (b) Shapiro PJ, Cotter WD, Schaefer WP, Labinger JA, Bercaw JE (1994) J Am Chem Soc 116:4623
250. Daniele S, Hubert-Pfalzgraf LG, Vaissermann J (1995) Polyhedron 14:327
251. Burns CJ, Andersen RA (1987) J Organomet Chem 325:31
252. Stults SD, Andersen RA, Zalkin A (1990) Organometallics 9:115
253. Anwander R, Palm C, Groeger O, Engelhardt G (1998) Organometallics 17:2027
254. van der Heijden H, Schaverien CJ, Orpen AG (1989) Organometallics 8:255
255. Guan J, Jin S, Lin Y, Shen Q (1992) Organometallics 11:2483
256. Hogerheide MP, Jastrzebski JTBH, Boersma J, Smeets WJJ, Spek AL, van Koten G (1994) Inorg Chem 33:4431
257. For an example, see: Haar MD, Stern CL, Marks TJ (1996) Organometallics 15:1765
258. Herrmann WA, Anwander R, Riepl H, Scherer W, Withaker CR (1993) Organometallics 12:4342
259. Evans WJ, Shreeve JL, Broomhall-Dillard RNR, Ziller JW (1995) J Organomet Chem 501:7
260. Schumann H, Genthe W, Bruncks N, Pickardt J (1982) Organometallics 1:1194
261. Deacon GB, Shen Q (1996) J Organomet Chem 511:1
262. Burns CJ, Andersen RA (1987) J Am Chem Soc 109:915
263. Casey CP, Hallenbeck SL, Pollock DW, Landis CR (1995) J Am Chem Soc 117:9770
264. Burns CJ, Andersen RA (1987) J Am Chem Soc 109:941
265. (a) Cotton FA, Schwotzer W (1986) J Am Chem Soc 108:4657; (b) Fan B, Shen Q, Lin Y (1989) J Organomet Chem 376:61
266. Nolan SP, Marks TJ (1989) J Am Chem Soc 111:8538
267. Evans WJ, Ulibarri TA, Ziller JW (1988) J Am Chem Soc 110:6877
268. (a) Evans WJ, Kociok-Köhn G, Ziller JW (1992) Angew Chem 104:1114; Angew Chem Int Ed Engl 31:1081; (b) Evans WJ, Kociok-Köhn G, Leong VS, Ziller JW (1992) Inorg Chem 31:3592
269. (a) Brookhart M, Green MLH (1983) J Organomet Chem 250:395; (b) Brookhart M, Green MLH, Wong L-L (1988) Prog Inorg Chem 36:1
270. Ziegler T, Folga E, Berces A (1993) J Am Chem Soc 115:636
271. Schaverien CJ, Nesbitt GJ (1992) J Chem Soc Dalton Trans 157
272. Piers WE, Bercaw JE (1990) J Am Chem Soc 112:9406
273. Burger BJ, Thompson ME, Cotter WD, Bercaw JE (1990) J Am Chem Soc 112:1566
274. For examples, see: (a) Jeske G, Lauke H, Mauermann H, Swepston PN, Schumann H, Marks TJ (1985) J Am Chem Soc 107:8091; (b) van der Heijden H, Pasman P, de Boer EJM, Schaverien CJ, Orpen AG (1989) Organometallics 8:1459
275. Klooster WT, Lu RS, Anwander R, Evans WJ, Koetzle TF, Bau R (1998) Angew Chem 110:1326; Angew Chem Int Ed Engl 37:1268

276. Barnhart DM, Clark DL, Gordon JC, Huffmann JC, Watkin JG, Zwick BD (1993) J Am Chem Soc 115:8461
277. Tilley TD, Andersen RA, Zalkin A (1982) J Am Chem Soc 104:3725
278. Evans WJ, Grate JW, Choi HW, Bloom I, Hunter WE, Atwood JL (1985) J Am Chem Soc 107:941
279. Anwander R (1998) Angew Chem 110:619; Angew Chem Int Ed Engl 37:599
280. Obora Y, Ohta T, Stern CL, Marks TJ (1997) J Am Chem Soc 119:3745
281. Evans WJ, Drummond DK (1988) Organometallics 7:797
282. Watson PL, Tulip TH, Williams I (1990) Organometallics 9:1999
283. Kretschmer WP, Teuben JH, Troyanov SI (1998) Angew Chem 110:92; Angew Chem Int Ed Eng 37:88
284. Poncelet O, Hubert-Pfalzgraf LG, Daran J-C, Astier R (1989) J Chem Soc Chem Commun 1846
285. Mehrotra RC, Singh A (1996) Chem Soc Rev 1
286. (a) Aime S, Crich SG, Gianolio E, Terreno E, Beltrami A, Uggeri F (1998) Eur J Inorg Chem 1283; (b) Aime S, Botta M, Fasano M, Terreno E (1998) Chem Soc Rev 27:19
287. Watson PL (1983) J Am Chem Soc 105:6491
288. Thompson ME, Baxter SM, Bulls AR, Burger BJ, Nolan MC, Santarsiero BD, Schaefer WP, Bercaw JE (1987) J Am Chem Soc 109:203
289. Booij M, Meetsma A, Teuben JH (1991) Organometallics 10:3246
290. Bradley DC, Hursthouse MB, Aspinall HC, Sales KD, Walker NPC, Hussain B (1989) J Chem Soc Dalton Trans 623
291. Watson PL (1982) J Am Chem Soc 104:337
292. (a) Cagne MR, Stern CL, Marks TJ (1992) J Am Chem Soc 114:275; (b) Li Y, Fu P-F, Marks TJ (1994) Organometallics 13:439
293. Evans WJ, Wayda AL, Hunter WE, Atwood JL (1981) J Chem Soc Chem Commun 706
294. Clair MA St, Santarsiero BD, Bercaw JE (1989) Organometallics 8:17
295. Evans WJ, Forrestal KJ, Ziller JW (1995) J Am Chem Soc 117:12635
296. Campion BK, Heyn RH, Tilley TD (1990) J Am Chem Soc 112:2011
297. Watson PL, Roe DC (1982) J Am Chem Soc 104:6471
298. Fryzuk MD, Haddad TS, Rettig SJ (1991) Organometallics 10:2026
299. Bradley DC, Chudzynska H, Hammond ME, Hursthouse MB, Motevalli M, Wu R (1992) Polyhedron 11:375
300. Poncelet O, Hubert-Pfalzgraf LG, Daran J-C (1990) Polyhedron 9:1305
301. Kagan HB, Sasaki M, Collin J (1988) Pure Appl Chem 60:1725
302. Evans WJ, Grate JW, Bloom I, Hunter WE, Atwood JL (1985) J Am Chem Soc 107:405
303. Zhang X, Loppnow GR, McDonald R, Takats J (1995) J Am Chem Soc 117:7828
304. Evans WJ (1987) In: Suslick KS (ed) High Energy Processes in Organometallic Chemistry, ACS Symposium Series 333, American Chemical Society, Washington, DC, p 278
305. (a) Evans WJ, Hughes LA, Drummond DK, Zhang H, Atwood JL (1986) J Am Chem Soc 108:1722; (b) Evans WJ, Drummond DK (1986) J Am Chem Soc 108:7440
306. Hou Z, Wakatsuki Y (1997) Chem Eur J 3:1005
307. Hou Z, Yamazaki H, Kobayashi K, Fujiwara Y, Taniguchi H (1992) Organometallics 11:2711
308. Makioka Y, Taniguchi Y, Fujiwara Y, Takaki K, Hou Z, Wakatsuki Y (1996) Organometallics 15:5476
309. Hou Z, Fujita A, Zhang Y, Miyano T, Yamazaki H, Wakatsuki Y (1998) J Am Chem Soc 120:754
310. Cassani MC, Lappert MF, Laschi F (1997) Chem Commun 1563
311. Evans WJ, Deming TJ, Ziller JW (1989) Organometallics 8:1581
312. Sen A, Stecher HA, Rheingold AL (1992) Inorg Chem 31:473
313. Sasai H, Arai T, Shibasaki M (1994) J Am Chem Soc 116:1571
314. (a) Yasuda H, Tamai H (1993) Prog Polym Sci 18:1097; (b) Yasuda H, Yamamoto H, Yokota K, Miyake S, Nakamura A (1992) J Am Chem Soc 114:4908

315. Boffa LS, Novak BM (1994) Macromolecules 27:6993
316. Evans WJ, Dominguez R, Hanusa TP (1986) Organometallics 5:1291
317. Evans WJ, Ulibarri TA, Ziller JW (1991) organometallics 10:134
318. Deelman B-J, Booij M, Meetsma A, Teuben JH, Kooijman H, Spek AL (1995) Organometallics 14:2306
319. Schumann H, Meese-Marktscheffel JA, Dietrich A, Görlitz FH (1992) 430:299
320. Herrmann WA, Anwander R, Kleine M, Öfele K, Riede J, Scherer W (1992) Chem Ber 125:2931
321. Yunlu K, Gradeff PS, Edelstein N, Kot W, Shalimoff G, Streib WE, Vaartstra BA, Caulton KG (1991) Inorg Chem 30:2317
322. Piguet C, Bernardinelli G, Hopfgartner G (1997) Chem Rev 97:2005
323. Kawa M, Fréchet JMJ (1998) Chem Mater 10:286

Lanthanide Triflate-Catalyzed Carbon-Carbon Bond-Forming Reactions in Organic Synthesis

Shū Kobayashi

Graduate School of Pharmaceutical Sciences, The University of Tokyo, Hongo, Bunkyo-ku, Tokyo 113-0033, Japan *E-mail: skobayas@mol.f.u-tokyo.ac.jp*

Versatile carbon-carbon bond-forming reactions using lanthanide triflates ($Ln(OTf)_3$) as catalysts are discussed. Lanthanide triflates are new types of Lewis acids different from typical Lewis acids such as $AlCl_3$, BF_3, $SnCl_4$, etc. While most Lewis acids are decomposed or deactivated in the presence of water, lanthanide triflates are stable and works as Lewis acids in water solutions. Many nitrogen-containing compounds such as imines and hydrazones are also successfully activated by using a small amount of $Ln(OTf)_3$. Lanthanide triflates are also excellent Lewis acid catalysts in organic solvents. A catalytic amount of $Ln(OTf)_3$ is enough to complete reactions in most cases. In addition, $Ln(OTf)_3$ can be recovered after reactions are completed and can be reused. Several chiral lanthanide catalysts for asymmetric Diels-Alder, aza Diels-Alder, and 1,3-dipolar cycloaddition reactions are also described.

Keywords: Lanthanide triflate, Lewis acid, Carbon-carbon bond formation, Asymmetric synthesis, Catalysis, Organic synthesis

1 **Introduction** 64

2 **Aldol Reactions** 65

2.1 Aldol Reactions in Aqueous Media 65
2.2 Recovery and Reuse of the Catalyst 72
2.3 Aldol Reactions in Organic Solvents 72
2.4 Aldol Reactions in Micellar Systems 75

3 **Diels-Alder Reactions** 77

4 **Allylation Reactions** 79

5 **Mannich-Type Reactions** 81

5.1 Reactions of Imines with Silyl Enolates 81
5.2 One-Pot Synthesis of β-Amino Esters from Aldehydes 83
5.3 Use of Acylhydrazones as Electrophiles
in Mannich-Type Reactions 86
5.4 Aqueous Mannich-Type Reaction 89

6 Aza Diels-Alder Reactions 91

6.1 Reactions of Imines with Dienes or Alkenes 91
6.2 Three-Components Coupling Reactions of Aldehydes, Amines, and Dienes or Alkenes 93
6.3 Reaction Mechanism 96

7 Asymmetric Diels-Alder Reactions 98

8 Asymmetric Aza Diels-Alder Reactions 106

9 Catalytic Enantioselective 1,3-Dipolar Cycloadditions 109

10 Conclusions 112

11 References and Notes 113

1 Introduction

Lewis acid-catalyzed carbon-carbon bond forming reactions have been of great interest in organic synthesis because of their unique reactivities, selectivities, and for the mild conditions used [1]. While various kinds of Lewis acid-promoted reactions have been developed and many have been applied in industry, these reactions must be carried out under strict anhydrous conditions. The presence of even a small amount of water stops the reaction because most Lewis acids immediately react with water rather than the substrates and decompose or deactivate, and this has restricted the use of Lewis acids in organic synthesis.

On the other hand, lanthanide compounds were expected to act as strong Lewis acids because of their hard character and to have strong affinity toward carbonyl oxygens [2]. Among these compounds, lanthanide triflates ($Ln(OTf)_3$) were expected to be one of the strongest Lewis acids because of the electron-withdrawing trifluoromethanesulfonyl group. Their hydrolysis was postulated to be slow, based on their hydration energies and hydrolysis constants [3]. In fact, while most metal triflates are prepared under strict anhydrous conditions, Lanthanide triflates are reported to be prepared in aqueous solution [4, 5]. The large radius and the specific coordination number of lanthanide(III) are also unique, and many investigations using $Ln(OTf)_3$ as Lewis acids have been performed.

In this paper, use of $Ln(OTf)_3$ as Lewis acid catalysts in carbon-carbon bond-forming reactions in both aqueous and organic solvents is overviewed.

2
Aldol Reactions

2.1
Aldol Reactions in Aqueous Media

The titanium tetrachloride-mediated aldol reaction of silyl enol ethers with aldehydes was first reported in 1973 [6]. The reaction is notably distinguished from the conventional aldol reactions carried out under basic conditions; it proceeds in a highly regioselective manner to afford cross aldols in high yields [7]. Since this pioneering effort, several efficient activators such as trityl salts [8], Cray montmorillonite [9], fluoride anions [10], etc. [11] have been developed to realize high yields and selectivities, and now the reaction is considered to be one of the most important carbon-carbon bond forming reactions in organic synthesis. These reactions are usually carried out under strictly anhydrous conditions. The presence of even a small amount of water causes lower yields, probably due to the rapid decomposition or deactivation of the promoters and the hydrolysis of the silyl enol ethers. Furthermore, the promoters cannot be recovered and reused because they decompose under usual quenching conditions.

On the other hand, the water-promoted aldol reaction of silyl enol ethers with aldehydes was reported in 1986 [12]. While the report that the aldol reactions proceeded without catalyst in water was unique, the yields and the substrate scope were not satisfactory. In 1991 the first example of Lewis acid catalysis in aqueous media was reported; that is the hydroxymethylation reaction of silyl enol ethers with commercial formaldehyde solution using $Ln(OTf)_3$. Formaldehyde is a versatile reagent as one of the most highly reactive C1 electrophiles in organic synthesis [13]. Dry gaseous formaldehyde required for many reactions has some disadvantages because it must be generated before use from solid polymer paraformaldehyde by way of thermal depolymerization and it itself-polymerizes easily [14]. On the other hand, commercial formaldehyde solution, which is an aqueous solution containing 37% formaldehyde and 8–10% methanol, is cheap, easy to handle, and stable even at room temperature. However, the use of this reagent is strongly restricted due to the existence of a large amount of water. For example, the titanium tetrachloride ($TiCl_4$)-promoted hydroxymethylation reaction of silyl enol ethers was carried out by using trioxane as a HCHO source under strict anhydrous conditions [6b, 15]. Formaldehyde water solution could not be used because $TiCl_4$ and the silyl enol ether reacted with water rather than HCHO in that water solution.

The effects of $Ln(OTf)_3$ in the reaction of 1-phenyl-1-trimethylsiloxypropene (**1**) with commercial formaldehyde solution were examined [16]. In most cases the reactions proceeded smoothly to give the corresponding adducts in high yields. The reactions were most effectively carried out in commercial formaldehyde solution-THF media under the influence of a catalytic amount of $Yb(OTf)_3$.

Several examples of the hydroxymethylation reaction of silyl enol ethers with commercial formaldehyde solution are shown in Table 1, and the following char-

Table 1. Reaction of silyl enol ethers with commercial formaldehyde solution catalyzed by $Yb(OTf)_3$

Entry	Silyl enol ether	Product	Yield/%
1	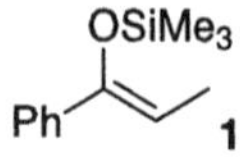 $OSiMe_3$ a) Ph **1**	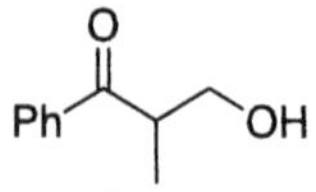 O Ph OH	94
2	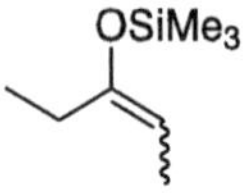 $OSiMe_3$ b)	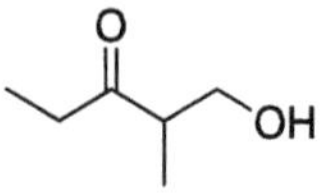 O OH	85
3	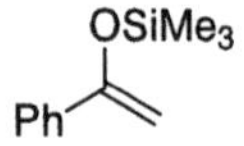 $OSiMe_3$ Ph	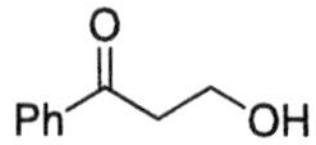 O Ph OH	77
4	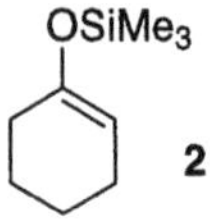 $OSiMe_3$ **2**	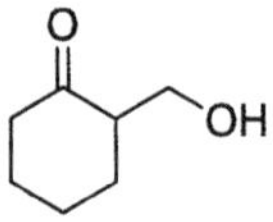 O OH	82
5	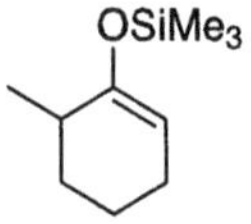 $OSiMe_3$	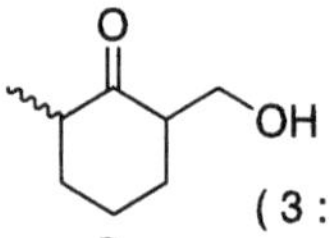 O OH (3 : 2)	86
6	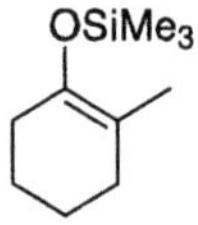 $OSiMe_3$	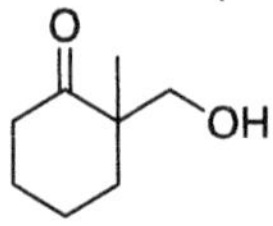 O OH	92
7	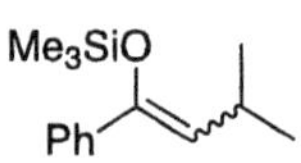 Me_3SiO a) Ph	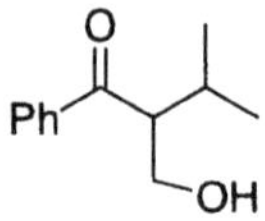 O Ph OH	92
8	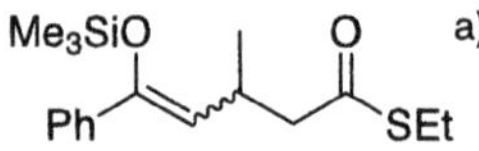 Me_3SiO O a) Ph SEt	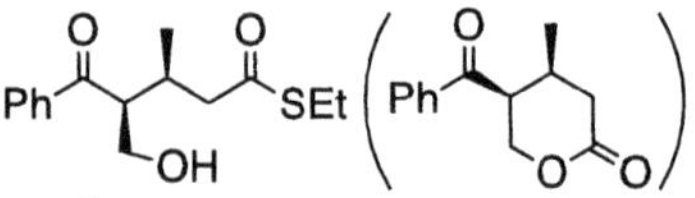 O O Ph SEt OH (O Ph O O) c)	88
9	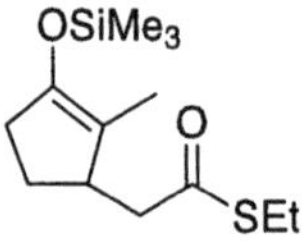 $OSiMe_3$ O SEt	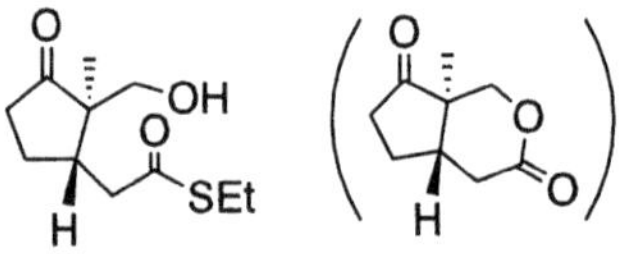 O OH O SEt H (O O O H) d)	83
10	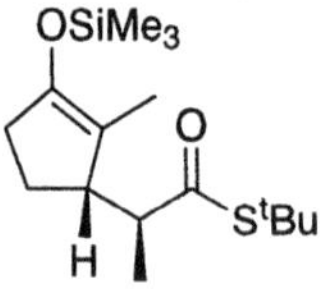 $OSiMe_3$ O S^tBu H	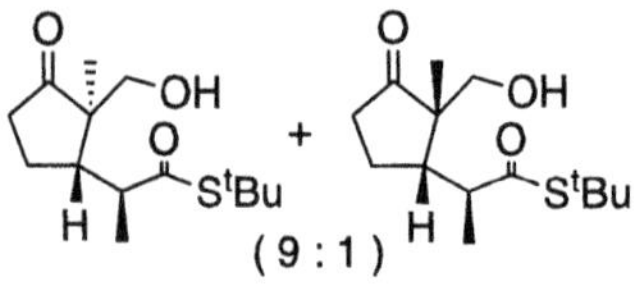 O OH O S^tBu H + O OH O S^tBu H (9 : 1)	90

[a] $Z/E = >98/2$.

[b] $Z/E = 1/4$.

[c] The mixture of the hydroxy thioester and the lactone (2:1) was obtained. The other diastereomers were not observed.

[d] The mixture of the hydroxy thioester and lactone (3:1) was obtained. Less than 3% yield of the other diastereomers was observed.

PhCHO + **2** (OSiMe$_3$) → Ln(OTf)$_3$ (10 mol%), H_2O-THF (1:4), rt, 20 h → Ph, OH, O

Table 2. Effect of Lanthanide Triflates

Ln(OTf)$_3$	Yield/%	Ln(OTf)$_3$	Yield/%
La(OTf)$_3$	8	Dy(OTf)$_3$	73
Pr(OTf)$_3$	28	Ho(OTf)$_3$	47
Nd(OTf)$_3$	83	Er(OTf)$_3$	52
Sm(OTf)$_3$	46	Tm(OTf)$_3$	20
Eu(OTf)$_3$	34	Yb(OTf)$_3$	91
Gd(OTf)$_3$	89	Lu(OTf)$_3$	88

acteristic features of this reaction are noted. (1) In every case, the reactions proceeded smoothly under extremely mild conditions (almost neutral) to give the corresponding hydroxymethylated adducts in high yields. Sterically hindered silyl enol ethers also worked well and the diastereoselectivities were high. (2) Di- and poly-hydroxymethylated products were not observed [17]. (3) The absence of equilibrium (double bond migration) in silyl enol ethers allowed for a regiospecific hydroxymethylation reaction. (4) Only a catalytic amount of Yb(OTf)$_3$ was required to complete the reaction. The amount of the catalyst was examined by taking the reaction of **1** with commercial formaldehyde solution as a model, and the reaction was found to be catalyzed by even 1 mol% of Yb(OTf)$_3$: 1 mol% (90% yield); 5 mol% (90% yield); 10 mol% (94% yield); 20 mol% (94% yield); 100 mol% (94% yield).

The use of Ln(OTf)$_3$ in the activation of aldehydes other than formaldehyde was also investigated [18]. The model reaction of 1-trimethylsiloxycyclohexene (**2**) with benzaldehyde under the influence of a catalytic amount of Yb(OTf)$_3$ (10 mol%) was examined. The reaction proceeded smoothly in H_2O-THF (1:4), but the yields were low when water or THF was used alone. Among several Ln(OTf)$_3$ screened, neodymium triflate (Nd(OTf)$_3$), gadolinium triflate (Gd(OTf)$_3$), Yb(OTf)$_3$, and lutetium triflate (Lu(OTf)$_3$) were quite effective, while the yield of the desired aldol adduct was lower in the presence of lanthanum triflate (La(OTf)$_3$), praseodymium triflate (Pr(OTf)$_3$) or thulium triflate (Tm(OTf)$_3$) (Table 2).

Lanthanide triflates were found to be effective for the activation of formaldehyde water solution. The effect of ytterbium salts was also investigated (Table 3) [19]. While the Yb salts with less nucleophilic counter anions such as OTf^- or ClO_4^- effectively catalyzed the reaction, only low yields of the product were obtained when the Yb salts with more nucleophilic counter anions such as Cl^-, OAc^-, NO_3^-, and SO_4^{2-} were employed. The Yb salts with less nucleophilic coun-

PhCHO + OSiMe$_3$ (cyclohexenyl, **2**) → Yb salt (10 mol%), H_2O-THF (1:4), rt, 19 h → Ph(OH)CH–(2-oxocyclohexyl)

Table 3. Effect of Yb salts

Yb salt	Yield/%
$Yb(OTf)_3$	91 [a)]
$Yb(ClO)_4$	88 [b)]
$YbCl_3$	3
$Yb(OAc)_3 \cdot 8H_2O$	14
$Yb(NO_3)_3 \cdot 5H_2O$	7
$Yb_2(SO_4)_3 \cdot 5H_2O$	trace

[a] *syn/anti* = 73/27
[b] 76/24

ter anions are more cationic and the high Lewis acidity promotes the desired reaction.

Several examples of the present aldol reaction of silyl enol ethers with aldehydes are listed in Table 4. In every case, the aldol adducts were obtained in high yields in the presence of a catalytic amount of $Yb(OTf)_3$, $Gd(OTf)_3$, or $Lu(OTf)_3$ in aqueous media. Diastereoselectivities were generally good to moderate. One feature in the present reaction is that water soluble aldehydes, for instance, acetaldehyde, acrolein, and chloroacetaldehyde, can be reacted with silyl enol ethers to afford the corresponding cross aldol adducts in high yields (entries 5–7). Some of these aldehydes are commercially supplied as water solutions and are appropriate for direct use. Phenylglyoxal monohydrate also worked well (entry 8). It is known that water often interferes with the aldol reactions of metal enolates with aldehydes and that in the cases where such water soluble aldehydes are employed, some troublesome purifications including dehydration are necessary. Moreover, salicylaldehyde (entry 9) and 2-pyridinecarboxaldehyde (entry 10) could be successfully employed. The former has a free hydroxy group which is incompatible with metal enolates or Lewis acids, and the latter is generally difficult to use under the influence of Lewis acids because of the coordination of the nitrogen atom to the Lewis acids resulting in the deactivation of the acids.

The aldol reactions of silyl enol ethers with aldehydes were also found to proceed smoothly in water-ethanol-toluene [20]. Some reactions proceeded much faster in this solvent system than in THF-water. Furthermore, the new solvent system realized continuous use of the catalyst by a very simple procedure.

Several water-organic solvent systems were examined in the model reaction of 1-phenyl-1-trimethylsiloxypropene (**1**) with 2-pyridinecarboxaldehyde un-

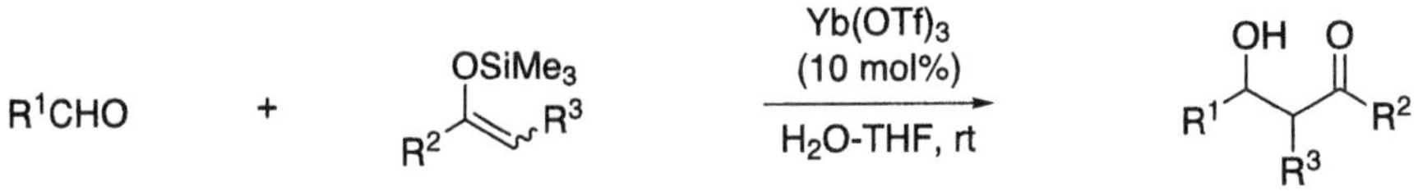

R^1CHO + $R^2C(OSiMe_3)=CHR^3$ $\xrightarrow[H_2O\text{-}THF,\ rt]{Yb(OTf)_3\ (10\ mol\%)}$ $R^1CH(OH)CH(R^3)C(=O)R^2$

Table 4. Lanthanide triflate-catalyzed aldol reactions in aqueous media

Entry	Aldehyde	Silyl Enol Ether	Yield/%
1	PhCHO	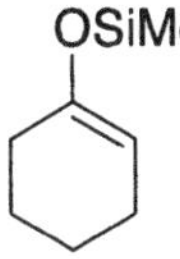 OSiMe$_3$ (cyclohexenyl) **2**	91 [a)]
2	PhCHO	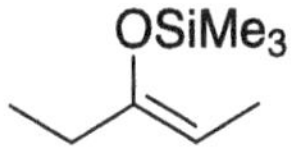 OSiMe$_3$	89 [b)]
3	PhCHO	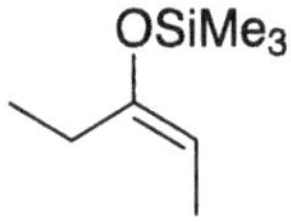 OSiMe$_3$	93 [c)]
4	PhCHO	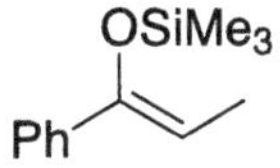 Ph, OSiMe$_3$ **1**	81 [d)]
5	CH_3CHO	**1**	93 [e, f)]
6	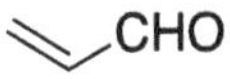 $CH_2=CHCHO$	**1**	82 [e, g)]
7	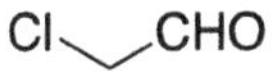 $ClCH_2CHO$	**1**	95 [h)]
8	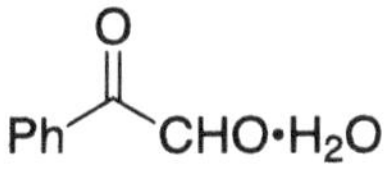 $PhCOCHO\cdot H_2O$	**1**	67 [i)]
9	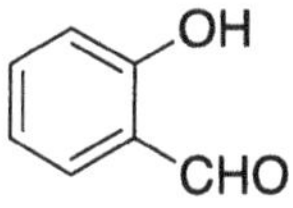 OH, CHO (salicylaldehyde)	**1**	81 [j, k)]
10	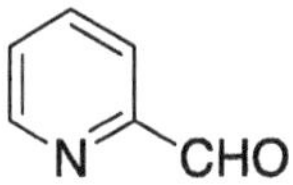 N, CHO (pyridine-2-carbaldehyde)	**1**	87 [j, l)]

[a] *syn/anti*=73/27.
[b] *syn/anti*=63/37.
[c] *syn/anti*=71/29.
[d] *syn/anti*=53/47.
[e] $Gd(OTf)_3$ was used instead of $Yb(OTf)_3$.
[f] *syn/anti*=46/54.
[g] *syn/anti*=60/40.
[h] *syn/anti*=45/55.
[i] *syn/anti*=27/73.
[j] $Lu(OTf)_3$ was used instead of $Yb(OTf)_3$.
[k] *syn/anti*=55/45.
[l] *syn/anti*=42/58.

CHO + $OSiMe_3$ Ph → $Yb(OTf)_3$ (10 mol%), solvent, rt, 4 h → OH O Ph

Table 5. Effect of solvents

Solvent		Yield / %	*syn*/*anti*
H_2O / Toluene (1 : 4)		0	—
H_2O / EtOH / Toluene	(1 : 3 : 4)	30	37 / 63
	(1 : 5 : 4)	41	39 / 61
	(1 : 7 : 4)	70	41 / 59
	(1 : 10 : 4)	96	40 / 60
H_2O / THF (1 : 4)		12	43 / 57

der the influence of 10 mol% $Yb(OTf)_3$ (Table 5). While the reaction proceeded sluggishly in a water-toluene system, the adduct was obtained in a good yield when ethanol was added to this system. The yield increased in accordance with the amount of ethanol, and it was noted that the reaction proceeded much faster in the water-ethanol-toluene system than in the original water-THF system.

Although the water-ethanol-toluene (1:7:4) system was one phase, it easily became two phases by adding toluene after the reaction was completed. The product was isolated from the organic layer by a usual work up. On the other hand, the catalyst remained in the aqueous layer, which was used directly in the next reaction without removing water. It is noteworthy that the yields of the second, third, and fourth runs were comparable to that of the first run (Eq. 1).

CHO + $OSiMe_3$ Ph (**1**) → $Yb(OTf)_3$ (10 mol%), H_2O/EtOH/toluene (1 : 7 : 4), rt, 8 h → OH O Ph (1)

1st run: 86% (syn /anti = 38/62)
2nd run: 82% (syn /anti = 38/62)
3rd run: 90% (syn /anti = 38/62)
4th run: 82% (syn /anti = 39/61) ■

Several examples of the present aldol reactions of silyl enol ethers with aldehydes in water-ethanol-toluene are listed in Table 6. 3-Pyridinecarboxaldehyde as well as 2-pyridinecarboxaldehyde, salicylaldehyde, and formaldehyde water

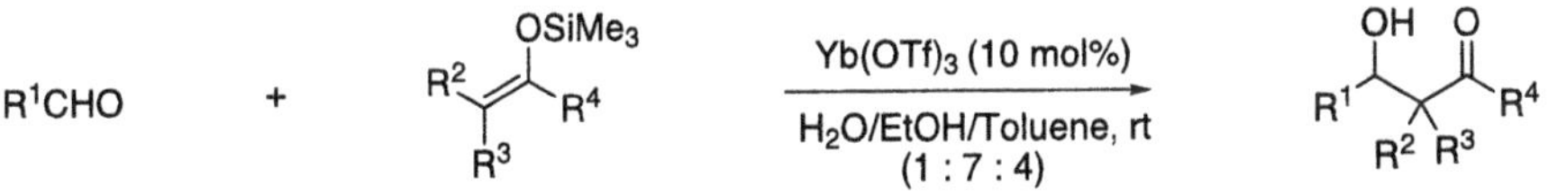

R^1CHO + R^2, R^3, $OSiMe_3$, R^4 — $Yb(OTf)_3$ (10 mol%), H_2O/EtOH/Toluene, rt (1 : 7 : 4) → OH O, R^1, R^4, R^2 R^3

Table 6. b(OTf)$_3$-catalyzed aldol reactions of silyl enol ethers with aldehydes in water-ethanol-toluene

Aldehyde	Silyl Enol Ether	Product	Yield / %
PhCHO	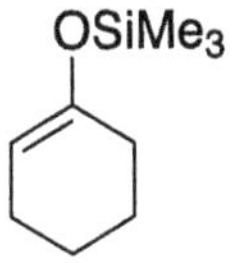 $OSiMe_3$ **2**	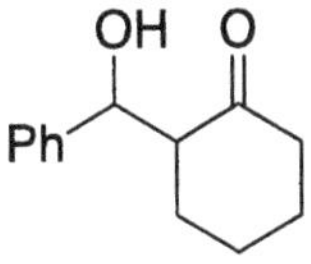 Ph, OH O	89 [a)]
PhCHO	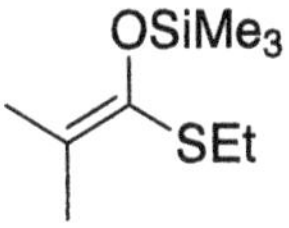 $OSiMe_3$, SEt	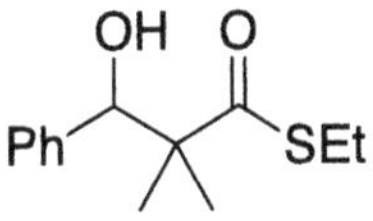 Ph, OH O, SEt	95
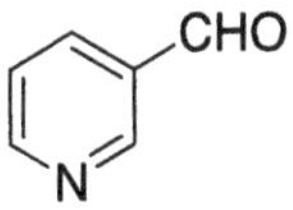 CHO, N	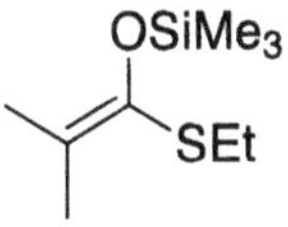 $OSiMe_3$, SEt	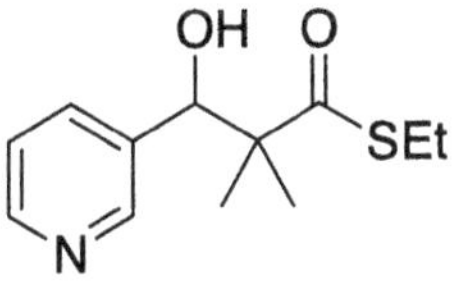 OH O, SEt, N	87
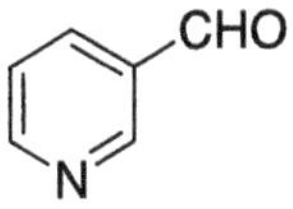 CHO, N	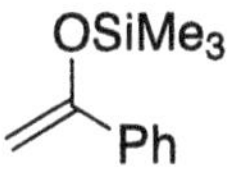 $OSiMe_3$, Ph	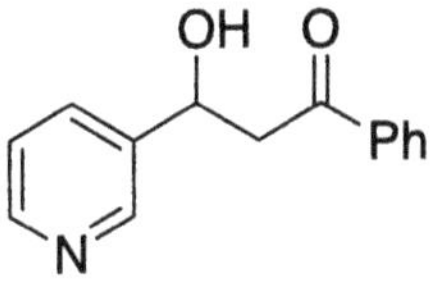 OH O, Ph, N	82
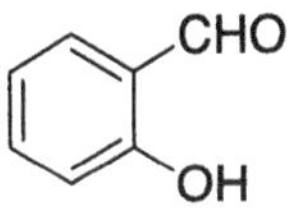 CHO, OH	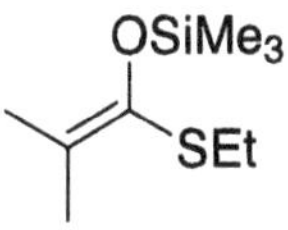 $OSiMe_3$, SEt	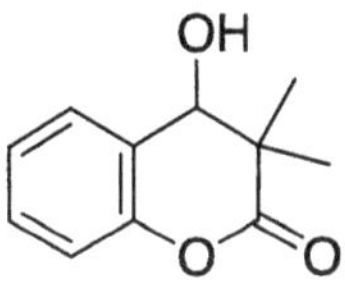 OH, O, O	96
HCHO aq.	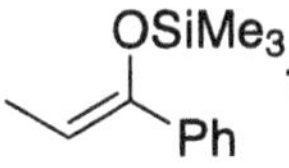 $OSiMe_3$ **1**, Ph	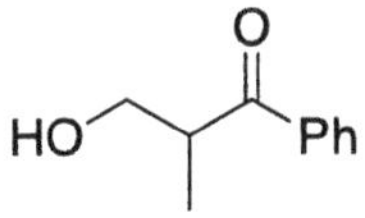 HO, O, Ph	90

[a] *syn/anti*=74/26.

solution worked well. As for silyl enol ethers, not only ketone enol ethers but also silyl enolates derived from thioesters were used. In every case, the adducts were obtained in high yields in the presence of 10 mol% $Yb(OTf)_3$.

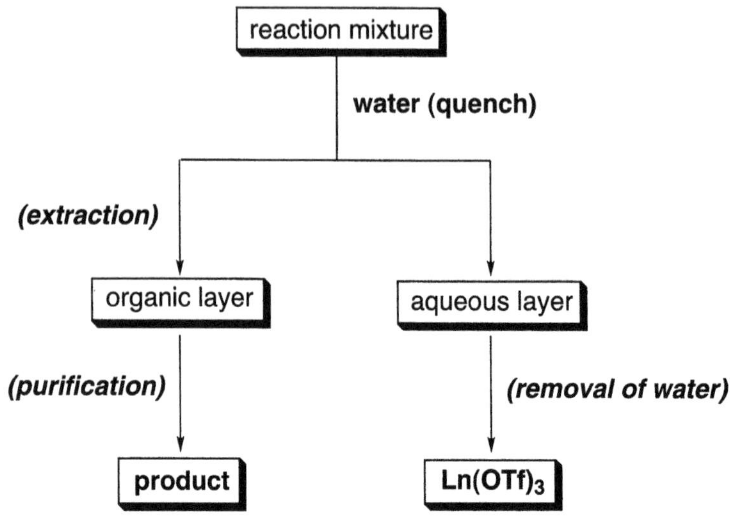

Scheme 1

2.2 Recovery and Reuse of the Catalyst

While continuous use of $Ln(OTf)_3$ is possible, it is also easy to recover $Ln(OTf)_3$ themselves. Lanthanide triflates are more soluble in water than in organic solvents such as dichloromethane. Almost 100% of $Ln(OTf)_3$ was quite easily recovered from the aqueous layer after the reaction was completed and it could be reused. For example, first use (20 mol% of $Yb(OTf)_3$) in the reaction of **1** with formaldehyde water solution (94% yield); second use (91% yield); third use (93% yield). The reactions are usually quenched with water and the products are extracted with an organic solvent (for example, dichloromethane). Lanthanide triflate is in aqueous layer and only removal of water gives the catalyst which can be used in the next reaction (Scheme 1). It is noteworthy that lanthanide triflates are expected to solve some severe environmental problems induced by Lewis acid-promoted reactions in industrial chemistry [21].

2.3 Aldol Reactions in Organic Solvents

Lanthanide triflates were found to be excellent Lewis acids not only in aqueous media but also in organic solvents. The reaction of ketene silyl acetal **3** with benzaldehyde proceeded smoothly in the presence of 10 mol% of $Yb(OTf)_3$ in dichloromethane at –78 °C, to afford the corresponding aldol-type adduct in a 94% yield. The same reaction at room temperature also went quite cleanly without side reactions and the desired adduct was obtained in a 95% yield. No adduct

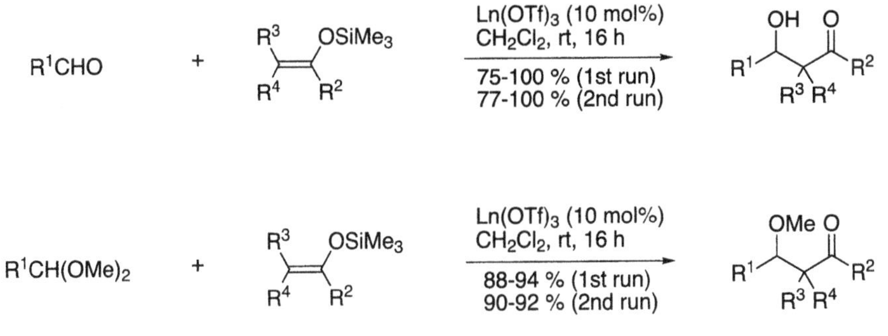

R^1CHO + $R^3R^4C=C(OSiMe_3)R^2$ → $Ln(OTf)_3$ (10 mol%), CH_2Cl_2, rt, 16 h; 75-100 % (1st run), 77-100 % (2nd run)

$R^1CH(OMe)_2$ + $R^3R^4C=C(OSiMe_3)R^2$ → $Ln(OTf)_3$ (10 mol%), CH_2Cl_2, rt, 16 h; 88-94 % (1st run), 90-92 % (2nd run)

Scheme 2

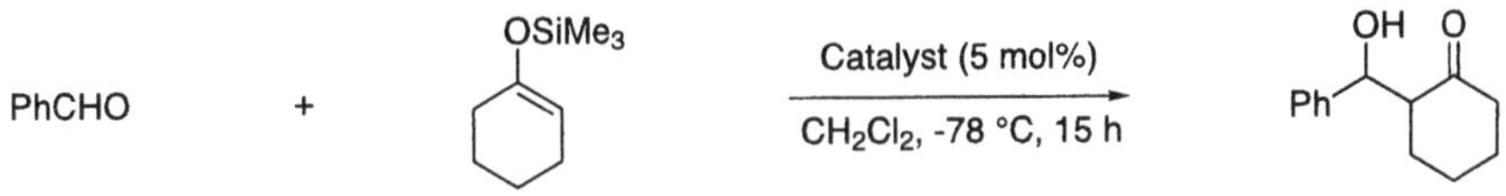

PhCHO + **2** → Catalyst (5 mol%), CH_2Cl_2, -78 °C, 15 h

Table 7. Effect of catalysts

Entry	Catalyst	Yield/%
1	$Sc(OTf)_3$	81
2	$Y(OTf)_3$	trace
3	$Yb(OTf)_3$	trace

was obtained in THF-water or toluene-ethanol-water, because hydrolysis of the ketene silyl acetal preceded the desired aldol reaction in such solvents. In other organic solvents such as toluene, THF, acetonitrile, and DMF, $Yb(OTf)_3$ worked well, and it was found that other $Ln(OTf)_3$ also catalyzed the above aldol reaction effectively (85–95% yields).

Silyl enolates derived from not only esters but also thioesters and ketones reacted with aldehydes to give the corresponding adducts in high yields (Scheme 2) [22]. Furthermore, acetals reacted smoothly with silyl enolates to afford the corresponding aldol-type adducts in high yields. It should be noted that the catalysts could be easily recovered from the aqueous layer after the reactions were quenched with water and could be reused, and that the yields of the second run were almost comparable to those of the first run in every case.

Scandium triflate ($Sc(OTf)_3$) was found to be an effective catalyst in the aldol reactions [23]. The activities of various triflate catalysts were evaluated in the aldol reaction of 1-trimethylsiloxycyclohexene (**2**) with benzaldehyde in dichloromethane (Table 7). While the reaction scarcely proceeded at –78 °C in the presence of $Yb(OTf)_3$ or $Y(OTf)_3$ [24], the aldol adduct was obtained in an 81%

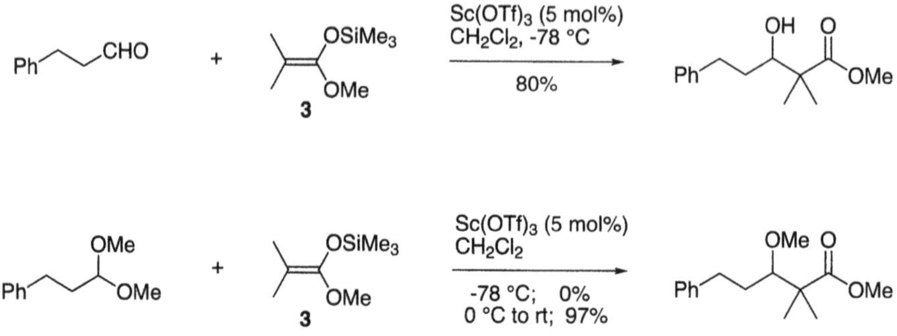

Scheme 3

yield in the presence of $Sc(OTf)_3$. Obviously $Sc(OTf)_3$ is more active than $Y(OTf)_3$ or $Yb(OTf)_3$ in this case.

Several examples of the $Sc(OTf)_3$-catalyzed aldol reactions of silyl enolates with aldehydes were examined, and it was found that silyl enolates derived from ketones, thioesters, and esters reacted smoothly with aldehydes in the presence of 5 mol% of $Sc(OTf)_3$ to afford the aldol adducts in high yields. $Sc(OTf)_3$ was also found to be an effective catalyst in the aldol-type reaction of silyl enolates with acetals. The reactions proceeded smoothly at −78 °C or room temperature to give the corresponding aldol-type adducts in high yields without side reaction products. It should be noted that aldehydes were more reactive than acetals [25]. For example, while 3-phenylpropionaldehyde reacted with the ketene silyl acetal of methyl 2-methylpropionate (**3**) at −78 °C to give the aldol adduct in an 80% yield, no reaction occurred at −78 °C in the reaction of the same ketene silyl acetal with 3-phenylpropionaldehyde dimethylacetal. The acetal reacted with the ketene silyl acetal at 0 °C to room temperature to give the aldol-type adduct in a 97% yield (Scheme 3).

$Sc(OTf)_3$ can behave as a Lewis acid catalyst even in aqueous media. $Sc(OTf)_3$ was stable in water and was effective in the aldol reactions of silyl enolates with aldehydes in aqueous media. The reactions of usual aromatic and aliphatic aldehydes such as benzaldehyde and 3-phenylpropionaldehyde with silyl enolates were carried out in both aqueous and organic solvents, and water-soluble formaldehyde and chloroacetaldehyde were directly treated as water solutions with silyl enolates to afford the aldol adducts in good yields. Moreover, the catalyst could be recovered almost quantitatively from the aqueous layer after the reaction was completed. The recovered catalyst was also effective in the second reaction, and the yield of the second run was comparable to that of the first run (Eq. 2).

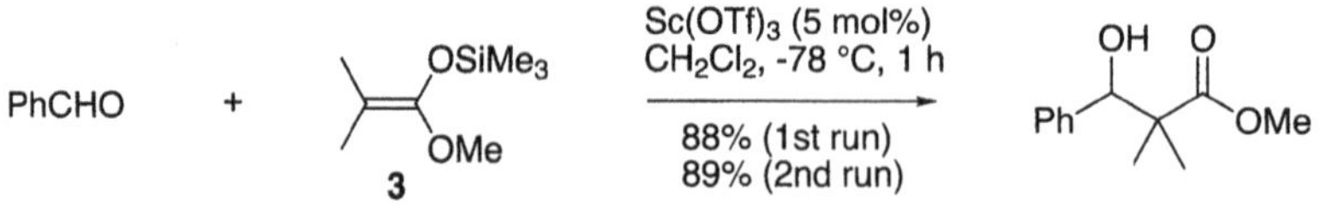

(2)

2.4 Aldol Reactions in Micellar Systems

$Sc(OTf)_3$-catalyzed aldol reactions of silyl enol ethers with aldehydes were successfully carried out in micellar systems. While the reactions proceeded sluggishly in pure water, remarkable enhancement of the reactivity was observed in the presence of a small amount of a surfactant. In these systems, versatile carbon-carbon bond-forming reactions proceeded in water without using any organic solvents.

Lewis acid catalysis in micellar systems was first found in the model reaction of **1** with benzaldehyde (Table 8) [26]. While the reaction proceeded sluggishly in the presence of 0.2 equiv. $Yb(OTf)_3$ in water, remarkable enhancement of the reactivity was observed when the reaction was carried out in the presence of 0.2 equiv. $Yb(OTf)_3$ in an aqueous solution of sodium dodecylsulfate (SDS, 0.2 equiv., 35 mmol/l), and the corresponding aldol adduct was obtained in a 50% yield. In the absence of the Lewis acid and the surfactant (water-promoted conditions), only 20% yield of the aldol adduct was isolated after 48 h, while a 33% yield of the aldol adduct was obtained after 48 h in the absence of the Lewis acid in an aqueous solution of SDS. The amounts of the surfactant also influ-

PhCHO + 1 (Ph–C(OSiMe$_3$)=CH–CH$_3$) →[cat. Ln(OTf)$_3$, surfactant, H$_2$O] PhCOCH(CH$_3$)CH(OH)Ph

Table 8. Effect of $Ln(OTf)_3$ and surfactants

$Ln(OTf)_3$/eq.	Surfactant/equiv.	Time/h	Yield/%
$Yb(OTf)_3$/0.2	——	48	17
$Yb(OTf)_3$/0.2	SDS/0.04	48	12
$Yb(OTf)_3$/0.2	SDS/0.1	48	19
$Yb(OTf)_3$/0.2	SDS/0.2	48	50
$Yb(OTf)_3$/0.2	SDS/1.0	48	22
$Sc(OTf)_3$/0.2	SDS/0.2	17	73
$Sc(OTf)_3$/0.1	SDS/0.2	4	88
$Sc(OTf)_3$/0.1	TritonX-100 [a)]/0.2	60	89
$Sc(OTf)_3$/0.1	CTAB/0.2	4	trace

a) $H_3C-C(CH_3)_2-CH_2-C(CH_3)_2-C_6H_4-(OCH_2CH_2)_xOH$

R^1CHO + silyl enol ether (OSiMe$_3$, R^2, R^3) → [Sc(OTf)$_3$ (0.1 eq.), SDS (0.2 eq.), H_2O, rt] → $R^2C(O)CH(R^3)CH(OH)R^1$

Table 9. Sc(OTf)$_3$-catalyzed aldol reactions in micellar systems

Aldehyde	Silyl Enol Ether	Yield/%
PhCHO	PhC(OSiMe$_3$)=CHCH$_3$ **1**	88 [a]
PhCH$_2$CH$_2$CHO	**1**	86 [b]
PhCH=CHCHO	**1**	88 [c]
HCHO	**1**	82 [d]
PhCHO	EtC(OSiMe$_3$)=CHCH$_3$	88 [e]
PhCHO	PhC(OSiMe$_3$)=CH$_2$	75 [f,g]
PhCHO	EtSC(OSiMe$_3$)=C(CH$_3$)$_2$	94
PhCHO	MeOC(OSiMe$_3$)=C(CH$_3$)$_2$ **3**	84 [g]

[a] *syn/anti*=50/50.
[b] *syn/anti*=45/55.
[c] *syn/anti*=41/59.
[d] Commercially available HCHO aq. (3 ml), **1** (0.5 mmol), Sc(OTf)$_3$ (0.1 mmol), and SDS (0.1 mmol) were combined.
[e] *syn/anti*=57/43.
[f] Sc(OTf)$_3$ (0.2 equiv.) was used.
[g] Additional silyl enolate (1.5 equiv.) was charged after 6 h.

enced the reactivity and the yield was improved when Sc(OTf)$_3$ was used as a Lewis acid catalyst. Judging from the critical micelle concentration, micelles would be formed in these reactions, and it is noteworthy that the Lewis acid-catalyzed reactions proceeded smoothly in micellar systems [27]. It was also found that the surfactants influenced the yield, and that TritonX-100 was effective in the aldol reaction (but required long reaction time), while only a trace amount of the adduct was detected when using cetyltrimethylammonium bromide (CTAB) as a surfactant [28].

Several examples of the $Sc(OTf)_3$-catalyzed aldol reactions in micellar systems are shown in Table 9. Not only aromatic, but also aliphatic and α,β-unsaturated aldehydes reacted with silyl enol ethers to afford the corresponding aldol adducts in high yields. Formaldehyde water solution also worked well. Ketene silyl acetal **3**, which is known to hydrolyze very easily even in the presence of a small amount of water, reacted with an aldehyde in the present micellar system to afford the corresponding aldol adduct in a high yield.

It should be noted that the reactions were successfully carried out in water without using any organic solvents. Use of the reusable scandium catalyst and water as a solvent would result in clean and environmentally friendly systems.

3 Diels-Alder Reactions

Although many Diels-Alder reactions have been carried out at higher reaction temperatures without catalysts, heat sensitive compounds in complex and multistep syntheses cannot be employed. While Lewis acid catalysts allow the reactions to proceed at room temperature with satisfactory yields, they are often accompanied by diene polymerization and excess amounts of the catalyst are often needed to catalyze carbonyl-containing dienophiles [29].

Lanthanide triflates were also found to be efficient catalysts in the Diels-Alder reactions of carbonyl-containing dienophiles with cyclopentadiene [30]. A catalytic amount of $Yb(OTf)_3$ was enough to promote the reactions to give the corresponding adducts in high yields, and the catalyst could be easily recovered and reused.

In the Diels-Alder reactions, $Sc(OTf)_3$ was clearly the most effective among $Ln(OTf)_3$ as a catalyst [31]. While in the presence of 10 mol% of $Y(OTf)_3$ or $Yb(OTf)_3$, only a trace amount of the adduct was obtained in the Diels-Alder reaction of MVK with isoprene, and the reaction proceeded quite smoothly to give the adduct in a 91% yield in the presence of 10 mol% of $Sc(OTf)_3$.

Several examples of the $Sc(OTf)_3$-catalyzed Diels-Alder reactions are shown in Table 10. In every case, the Diels-Alder adducts were obtained in high yields with *endo* selectivities.

The present Diels-Alder reactions proceeded even in aqueous media (Eq. 3) [32]. Thus, naphthoquinone reacted with cyclopentadiene in THF-H_2O (9:1) at room temperature to give the corresponding adduct in a 93% yield (*endo*/*exo*= 100/0).

$Sc(OTf)_3$ (10 mol%)
THF : H_2O (9 : 1)
93% yield, *endo*/*exo* = 100/ 0 (3)

Recovery and reuse of the catalyst were also possible in this reaction. After the reaction was completed, the aqueous layer was concentrated to give the catalyst. The recovered catalyst was effective in subsequent Diels-Alder reactions, and it

Table 10. $Sc(OTf)_3$-catalyzed Diels-Alder reactions

Dienophile	Diene	Product	Yield /%	endo/ exo
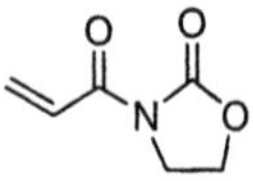		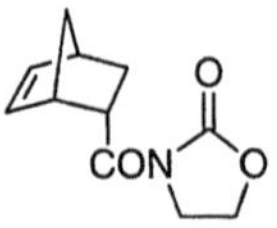	95	87 / 13
		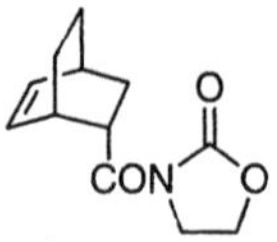	89	100 / 0
	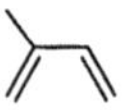	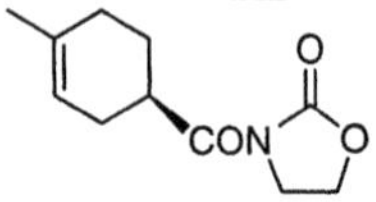	90	—
		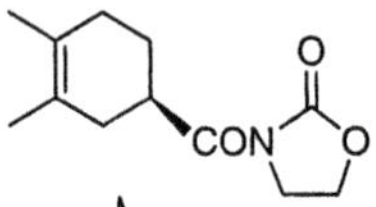	86	—
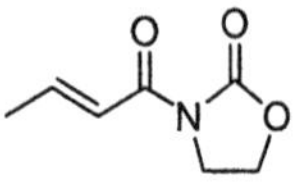		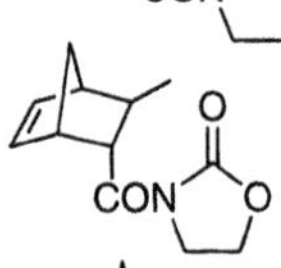	97	84 / 16
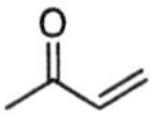			96	89 / 11
			83	>95 / 5
		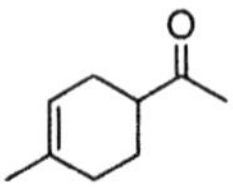	91	—
		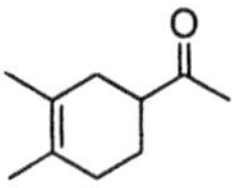	73	—
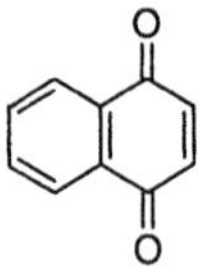		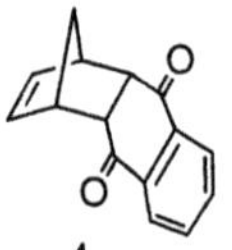	83	100 / 0
		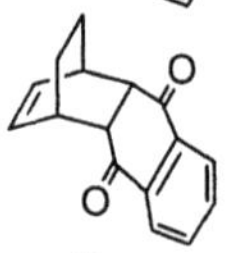	89	94 / 3
		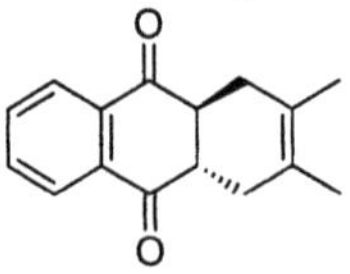	92	—

should be noted that the yields of the second and even the third runs were comparable to that of the first run.

4
Allylation Reactions

Synthesis of homoallylic alcohols by the reaction of allyl organometallics with carbonyl compounds is one of the most important processes in organic synthesis [33]. The allylation reactions of carbonyl compounds were found to proceed smoothly under the influence of 5 mol% of $Sc(OTf)_3$ [34] by using tetraallyltin [35] as an allylating reagent [36]. Several examples are shown in Table 11. The reactions proceeded smoothly in the presence of only a catalytic amount of $Sc(OTf)_3$ under extremely mild conditions [8], and the adducts, homoallylic alcohols, were obtained in high yields. Ketones could also be used in the reaction (entries 4, 5). In most cases, the reactions were successfully carried out in aqueous media. It is noteworthy that unprotected sugars reacted directly to give the adducts in high yields (entries 7–9). The allylated adducts are intermediates for the synthesis of higher sugars [37]. Moreover, an aldehyde containing water of crystallization such as phenylglyoxal monohydrate reacted with tetraallyltin to give the di-allylated adduct in high yield (entry 10). Under the present reaction conditions, salicylaldehyde and 2-pyridinecarboxaldehyde reacted with tetraallyltin to afford the homoallylic alcohols in good yields (entries 11, 12). Under general Lewis acid conditions, these compounds react with the Lewis acids rather than the nucleophile. Furthermore, several kinds of solvents could be used. The reactions also proceeded under non-aqueous conditions. Water-sensitive substrates under Lewis acid conditions could be reacted in an appropriate organic solvent (entries 13, 14).

$Yb(OTf)_3$ is also effective in the present allylation reactions. For example, 3-phenylpropionaldehyde reacted with tetraallyltin in the presence of 5 mol% of $Yb(OTf)_3$ to afford the adduct in a 90% yield.

The water-ethanol-toluene system could be successfully applied to the present allylation reactions. An example of the allylation reaction of tetraallyltin with aldehyde is shown in Eq. (4), and in this case continuous use of the catalyst was also realized.

Ph⌒⌒CHO + (allyl)$_4$Sn —[$Yb(OTf)_3$ (10 mol%); H_2O/EtOH/toluene (1 : 7 : 4), rt, 8 h]→ Ph⌒⌒CH(OH)⌒CH=CH$_2$ (4)

1st run: 90%; 2nd run: 95%; 3rd run: 96%; 4th run: 89%

Furthermore, the $Sc(OTf)_3$-catalyzed allylation reactions of aldehydes with tetraallyltin proceeded smoothly in micellar systems to afford the correspond-

Table 11. $Sc(OTf)_3$-catalyzed allylation reactions of carbonyl compounds with tetraallyltin[a)]

Entry	Carbonyl Compound	Product	Solvent	Yield/%
1	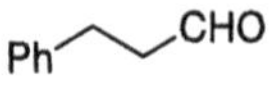 Ph CHO	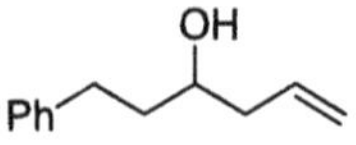 OH Ph	H_2O : THF (1 : 9)	92
			H_2O : EtOH (1 : 9)	96
			H_2O : CH_3CN (1 : 9)	96
			EtOH	86
			CH_3CN	94
2	PhCHO	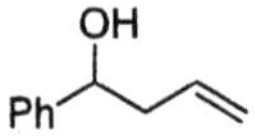 OH Ph	H_2O : THF (1 : 9)	94
			CH_3CN	82
3	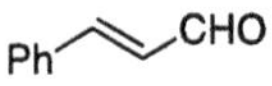 Ph CHO	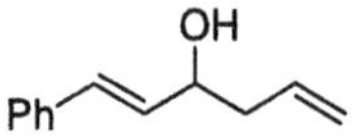 OH Ph	H_2O : THF (1 : 9)	98
			CH_3CN	94
4	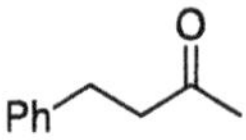 O Ph	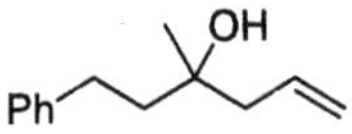 OH Ph	CH_2Cl_2	78
5	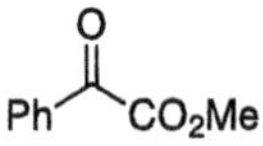 O Ph CO_2Me	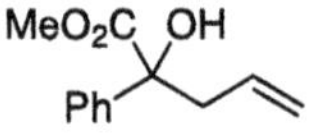 MeO_2C OH Ph	H_2O : THF (1 : 9)	87
6	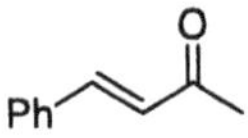 O Ph	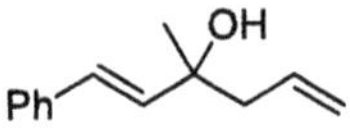 OH Ph	CH_2Cl_2	82
7	D-arabinose	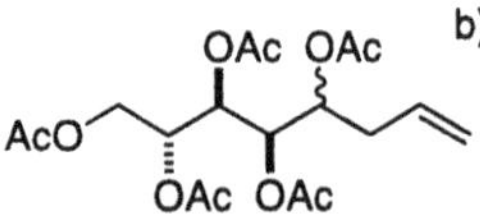 OAc OAc AcO OAc OAc [b)]	H_2O : THF (1 : 4)	81[c)]
			H_2O : EtOH (1 : 4)	89[d)]
			H_2O : CH_3CN (1 : 9)	93[e)]
8	2-deoxy-D-ribose	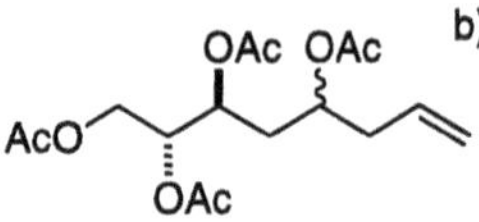 OAc OAc AcO OAc [b)]	H_2O : THF (1 : 9)	89[f)]
9	2-deoxy-D-glucose	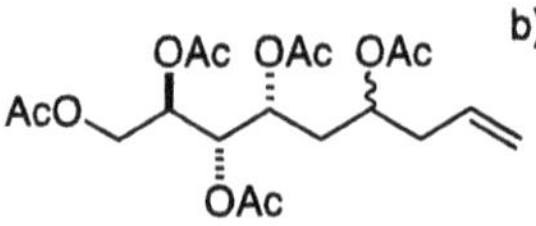 OAc OAc OAc AcO OAc [b)]	H_2O : THF (1 : 9)	88[f)]
10	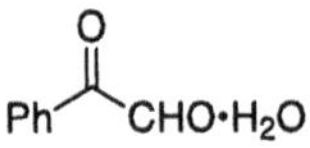 O Ph CHO·H_2O	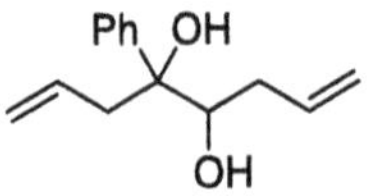 Ph OH OH	CH_3CN	78[g)]

Table 11. (continued)

Entry	Carbonyl Compound	Product	Solvent	Yield/%
11	2-HO-C$_6$H$_4$-CHO	2-HO-C$_6$H$_4$-CH(OH)CH$_2$CH=CH$_2$	H_2O : THF (1 : 9)	quant.
			CH_3CN	90
12	2-pyridyl-CHO	2-pyridyl-CH(OH)CH$_2$CH=CH$_2$	H_2O : THF (1 : 9)	quant.
			CH_3CN	84
13	2-(MeOCH$_2$O)-C$_6$H$_4$-CHO	2-(MeOCH$_2$O)-C$_6$H$_4$-CH(OH)CH$_2$CH=CH$_2$	CH_3CN	87
14	4-(EtO)$_2$CH-C$_6$H$_4$-CHO	4-(EtO)$_2$CH-C$_6$H$_4$-CH(OH)CH$_2$CH=CH$_2$	CH_3CN	90

[a] Carried out at 25 °C except for entries 8 and 9 (60 °C).
[b] The products were isolated after acetlyation.
[c] *syn/anti*=28/72.
[d] *syn/anti*=27/73.
[e] *syn/anti*=26/74.
[f] *syn/anti*=50/50.
[g] Diastereomer ratio=88/12. Relative configuration assignment was not made.

ing homoallylic alcohols in high yields. The reactions were successfully carried out in the presence of a small amount of a surfactant in water without using any organic solvents.

5 Mannich-Type Reactions

5.1 Reactions of Imines with Silyl Enolates

The Mannich and related reactions provide one of the most fundamental and useful methods for the synthesis of β-amino ketones or esters. Although the classical protocols include some severe side reactions, new modifications using preformed iminium salts and imines have been developed [38]. Among them, reactions of imines with enolate components, especially silyl enolates, provide useful and promising methods leading to β-amino ketones or esters. The first report using a stoichiometric amount of $TiCl_4$ as a promoter appeared in 1977 [39], and since then some efficient catalysts have been developed [40].

N-R2 / R1 H + OSiMe3 / R3 R5 / R4 → Ln(OTf)3 (5 mol%), CH_2Cl_2, 0 °C → R2 NH O / R1 R5 / R3 R4

Table 12. Reactions of imines with silyl enolates

Imine	Silyl enolate	Ln	Yield/%
N, Ph, Ph, H	OSiMe3, OMe **3**	Yb Y	97 [a)] 81
	OSiMe3, SEt	Yb	95
N, Ph, Ph, H	**3**	Yb Y	86 78
	OSiMe3, SEt	Yb Sc	65 80
	OSiMe3, OBn	Yb	80 [b)]
N, Ph, O, H	**3**	Yb	95
	OSiMe3, OBn	Yb	67 [c)]
N, Ph, O, H	**3**	Yb	88
N, Ts, Ph, H	**3**	Yb	88
N, Ph, Ph, O, H	**3**	Yb	60
N, Ph, C_4H_9, H	**3**	Yb	47

[a] Second use=96% yield; third use=98% yield.
[b] *syn/anti*=18/82.
[c] *syn/anti*=21/79.

In aqueous media, water coordinates lanthanide triflates under equilibrium conditions, and thus activation of carbonyl compounds using a catalytic amount of the Lewis acid has been performed. It was expected that, based on the same consideration, the catalytic activation of imines would be possible by using lanthanide triflates.

The reactions of imines with silyl enolates were tested in the presence of 5 mol% of $Ln(OTf)_3$, and selected examples are shown in Table 12 [41]. In most cases the reactions proceeded smoothly in the presence of 5 mol% of $Yb(OTf)_3$ (a representative lanthanide triflate) to afford the corresponding β–amino ester derivatives in good to high yields. Yttrium triflate ($Y(OTf)_3$) was also effective, and the yield was improved when $Sc(OTf)_3$ was used instead of $Yb(OTf)_3$ as a catalyst. Not only silyl enolates derived from esters, but also that derived from a thioester, worked well to give the desired β–amino esters and thioester in high yields. In the reactions of the silyl enolate derived from benzyl propionate, *anti* adducts were obtained in good selectivities. In addition, the catalyst could be recovered after the reaction was completed and could be reused.

5.2 One-Pot Synthesis of β-Amino Esters from Aldehydes

While the Lewis acid-catalyzed reactions of imines with silyl enolates are one of the most efficient methods for the preparation of β-amino esters, many imines are hygroscopic, unstable at high temperatures, and difficult to purify by distillation or column chromatography. It is desirable from a synthetic point of view that imines, generated in situ from aldehydes and amines, immediately react with silyl enolates and provide β-amino esters in a one-pot reaction. However, most Lewis acids cannot be used in this reaction because they decompose or deactivate in the presence of the amines and water that exist during imine formation. Due to the unique properties of $Ln(OTf)_3$, their use as catalysts for the above one-pot preparation of β-amino esters from aldehydes was planned.

A general scheme of the one-pot synthesis of β-amino esters from aldehydes is shown in Scheme 4 [42]. In the presence of a catalytic amount of $Yb(OTf)_3$ and an additive (a dehydrating reagent such as MS 4 A or $MgSO_4$), an aldehyde was treated with an amine and then with a silyl enolate in the same vessel. Several examples are shown in Tables 13 and 14, and the following characteristic features of this reaction are noted.

1. In every case, β-amino esters were obtained in high yields. Silyl enolates derived from esters as well as thioesters reacted smoothly to give the adducts.

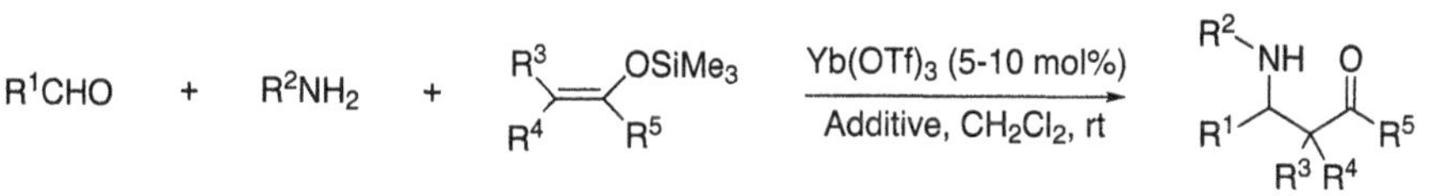

R^1CHO + R^2NH_2 + $R^3R^4C{=}C(OSiMe_3)R^5$ $\xrightarrow[\text{Additive, } CH_2Cl_2\text{, rt}]{Yb(OTf)_3\ (5\text{-}10\ mol\%)}$ $R^1CH(NHR^2)CR^3R^4C(O)R^5$

Scheme 4

Table 13. One-pot synthesis of β-amino esters from aldehydes

Entry	R^1	R^2	Silyl enolate	Additive [a]	Yield/%
1	Ph	Ph	OSiMe3 / OMe **3**	MS4A $MgSO_4$	90 89
2	Ph	Bn	**3**	MS4A	85
3	Ph	p-MeOPh	**3**	$MgSO_4$	91 [b]
4	Ph	o-MeOPh	**3**	MS4A	96
5	Ph	Ph	OSiMe3 / SEt	MS4A	90
6	Ph	Bn	OSiMe3 / SEt	MS4A	62 [b]
7	Ph	p-MeOPh	OSiMe3 / SEt	MS4A	79 84,[b] 87 [b,c]
8	Ph	C_4H_9	**3**	MS4A	89
9	PhCO [d]	Ph	**3**	$MgSO_4$	82
10	PhCO [d]	Ph	OSiMe3 / Ph	$MgSO_4$	87
11	PhCH=CH	p-MeOPh	**3**	$MgSO_4$	92 [e]
12	$Ph(CH_2)_2$	Bn	**3**	$MgSO_4$	83 [f]
13	C_4H_9	Bn	**3**	$MgSO_4$	77 [f]
14	C_8H_{17}	Bn	**3**	$MgSO_4$	81 [f]
15	C_8H_{17}	Ph_2CH	**3**	$MgSO_4$	89 [g]

[a] MS4A or $MgSO_4$ was used. Almost comparable yields were obtained in each case.
[b] CH_3CN was used as a solvent.
[c] $Sc(OTf)_3$ was used instead of $Yb(OTf)_3$.
[d] Monohydrate.
[e] C_2H_5CN, –78 °C.
[f] –78 to 0 °C.
[g] 0 °C.

No adducts between aldehydes and the silyl enolates were observed in any reaction.

2. A silyl enol ether derived from a ketone also worked well to afford the β-amino ketone in a high yield (Table 13, entry 10).
3. Aliphatic aldehydes reacted with amines and silyl enolates to give the corresponding β-amino esters in high yields. In some reactions of imines, it is known that aliphatic enolizable imines prepared from aliphatic aldehydes gave poor results.

R^1CHO + R^2NH_2 + R^3 / $OSiMe_3$ / R^4 $\xrightarrow[\text{MS4A, } C_2H_5CN\text{, -78 °C}]{Yb(OTf)_3 \text{ (5-10 mol\%)}}$ R^2 NH O / R^1 R^4 / R^3

Table 14. Diastereoselective one-pot synthesis of β-amino esters from aldehydes

Entry	R^1	R^2	Silyl enolate	Yield/%	*syn* / *anti*[a]
1	Ph	Bn	$OSiMe_3$ / OMe	90	1 / 13.3
2	Ph	Bn	OTBS / TBSO / OMe	78	1 / 9.0
3	$Ph(CH_2)_2$	Ph_2CH	BnO / $OSiMe_3$ / OMe	88	8.1 / 1
4	C_4H_9	Ph_2CH	BnO / $OSiMe_3$ / OMe	90	8.1 / 1
5	$(CH_3)_2CHCH_2$	Ph_2CH	BnO / $OSiMe_3$ / OMe	86	7.3 / 1

[a] Determined by ^1H NMR analysis.

4. Phenylglyoxal monohydrate also worked well in this reaction. The imine derived from phenylglyoxal is unstable and a troublesome treatment is known to be required for its use [43].
5. The catalyst could be recovered after the reaction was completed and could be reused (first run, 91%; second run, 92%, in the reaction of benzaldehyde, *p*-anisidine, and silyl enolate **3** (Table 13, entry 3)).
6. As for the diastereoselectivity of this reaction, good results were obtained after examination of the reaction conditions. While *anti* adducts were produced preferentially in the reactions of benzaldehyde, *syn* adducts were obtained with high selectivities in the reactions of aliphatic aldehydes (Table 14).
7. The high yields of the present one-pot reactions depend on the unique properties of $Ln(OTf)_3$ as the Lewis acid catalysts. Although $TiCl_4$ and TMSOTf are known to be effective for the activation of imines [39, 44], the use of even stoichiometric amounts of $TiCl_4$ and TMSOTf instead of $Ln(OTf)_3$ in the present one-pot reactions gave only trace amounts of the product in both cases (Table 15).
8. One-pot preparation of a β-lactam from an aldehyde, an amine, and a silyl enolate has been achieved based on the present reaction (Scheme 5). The reaction of the aldehyde, the amine, and **2** was carried out under the standard conditions, and $Hg(OCOCF_3)_2$ was then added to the same pot. The desired β-lactam was isolated in a 78% yield.

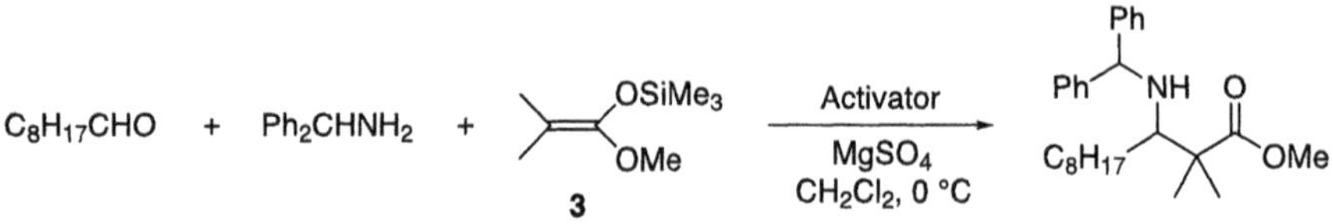

$C_8H_{17}CHO$ + Ph_2CHNH_2 + **3** (OSiMe₃/OMe ketene silyl acetal) → Activator, $MgSO_4$, CH_2Cl_2, 0 °C

Table 15. Effect of catalysts

Entry	Activator		Yield/%
1	$Yb(OTf)_3$	10 mol%	89
2	$TiCl_4$	100 mol%	trace
3	TMSOTf	100 mol%	trace

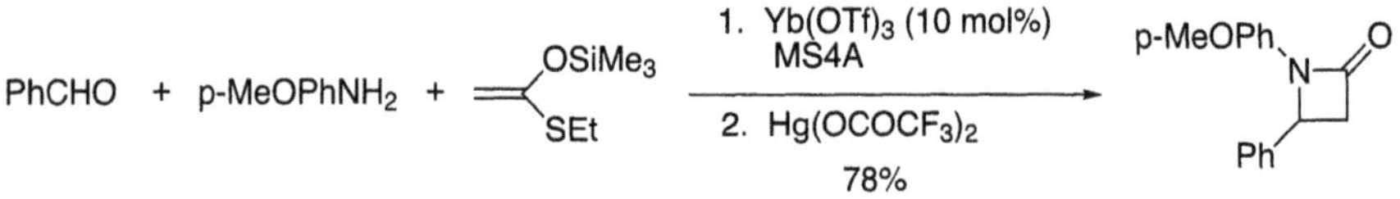

PhCHO + p-MeOPhNH₂ + (OSiMe₃/SEt) → 1. $Yb(OTf)_3$ (10 mol%) MS4A; 2. $Hg(OCOCF_3)_2$; 78%

Scheme 5

Thus, the one-pot synthesis of β-amino esters from aldehydes has been achieved by using lanthanide triflate catalysis. The high efficiency using simple starting materials and a catalytic amount of a reusable catalyst is especially noteworthy.

5.3 Use of Acylhydrazones as Electrophiles in Mannich-Type Reactions

Hydrazones are aldehyde and ketone equivalents as well as imines. Their stability is much higher than imines and actually hydrazones derived from aliphatic aldehydes are often crystalline and can be isolated and stored at room temperature. However, their reactivity as electrophiles is known to be low, and there have been many fewer reports on the reactions of hydrazones with nucleophiles than those of imines [45, 46].

While 3-phenylpropionaldehyde phenylhydrazone did not react with ketene silyl acetal **3b** (R^2=*t*Bu) derived from methyl isobutyrate at all, 3-phenylpropionaldehyde acylhydrazones reacted with **3b** in the presence of a catalytic amount of $Sc(OTf)_3$. Among the acylhydrazones tested, 4-trifluoromethylbenzoylhydrazone (**4a**, R^1=CF_3) gave the best yield (Table 16). It is noteworthy that the electronic effect of the benzoyl moieties influenced the yields dramatically. While hydrazones with electron-donating groups gave lower yields, higher yields were

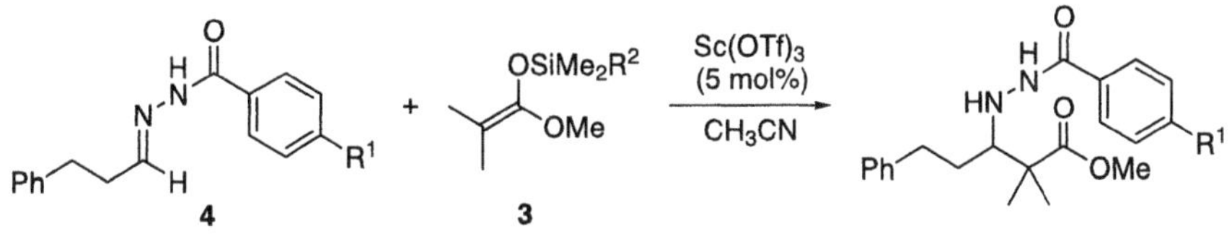

Table 16. Effect of R^1 and Lewis acids

Entry	Lewis Acid	R^1	R^2	Temp/°C	Yield/%
1	$Sc(OTf)_3$	H	^{t}Bu	-20	37
2	$Sc(OTf)_3$	MeO	^{t}Bu	-20	7
3	$Sc(OTf)_3$	NO_2	^{t}Bu	-20	77
4	$Sc(OTf)_3$	CF_3	^{t}Bu	-20	88
5	$Sc(OTf)_3$	CF_3	Me	-20	97 (75[a,b])
6	BF_3•OEt_2	CF_3	Me	0	4[a]
7	$TiCl_4$	CF_3	Me	0	13[a]
8	$SnCl_4$	CF_3	Me	0	46[a]

[a] Dichloromethane was used as a solvent.
[b] 0 °C.

obtained using hydrazones with electron-withdrawing groups. As for Lewis acids, $Sc(OTf)_3$ gave an excellent yield [47–49], and much lower yields were obtained by using typical Lewis acids such as $TiCl_4$, $SnCl_4$, and $BF_3.OEt_2$ [50].

Several examples of the reactions of 4-trifluoromethylbenzoylhydrazones with silyl enolates are shown in Table 17. Hydrazones derived from aromatic, aliphatic, and α,β-unsaturated aldehydes reacted with silyl enolates smoothly to afford the corresponding β-*N*′-acylhydrazinocarbonyl compounds in high yields. It is noted that several aliphatic hydrazones, readily prepared from aliphatic aldehydes, reacted with silyl enolates smoothly to afford the corresponding adducts in high yields [51]. All aliphatic acylhydrazones tested were crystalline and could be stored at room temperature. As for silyl enolates, the enolates derived from both esters and thioesters worked well. 1-Phenyl-1-trimethylsiloxyethene (a ketone-derived silyl enolate) also reacted with an aliphatic acylhydrazone to afford the corresponding adduct in a good yield.

Reductive cleavage of the nitrogen-nitrogen bond of the hydrazino compound (**6a**) was successfully carried out using Raney Ni under H_2 atmosphere (Scheme 6) [52]. Thus, adduct **5** was treated with a catalytic amount of Raney Ni (W-3) under H_2 (1 atm) at ambient temperature. After cleavage of the nitrogen-nitrogen bond, the resulting amine was protected as its *t*-butoxycarbonyl (Boc) group (**6**). It was found that β-lactam **7** was obtained by treatment of **5** with *n*-BuLi at –78 °C [53], while pyrazolone **8** was produced in the presence of NaOMe

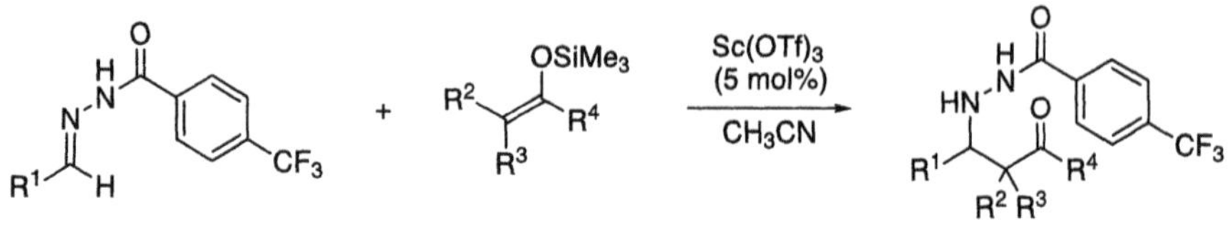

Table 17. Reactions of acylhydrazones with silyl enolates

Entry	R^1	R^2	R^3	R^4	Temp/°C	Yield/%
1	$Ph(CH_2)_2$	Me	Me	OMe	-20	97
2	$(CH_3)_2CHCH_2$	Me	Me	OMe	rt	88
3	$CH_3(CH_2)_5$	Me	Me	OMe	0	95
4	Ph	Me	Me	OMe	rt	73
5	$CH_3CH{=}CH$	Me	Me	OMe	0-rt	80
6	$PhCH{=}CH$	Me	Me	OMe	rt	75
7	$Ph(CH_2)_2$	H	H	SEt	0	79
8	$CH_3(CH_2)_5$	H	H	SEt	0	80
9	$(CH_3)_2CHCH_2$	H	H	SEt	0-rt	68
10	c-C_6H_{11}	H	H	SEt	0	68
11	$(CH_3)_2CHCH_2$	Me	Me	SEt	0-rt	83
12	$Ph(CH_2)_2$	OBn	H	O^iPr	-78	90[a,b]
13	$Ph(CH_2)_2$	H	Me	OPh	-45-0	95[a,c]
14	$Ph(CH_2)_2$	H	H	Ph	rt	66

[a] Propionitrile was used as a solvent.
[b] *syn/anti*=75/25.
[c] *syn/anti*=64/36.

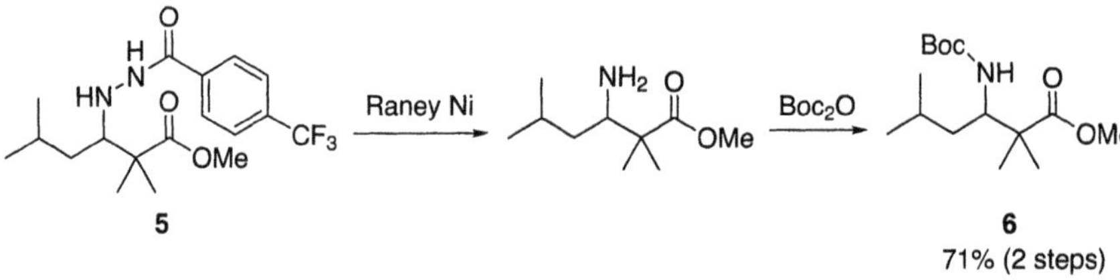

5 Raney Ni Boc₂O 6 71% (2 steps)

Scheme 6

at 70°C (Scheme 7). Since isomerization from **7** to **8** was observed under these conditions (NaOMe), **7** and **8** were expected to be kinetic and thermodynamic products, respectively. Moreover, pyrazolidinone **9** was obtained in the presence of samarium diiodide (SmI_2) [54] in THF-MeOH at 45°C.

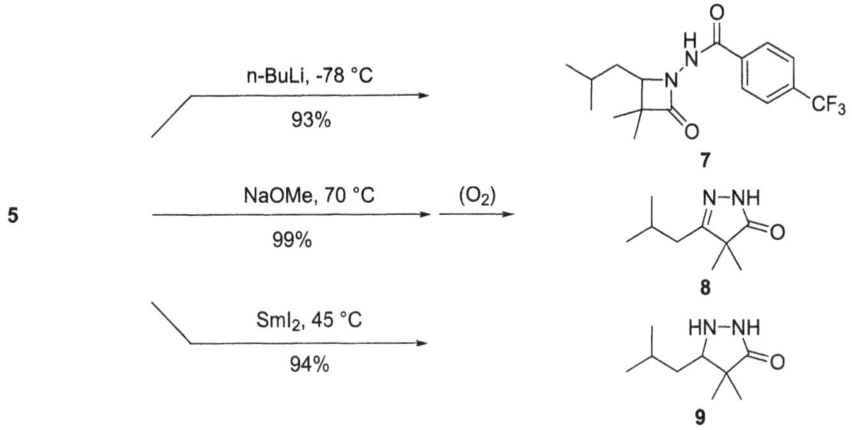

5

n-BuLi, -78 °C
93%
7

NaOMe, 70 °C
99%
(O_2)
8

SmI_2, 45 °C
94%
9

Scheme 7

5.4
Aqueous Mannich-Type Reaction

As mentioned in the previous sections, silyl enolates are excellent enolate components in the Mannich-type reactions with imines. Alternatively, it was found that vinyl ethers also reacted with imines smoothly in the presence of a catalytic amount of $Ln(OTf)_3$ to afford the corresponding β-amino ketones. In addition, the reactions proceeded smoothly by the combination of aldehydes, amines, and vinyl ethers in aqueous media [55].

A general scheme for the new Mannich-type reaction is shown below (Scheme 8); the procedure is very simple. In the presence of 10 mol% of $Yb(OTf)_3$, an aldehyde, an amine, and a vinyl ether were combined in a solution of THF-water (9:1) at room temperature to afford a β-amino ketone.

Selected examples of the present reaction are shown in Table 18. In all cases, β-amino ketones were obtained in good yields. Several characteristic features are noteworthy in this reaction. The procedure is very simple, consisting of simply mixing an aldehyde, an amine, a vinyl ether, and a small amount of $Ln(OTf)_3$ in aqueous solution. The catalyst could be recovered after the reaction was completed and could be reused (first run, 93%; second run, 83%; third run, 87%, in the reaction of phenylglyoxal•monohydrate, *p*-chloroaniline, and 2-methoxypropene). Commercially available formaldehyde and chloroacetaldehyde water solutions were used directly and the corresponding β-amino ketones were obtained in good yields. Phenylglyoxal•monohydrate, methyl glyoxylate, an aliphatic aldehyde, and an α,β-unsaturated ketone also worked well to give the corresponding β-amino esters in high yields. In some Mannich reactions with preformed iminium salts and imines, it is known that yields are often low because of the instability of the imines derived from these aldehydes or troublesome treatments are known to be required for their use [43, 56].

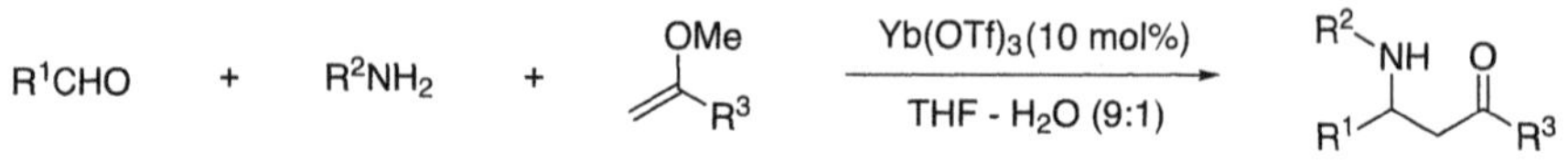

$R^1CHO + R^2NH_2 +$ (vinyl ether with OMe and R^3) $\xrightarrow[\text{THF - H}_2\text{O (9:1)}]{\text{Yb(OTf)}_3\text{(10 mol\%)}}$ β-amino ketone (R^1, R^2NH, R^3)

Scheme 8

Table 18. Synthesis of β-amino ketones in aqueous media

R^1	R^2	R^3	Yield/%
H	p-ClPh	Me	92
H	p-Ans	Me	76
H	p-Ans	Ph	quant.
Ph	p-ClPh	Me	90
Ph	p-Ans	Me	74
$Ph(CH_2)_2$	p-ClPh	Me	55
$ClCH_2$	p-ClPh	Me	59
PhCH=CH	p-ClPh	Me	73
PhCO	p-ClPh	Me	93
PhCO	Ph	Me	90
PhCO	p-Ans	Me	75
PhCO	p-Ans	Ph	85
MeO_2C	p-Ans	Me	67
MeO_2C	p-Ans	Ph	58

p-Ans=*p*-Anisidine

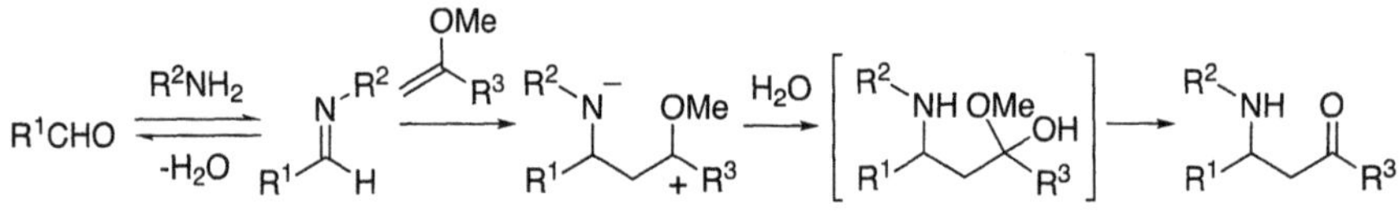

$R^1CHO \underset{-H_2O}{\overset{R^2NH_2}{\rightleftharpoons}}$ imine (R^1CH=NR^2) → (+ vinyl ether, OMe, R^3) → R^2N^- / OMe / $+R^3$ intermediate $\xrightarrow{H_2O}$ [R^2NH, OMe, OH, R^3] → β-amino ketone (R^1, R^2NH, R^3)

Scheme 9

A possible mechanism for the present reaction is shown in Scheme 9. It should be noted that dehydration accompanied by imine formation and successive addition of a vinyl ether proceed smoothly in aqueous solution and that the first aqueous Mannich-type reaction catalyzed by $Ln(OTf)_3$ has been developed [57]. Use of $Ln(OTf)_3$, a water-tolerant Lewis acid, is key and essential in this reaction.

6 Aza Diels-Alder Reactions

6.1 Reactions of Imines with Dienes or Alkenes

The imino Diels-Alder reaction is among the most powerful synthetic tools for constructing N-containing six-membered heterocycles, such as pyridines and quinolines [58]. Although Lewis acids often promote these reactions, more than stoichiometric amounts of the acids are required due to the strong coordination of the acids to nitrogen atoms [58]. Use of $Ln(OTf)_3$ as a catalyst was investigated in this reaction.

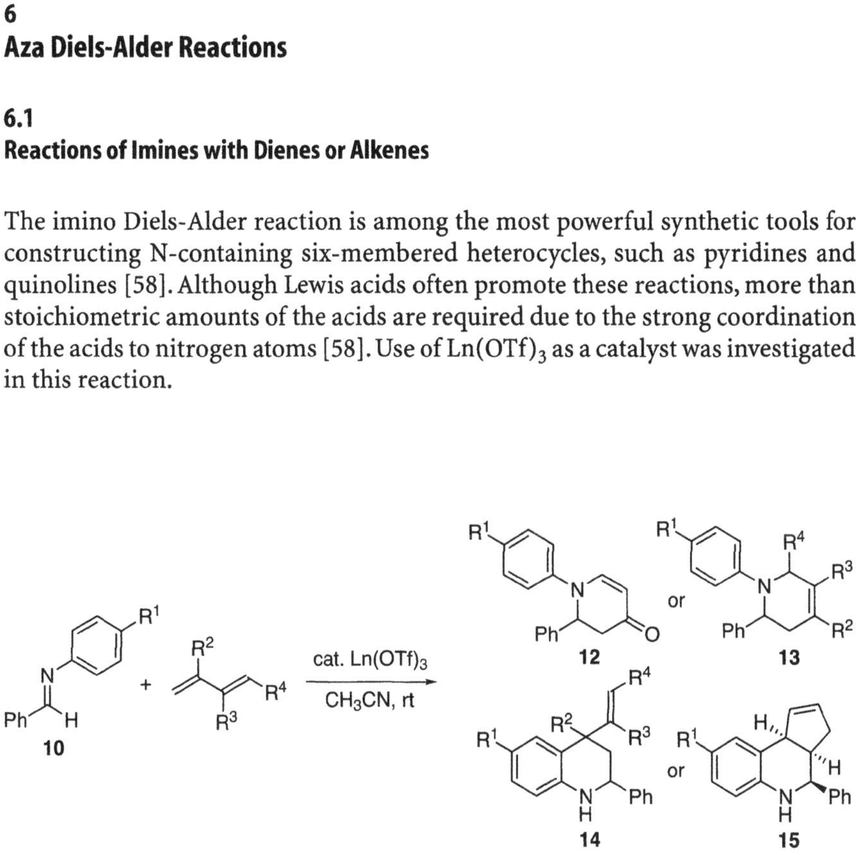

Table 19. Syntheses of pyridine dereivatives catalyzed by $Ln(OTf)_3$

R^1	Diene	$Ln(OTf)_3$ (mol%)	Product	Yield/%
H (**10a**)	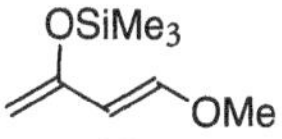 $OSiMe_3$, OMe **11**	$Yb(OTf)_3$ (10)	**12a**	93 (99)[a]
OMe (**10b**)			**12b**	82
Cl (**10c**)			**12c**	92
H (**10a**)	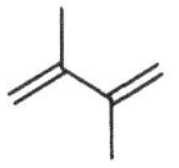	$Sc(OTf)_3$ (20)	**13a+14a**	54[b]
OMe (**10b**)			**13b+14b**	71[c]
Cl (**10c**)			**13c+14c**	50[d]
H (**10a**)		$Yb(OTf)_3$ (10)	**15a**	69 (91)[e,f]
OMe (**10b**)			**15b**	38[f]
Cl (**10c**)			**15c**	85[f]

[a] 10 mol% $Sc(OTf)_3$ was used. The reaction was carried out at 0 °C.
[b] **13a** 37%, **14a** 17%.
[c] **13b** 8%, **14b** 63%.
[d] **13c** 37%, **14c** 13%.
[e] 20 mol% $Sc(OTf)_3$ was used.
[f] *cis/trans*=94/6.

Table 20. Syntheses of quinoline derivatives catalyzed by $Ln(OTf)_3$

R^1	R^2		Alkene	Product	Yield/%	*cis/trans*
H	H	(10A)	SPh	16A	75	57/43
OMe	H	(10B)		16B	0	—
Cl	H	(10C)		16C	quant.	nd[a]
H	OMe	(10D)		16D	70	nd[a]
H	H	(10A)	OEt	17A	96	nd[a]
OMe	H	(10B)		17B	77	67/33
Cl	H	(10C)		17C	95	nd[a]
OMe	H	(10B)	Ph, $OSiMe_3$	18	quant.	83/17[b]

[a] Not determined.
[b] Relative configuration assignment was not made.

In the presence of 10 mol% of ytterbium triflate ($Yb(OTf)_3$, a representative lanthanide triflate), *N*-benzylideneaniline (**10a**) was treated with 2-trimethylsiloxy-4-methoxy-1,3-butadiene (Danishefsky's diene, **11**) [59] in acetonitrile at room temperature. The imino Diels-Alder reaction proceeded smoothly to afford the corresponding tetrahydropyridine derivative in a 93% yield (Table 19). The adduct was obtained quantitatively when $Sc(OTf)_3$ was used as a catalyst. Imines **10b** and **10c** also reacted smoothly with **11** to give the corresponding adducts in high yields. The reaction of **10a** with cyclopentadiene was performed under the same reaction conditions. It was found that the reaction course changed in this case and that a tetrahydroquinoline derivative was obtained in a 69% yield. In this reaction, the imine worked as an azadiene toward one of the double bonds of cyclopentadiene as a dienophile [43, 60]. In the reactions of 2,3-dimethylbutadiene, mixtures of tetrahydropyridine and tetrahydroquinoline derivatives were obtained.

Other examples and effects of $Ln(OTf)_3$ are shown in Tables 20 and 21, respectively [61]. A vinyl sulfide, a vinyl ether, and a silyl enol ether worked well as dienophiles to afford tetrahydroquinoline derivatives in high yields [62, 63]. As for the $Ln(OTf)_3$, heavy lanthanides such as Er, Tm, and Yb gave better results.

10a + cyclopentadiene → $Ln(OTf)_3$ (20mol %), CH_3CN, rt, 20h → 15a

Table 21. Effects of $Ln(OTf)_3$ (1)

Ln	Yield/%
Sc	94
Y	60
La	45
Pr	57
Nd	60
Sm	63
Eu	63
Gd	76
Dy	63
Ho	64
Er	97
Tm	92
Yb	85
Lu	72

6.2
Three-Components Coupling Reactions of Aldehydes, Amines, and Dienes or Alkenes

One synthetic problem in the imino Diels-Alder reactions is the imines' stability under the influence of Lewis acids. It is desirable that the imines activated by Lewis acids are immediately trapped by dienes or dienophiles [57]. In 1989, Sisko and Weinreb reported a convenient procedure for the imino Diels-Alder reaction of an aldehyde and a 1,3-diene with *N*-sulfinyl *p*-toluenesulfonamide via *N*-sulfonyl imine produced in situ, by using a stoichiometric amount of $BF_3.OEt_2$ as a promoter [64].

Bearing in mind the usefulness and efficiency of one-pot procedures, three-component coupling reactions between aldehydes, amines, and alkenes via imine formation and imino Diels-Alder reactions were examined by using $Ln(OTf)_3$ as a catalyst.

In the presence of 10 mol% of ytterbium triflate and magnesium sulfate, benzaldehyde was treated with aniline and **11** successively in acetonitrile at room

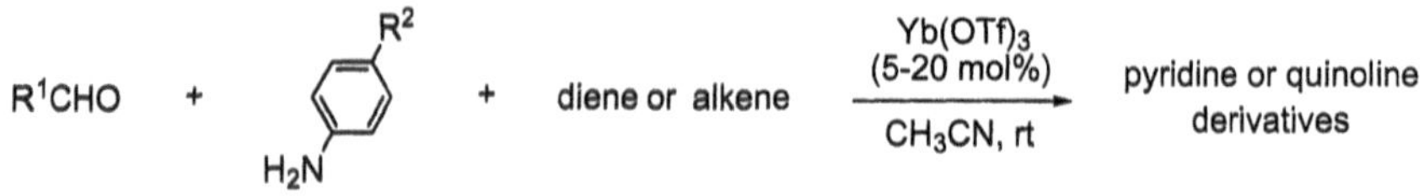

R^1CHO + (4-R^2-aniline, H_2N) + diene or alkene $\xrightarrow[CH_3CN,\ rt]{Yb(OTf)_3\ (5\text{-}20\ mol\%)}$ pyridine or quinoline derivatives

Table 22. One-pot synthesis of pyridine and quinoline derivatives

R^1	R^2	Diene or Alkene	Product	Yield/%	*cis/trans*
Ph	H	**11**	**12a**	80 (83)[a]	—
		 cyclopentadiene	**15a**	56	94/ 6
		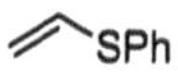 vinyl-SPh	**16a**	70	nd[b]
	Cl		**16c**	quant.	nd[b]
	H	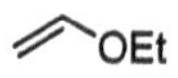 vinyl-OEt	**17a**	60	79/21
PhCO	H	**11**	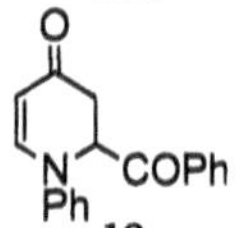 **19**	76	—
	H	 cyclopentadiene	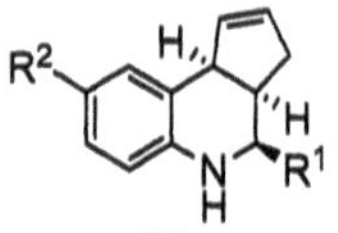 **20a**	94 (97)[c]	96/ 4 (96/ 4)
	OMe		**20b**	94	94/ 6
	Cl		**20c**	quant.	96/ 4
MeO_2C	H		**21a**	82	99/ 1
	Cl		**21c**	84	99/ 1
	Cl	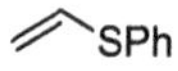 vinyl-SPh	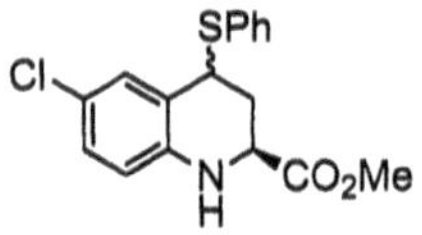 **22**	65	nd[b]
H[d]	Cl	 cyclopentadiene	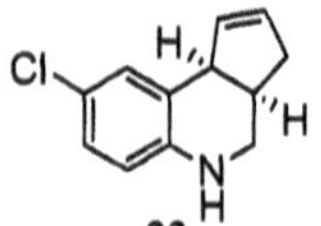 **23**	90[c]	—

[a] $Sc(OTf)_3$ (10 mol%) was used.
[b] Not determined.
[c] The reactions were carried out in aqueous solution (H_2O:EtOH:toluene=1:9:4).
[d] Commercial formaldehyde water solution was used.

PhCOCHO + [p-anisidine: NH_2, MeO] + [cyclopentadiene] $\xrightarrow[CH_3CN,\ rt,\ 20h]{Ln(OTf)_3\ (10\ mol\%)}$ [MeO, H, H, N, H, COPh] **20b**

Table 23. Effects of $Ln(OTf)_3$ (2)

Ln	Yield/%
Sc	63
Y	77
La	88
Pr	75
Nd	97
Sm	91
Eu	87
Gd	91
Dy	87
Ho	76
Er	84
Tm	84
Yb	94
Lu	80

temperature. The three-component coupling reaction proceeded smoothly to afford the corresponding tetrahydropyridine derivative in an 80% yield. It is noteworthy that $Yb(OTf)_3$ kept its activity and effectively catalyzed the reaction even in the presence of water and the amine [65]. Use of $Sc(OTf)_3$ slightly improved the yield. Other examples of the three-component coupling reaction are shown in Table 22. In the reaction between benzaldehyde, anisidine, and cyclopentadiene under the same reaction conditions, the reaction course changed and the tetrahydroquinoline derivative was obtained in a 56% yield. A vinyl sulfide, a vinyl ether, and a silyl enol ether worked well as dienophiles to afford tetrahydroquinoline derivatives in high yields. Phenylglyoxal monohydrate reacted with amines and **11** or cyclopentadiene to give the corresponding tetrahydropyridine or quinoline derivatives in high yields. As mentioned in the previous section, the imine derived from phenylglyoxal is known to be highly hygroscopic and its purification by distillation or chromatography is very difficult due to its instability

[43]. Moreover, the three-component coupling reactions proceeded smoothly in aqueous solution, and commercial formaldehyde water solution could be used directly. Most lanthanide triflates tested were effective in the three-component coupling reactions (Table 23). These reactions provide very useful routes for the synthesis of pyridine and quinoline derivatives.

6.3 Reaction Mechanism

In the reactions of **10a–10c** with cyclopentadiene, a vinyl sulfide, or a vinyl ether (**10a–c** work as azadienes), **10c** gave the best yields, while the yields using **10b** were lowest. The HOMO and LUMO energies and coefficients of **10a–c** and protonated **10a–c** are summarized in Table 24. These data do not correspond to the differences in reactivity between **10a–c** if the reactions are postulated to proceed via concerted [4+2] cycloaddition. On the other hand, the high reactivity of **4c** toward electrophiles compared to **10a** and **10b** may be accepted by assuming a stepwise mechanism.

The reaction of **10a** with 2-methoxypropene was tested in the presence of $Yb(OTf)_3$ (10 mol%). The main product was tetrahydroquinoline derivative **24a**, and small amounts of quinoline **25a** and β-amino ketone dimethylacetal **26a** were also obtained (Eq. 5). On the other hand, the three components coupling reaction between benzaldehyde, aniline, and 2-methoxypropene gave only a small amount of tetrahydroquinoline derivative **24a**, and the main products in this case were β-amino ketone **27a** and its dimethylacetal **26a** (Eq. 6). Similar results were obtained in the reaction of **4b** with 2-methoxypropene and the three

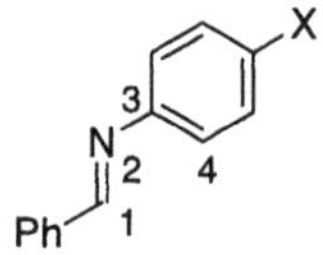

Table 24. HOMO and LUMO energies and coefficient of **10a–c**[a]

	HOMO /eV	Coefficient C1	N2	C3	C4	HOMO /eV	Coefficient C1	N2	C3	C4
10a	-8.97	0.32	0.35	-0.34	-0.24	-0.69	0.43	-0.38	-0.30	0.27
10b	-8.68	0.32	0.29	-0.40	-0.24	-0.64	0.43	-0.38	-0.28	0.28
10c	-8.93	0.30	0.31	-0.35	-0.22	-0.85	0.43	-0.35	-0.33	0.27
10a - H$^+$	-13.19	0.15	0.31	-0.39	-0.19	-5.75	0.63	-0.48	-0.09	0.21
10b - H$^+$	-12.53	0.20	0.18	-0.46	-0.08	-5.58	0.62	-0.49	-0.07	0.21
10c - H$^+$	-12.58	0.16	0.15	-0.37	-0.10	-5.77	0.63	-0.47	-0.19	0.21

[a] Calculated with MOPAC Ver. 6.01 using the PM3 Hamiltonian. MOPAC Ver. 7: Stewart JJP (1989) QCPE Bull 10:9. Revised as Ver. 6.01 by Tsuneo Hirano, University of Tokyo, for HITAC and UNIX machines (1989) JCPE Newslett 10:1.

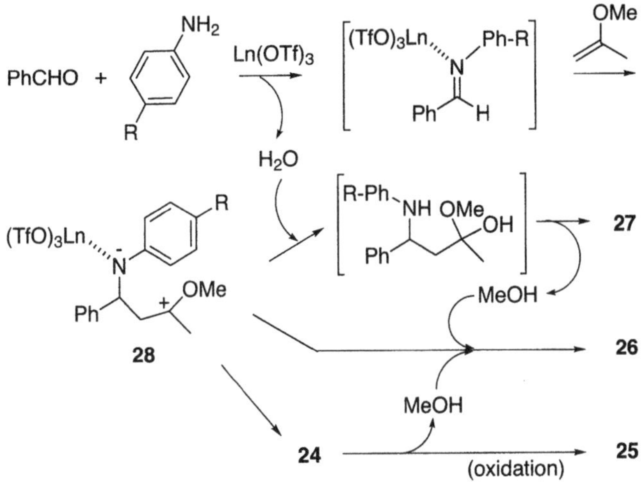

Scheme 10

components coupling reaction between benzaldehyde, anisidine, and 2-methoxypropene (Eqs. 7 and 8).

A possible mechanism of these reactions is shown in Scheme 10. Intermediate **28** is quenched by water and methanol generated in situ to afford **26** and **27**, respectively. While **24** is predominantly formed from **28** under anhydrous conditions, formation of **26** and **27** predominated in the presence of even a small amount of water. It is noted that these results suggest a stepwise mechanism in these types of imino Diels-Alder reactions [66].

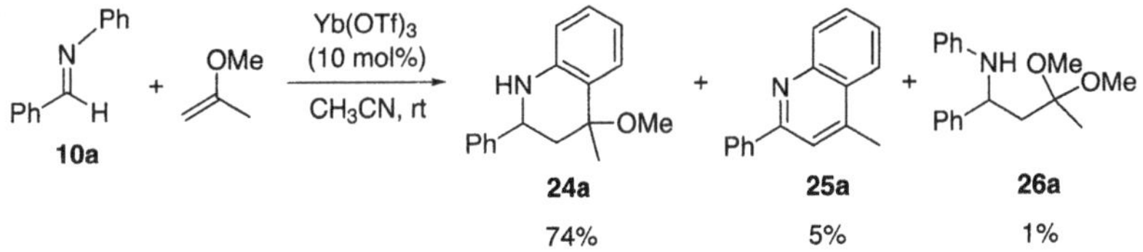

(5)

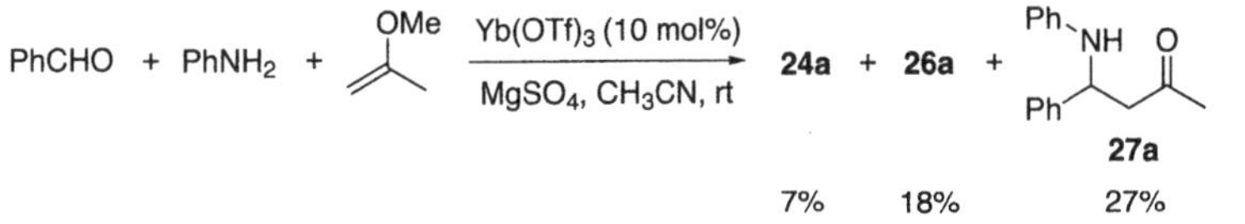

(6)

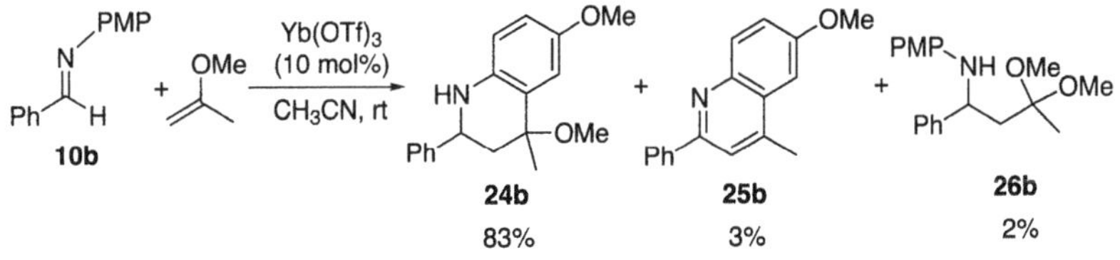

(7)

PMP = *p*-MeO-Ph

PhCHO + (p-anisidine, NH_2 / MeO) + (2-methoxypropene, OMe) $\xrightarrow[\text{MgSO}_4,\ \text{CH}_3\text{CN, rt}]{\text{Yb(OTf)}_3\ (10\ \text{mol\%})}$ **26b** + **27b** (8)

PMP = *p*-MeO-Ph

26b 34%, **27b** 37%

Thus, a new type of Lewis acid, lanthanide triflates, is quite effective for the catalytic activation of imines, and has achieved imino Diels-Alder reactions of imines with dienes or alkenes. The unique reactivities of imines which work as both dienophiles and azadienes under certain conditions were also revealed. Three-component coupling reactions between aldehydes, amines, and dienes or alkenes were successfully carried out by using $Ln(OTf)_3$ as catalysts to afford pyridine and quinoline derivatives in high yields. The triflates were stable and kept their activity even in the presence of water and amines. According to these reactions, many substituted pyridines and quinolines can be prepared directly from aldehydes, amines, and dienes or alkenes. A stepwise reaction mechanism in these reactions was suggested from the experimental results.

7 Asymmetric Diels-Alder Reactions

Recently, some efficient asymmetric Diels-Alder reactions catalyzed by chiral Lewis acids have been reported [67]. The chiral Lewis acids employed in these reactions are generally based on traditional acids such as titanium, boron, or aluminum reagents, and they are well modified to realize high enantioselectivities. Although lanthanide compounds were expected to be Lewis acid reagents, only a few asymmetric reactions catalyzed by chiral lanthanide Lewis acids were reported. Pioneering work by Danishefsky et al. demonstrated that $Eu(hfc)_3$ (an NMR shift reagent) catalyzed hetero-Diels-Alder reactions of aldehydes with siloxydienes, but enantiomeric excesses were moderate [68].

As shown in the previous section, lanthanide triflates were found to be good catalysts in the Diels-Alder reaction of various dienophiles with cyclic and acyclic dienes. The reactions proceeded smoothly in the presence of a catalytic amount of the triflate to give the corresponding adducts in high yields. Moreover, the catalyst was stable in water and was easily recovered from the aqueous layer after the reaction was completed, and could be reused. These unique properties were considered to be dependent on the specific characters of the lanthanides(III) [69], and design of chiral lanthanide triflates which could work as an efficient catalyst in the asymmetric Diels-Alder reaction was performed.

First, $Yb(OTf)_3$ was chosen as a representative of the lanthanide triflates. The chiral Yb triflate was prepared in situ from $Yb(OTf)_3$, (*R*)-(+)-binaphthol, and a tertiary amine (Eq. 9) [70], and a model reaction of 3-(2-butenoyl)-1,3-oxazolidin-2-one (**6**) with cyclopentadiene was examined [71].

$$Yb(OTf)_3 \xrightarrow[\text{MS 4A}]{\text{1.2 eq (binaphthol)}} \xrightarrow[0\ ^\circ C,\ 30\ \text{min}]{2.4\ \text{eq. amine}} \text{"chiral Yb triflate"} \quad (9)$$

Thus, in the presence of a chiral Yb triflate prepared from $Yb(OTf)_3$, (*R*)-(+)-binaphthol, and triethylamine at 0 °C for 0.5 h in dichloromethane, **29** reacted with cyclopentadiene at room temperature to afford the Diels-Alder adduct in an 87% yield (*endo*/*exo*=76/24) and the enantiomeric excess of the *endo* adduct was shown to be 33%. After screening several reaction conditions, it was found that the amine employed at the stage of the preparation of the chiral ytterbium triflate strongly influenced the diastereo- and enantioselectivities. In general, bulky amines gave better results and 70%, 75%, and 71% ees were observed when diisopropylethylamine, *cis*-2,6-dimethylpiperidine, and *cis*-1,2,6-trimethylpiperidine were used, respectively. In addition, a better result was obtained when the amine was combined with molecular sieves 4 Å (*cis*-1,2,6-trimethylpiperidine, 91% yield, *endo*/*exo*=86/14, *endo*=90% ee), and the enantiomeric excess was further improved to 95% when the reaction was carried out at 0 °C [72].

At this stage, although the reaction conditions were optimized, aging of the catalyst was found to take place. High selectivities (77% yield, *endo*/*exo*=89/11, *endo*=95% ee) were obtained when the diene and the dienophile were added just after $Yb(OTf)_3$, (*R*)-(+)-binaphthol, and a tertiary amine were stirred at 0 °C for 0.5 h in dichloromethane (the original catalyst system). On the other hand, the selectivities became lower in accordance with the stirring time of the catalyst solution and the temperature. These results seemed to be ascribed to the aging of the catalyst, but the best result (77% yield, *endo*/*exo*=89/11, *endo*=95% ee) was obtained when the mixture (the substrates and 20 mol% of the catalyst) was stirred at 0 °C for 20 h. It was suggested from this result that the substrates or the product stabilized the catalyst. The effect of the substrates or the product on the stabilization of the catalyst was then examined, and the dienophile (**29**) was found to be effective in preventing the catalyst from aging. When 20 mol% of the original catalyst system and **29** (additive) were stirred at 0 °C for 5.5 h in dichloromethane, the product was obtained in a 66% yield, *endo*/*exo*=87/13, and the enantiomeric excess of the *endo* adduct was 88%.

Moreover, after screening several additives other than **29**, it was found that some additives were effective not only in stabilizing the catalyst but also *in controlling the enantiofacial selectivities in the Diels-Alder reaction.* Selected examples are shown in Table 25. When 3-acetyl-1,3-oxazolidin-2-one (**30**) was combined with the original catalyst system (to form catalyst A), the *endo* adduct was

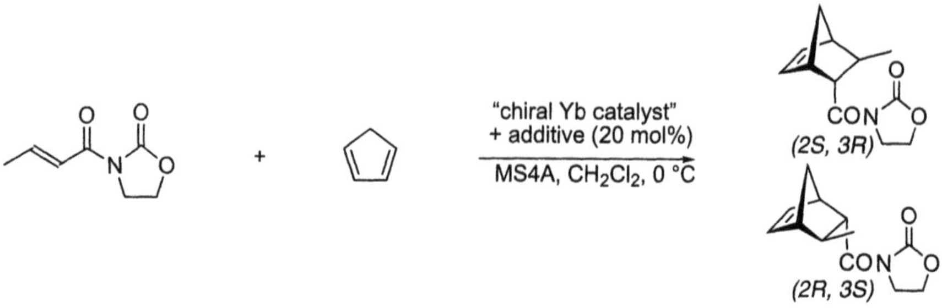

"chiral Yb catalyst"
+ additive (20 mol%)
MS4A, CH_2Cl_2, 0 °C

(2S, 3R)

(2R, 3S)

Table 25. Effect of additives

Additive	Yield/%	*endo/exo*	2*S*,3*R* /2*R*,3*S*	(ee (%))[a]
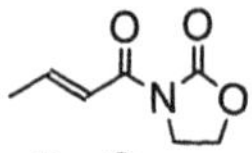 29	66	87/13	**94.0**/ 6.0	(88)
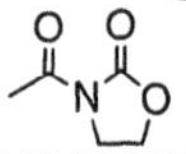 30	77	89/11	**96.5**/ 3.5	(93)
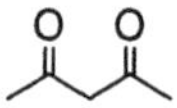	80	88/12	22.5/**77.5**	(55)
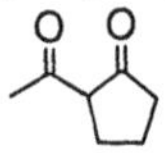	36	81/19	19.0/**81.0**	(62)
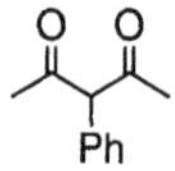	69	88/12	15.5/**84.5**	(69)
Ph	83	93/ 7	9.5/**90.5**	(81)[b]

[a] Enantiomer ratios of *endo* adducts.
[b] 1,2,2,6,6-Pentamethylpiperidine was used instead of *cis*-1,2,6-trimethylpiperidine. $Yb(OTf)_3$, MS4A, and the additive were stirred in dichloromethane at 40 °C for 3 h.

obtained in 93% ee and the absolute configuration of the product was *2 S, 3R*. On the other hand, when acetyl acetone derivatives were mixed with the catalyst, *reverse enantiofacial selectivities were observed.* The *endo* adduct with an absolute configuration of *2R, 3S* was obtained in 81% ee when 3-phenylacetylacetone (PAA) was used as an additive (catalyst B). In these cases, the chiral source was the same (*R*)-(+)-binaphthol. Therefore, *the enantioselectivities were controlled by the achiral ligands, 3-acetyl-1,3-oxazolidin-2-one and PAA* [73].

As shown in Table 26, the same selectivities were observed in the reactions of other 3-acyl-1,3-oxazolidin-2-ones. Thus, by using the same chiral source ((*R*)-(+)-binaphthol), both enantiomers of the Diels-Alder adducts between 3-acyl-1,3-oxazolidin-2-ones and cyclopentadiene were prepared. Traditional methods have required both enantiomers of chiral sources in order to prepare both enantiomers stereoselectively [74], but the counterparts of some chiral sources are of poor quality or are hard to obtain (for example, sugars, amino acids, alkaloids, etc.). *It is noted that the chiral catalysts with reverse enantiofacial selectivities could be prepared by using the same chiral source and a choice of achiral ligands.*

Table 26. Synthesis of both enantiomers of the Diels-Alder adducts between cyclopentadiene and dienophiles by use of catalysts A and B

Dienophile	Catalyst A			
	Yield/%	*endo/exo*	2*S*,3*R*/2*R*,3*S*	(ee (%))[b]
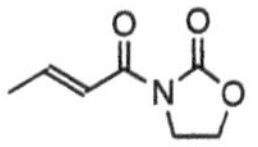	77	89/11	**96.5**/ 3.5	(93)
	77	89/11	**97.5**/ 2.5	(95)[c]
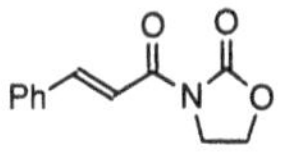 Ph	40	81/19	**91.5**/ 8.5[d]	(83)
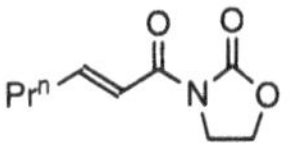 Pr^n	34	80/20	**93.0**/ 7.0	(86)
	81	80/20	**91.5**/ 8.5	(83)[c]

Dienophile	Catalyst B[a]			
	Yield/%	*endo/exo*	2*S*,3*R*/2*R*,3*S*	(ee (%))[b]
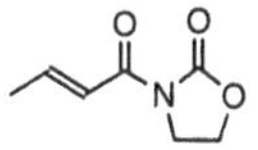	83	93/ 7	9.5/**90.5**	(81)
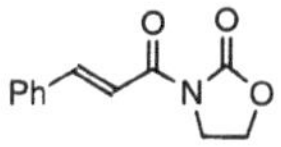 Ph	60	89/11	10.5/**89.5**[d]	(79)
	51	89/11	8.5/**91.5**[d]	(83)
	51	89/11	5.5/**94.5**[d]	(89)[e]
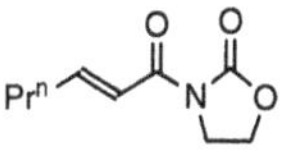 Pr^n	81	91/ 9	10.0/**90.0**	(80)
	85	91/ 9	9.0/**91.0**	(82)[e]
	60	91/ 9	7.5/**92.5**	(85)[f]

Catalyst A: $Yb(OTf)_3$+(*R*)-(+)-binaphthol+*cis*-1,2,6-trimethylpiperidine+MS4A+3-acetyl-1,3-oxazolidin-2-one (**30**)
Catalyst B: $Yb(OTf)_3$+(*R*)-(+)-binaphthol+*cis*-1,2,6-trimethylpiperidine+MS4A+3-phenylacetylacetone (PAA)
[a] 1,2,2,6,6-Pentamethylpiperidine was used instead of 1,2,6-trimethylpiperidine.
[b] Enantiomer ratios of *endo* adducts.
[c] Without additive.
[d] *2R,3R/2S,3S*.
[e] $Tm(OTf)_3$.
[f] $Er(OTf)_3$ was used instead of $Yb(OTf)_3$.

These exciting selectivities are believed to be strongly dependent on the specific coordination number of Yb(III) [69] (Scheme 11). Two binding sites for the ligands are now postulated in the Yb catalysts. **29** or **30** coordinates in site A under equilibrium conditions to stabilize the original catalyst system. When **29** coordinates Yb(III), cyclopentadiene attacks from the *si* face of **29** (site A favors *si* face attack). On the other hand, in catalyst B (the original catalyst system and PAA), site A is occupied by PAA [75]. Since another coordination site still remains in the Yb(III) catalyst owing to the specific coordination numbers, **29** coordinates at site B and cyclopentadiene attacks from the *re* face (site B favors *re* face attack).

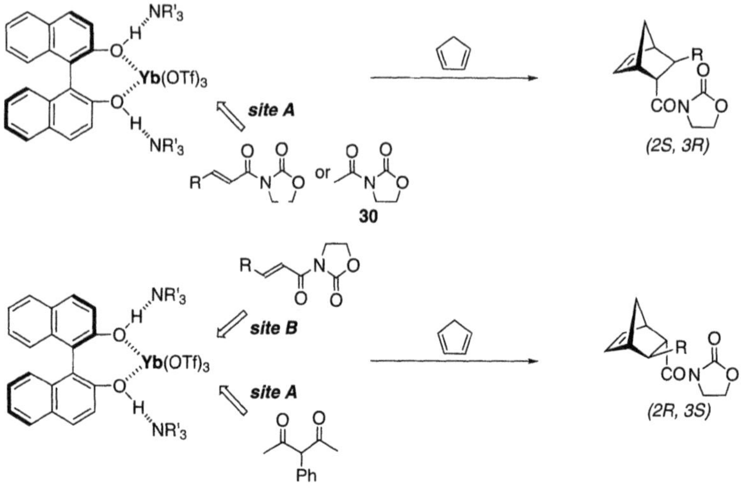

Scheme 11

Table 27. Effect of $Ln(OTf)_3$

Ln	Catalyst A Yield/%	*endo/exo*	2*S*,3*R*/2*R*,3*S*	(ee(%))[a]
Lu	60	89/11	96.5/3.5	(93)
Yb	77	89/11	96.5/3.5	(93)
Tm	46	86/14	87.5/12.5	(75)
Er	24	83/17	84.5/15.5	(69)
Ho	12	73/27	62.5/37.5	(25)
Y	6	70/30	60.0/40.0	(20)
Gd	0	—	—	—

Ln	Catalyst B Yield/%	*endo/exo*	2*S*,3*R*/2*R*,3*S*	(ee (%))[a]
Lu	30	89/11	24.5/75.5	(51)
Yb	88	92/ 8	15.0/85.0	(70)
Tm	72	91/ 9	13.0/87.0	(74)
Er	59	90/10	13.0/87.0	(74)
Ho	70	84/16	21.0/79.0	(58)
Y	85	91/ 9	19.5/80.5	(61)
Gd	61	85/15	28.5/71.5	(43)

[a] Enantiomer ratios of *endo* adducts.

The effect of other lanthanide triflates was also examined. As shown in Table 27, lanthanide elements strongly influenced the yields and selectivities. A slight difference between the two catalyst systems (catalysts A and B) on the ef-

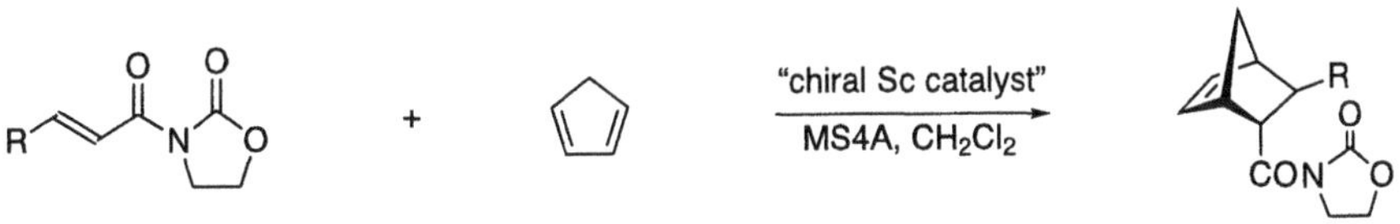

"chiral Sc catalyst"
MS4A, CH_2Cl_2

Table 28. Enantioselective Diels-Alder reactions using a chiral scandium catalyst

Dienophile	Catalyst/mol%	Yield/%	*endo/exo*	ee/% (*endo*)
3-(2-butenoyl)-1,3-oxazolidin-2-one	20	94	89/11	92 (2*S*,3*R*)
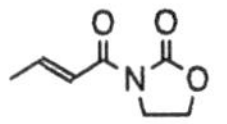	10	84	86/14	96 (2*S*,3*R*)
	5	84	87/13	93 (2*S*,3*R*)
	3	83	87/13	92 (2*S*,3*R*)
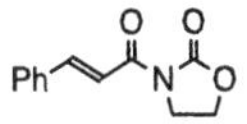 Ph (3-cinnamoyl-1,3-oxazolidin-2-one)	20	99	89/11	93 (2R,3*R*)
	10	96	90/10	97 (2R,3*R*)
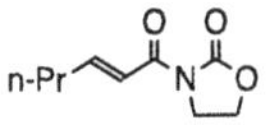 n-Pr (3-(2-hexenoyl)-1,3-oxazolidin-2-one)	20	95	78/22	74 (2*S*,3*R*)
	10	86	78/22	75 (2*S*,3*R*)

fect of the lanthanide elements was also observed. In catalyst A, lutetium triflate ($Lu(OTf)_3$) was also effective in generating the *endo* Diels-Alder adduct in 93% ee. The yields and selectivities diminished rapidly in accordance with the enlargement of the ionic radii. In catalyst B, on the other hand, the best results were obtained when thulium triflate ($Tm(OTf)_3$) or erbium triflate ($Er(OTf)_3$) was employed. Deviations to either larger or smaller ionic radii resulted in decreased selectivities, although the Diels-Alder adduct was obtained in an 85% yield with good selectivities (*endo*/*exo*=92/8, *endo* isomer=61% ee) even when holmium triflate ($Ho(OTf)_3$) was used.

Although $Sc(OTf)_3$ has slightly different properties compared with other lanthanide triflates, the chiral Sc catalyst could be prepared from $Sc(OTf)_3$, (*R*)-(+)-binaphthol, and a tertiary amine in dichloromethane [76]. The catalyst was also found to be effective in the Diels-Alder reactions of acyl-1,3-oxazolidin-2-ones with dienes. The amines employed in the preparation of the catalyst influenced the enantioselectivities strongly. For example, in the Diels-Alder reaction of 3-(2-butenoyl)-1,3-oxazolidin-2-one with cyclopentadiene (CH_2Cl_2, 0 °C), the enantiomeric excesses of the *endo* adduct depended crucially on the amines employed; aniline, 14% ee; lutidine, 46% ee; triethylamine, 51% ee; 2,2,6,6-tetramethylpiperidine, 51% ee; diisopropylethylamine, 69% ee; 2,6-dimethylpiperidine, 69% ee; 1,2,2,6,6-pentamethylpiperidine, 72% ee; and *cis*-1,2,6-trimethylpiperidine, 84% ee.

Several examples of the chiral Sc(III)-catalyzed Diels-Alder reactions are shown in Table 28. The highest enantioselectivities were observed when *cis*-1,2,6-trimethylpiperidine was employed as an amine. 3-(2-Butenoyl)-, 3-cinnamoyl-, and 3-(2-hexenoyl)-1,3-oxazolidin-2-ones reacted with cyclopentadi-

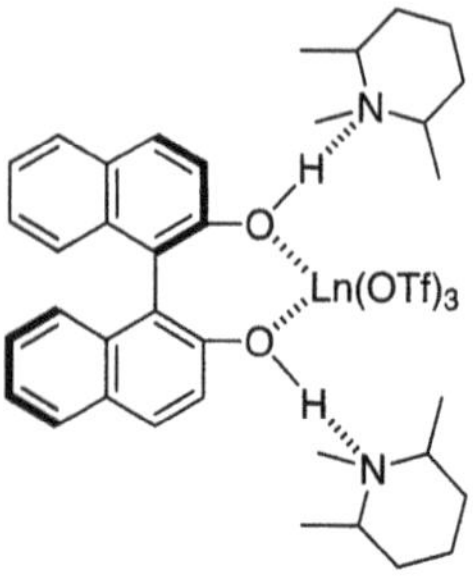

Scheme 12

ene smoothly in the presence of the chiral Sc catalyst to afford the corresponding Diels-Alder adducts in high yields and high selectivities. It should be noted that even 3 mol% of the catalyst was enough to complete the reaction and the *endo* adduct was obtained in a 92% ee.

The catalyst was also found to be effective for the Diels-Alder reactions of an acrylic acid derivative [77]. 3-Acryloyl-1,3-oxazolidin-2-one reacted with 2,3-dimethylbutadiene to afford the corresponding Diels-Alder adduct in a 78% yield and a 73% ee, whereas the reaction of 3-acryloyl-1,3-oxazolidin-2-one with cyclohexadiene gave a 72% ee for the *endo* adduct (88% yield, *endo*/*exo*=100/0).

Similar to the chiral Yb catalyst, aging was observed in the chiral Sc catalyst. It was also found that **30** or 3-benzoyl-1,3-oxazolidin-2-one was a good additive for stabilization of the catalyst, but that reverse enantioselectivities by additives were not observed. This can be explained by the coordination numbers of Yb(III) and Sc(III); while Sc(III) has up to seven ligands, specific coordination numbers of Yb(III) allow up to twelve ligands [69, 78].

As for the chiral ytterbium and scandium catalysts, the following structures were postulated. The unique structure shown in Scheme 12 was indicated by ^{13}C NMR and IR spectra. The most characteristic point of the catalysts was the existence of hydrogen bonds between the phenolic hydrogens of (*R*)-(+)-binaphthol and the nitrogens of the tertiary amines. The ^{13}C NMR spectra indicated these interactions, and the existence of the hydrogen bonds was confirmed by the IR spectra [79]. The coordination form of these catalysts may be similar to that of the lanthanide(III)-water or -alcohol complex [78]. It is noted that the structure is quite different from those of conventional chiral Lewis acids based on aluminum [80], boron [81], or titanium [82]. In the present chiral catalysts, the axial chirality of (*R*)-(+)-binaphthol is transferred via the hydrogen bonds to the amine parts, which shield one side of the dienophile effectively [83]. This is consistent with the experimental results showing that amines employed in the preparation of the chiral catalysts strongly influenced the selectivities and that bulky amines gave better selectivities.

Although the transition states of the chiral lanthanide(III)-catalyzed reactions are rather complicated due to the specific coordination number and stereochemistry of lanthanide(III), the sense of asymmetric induction in the chiral

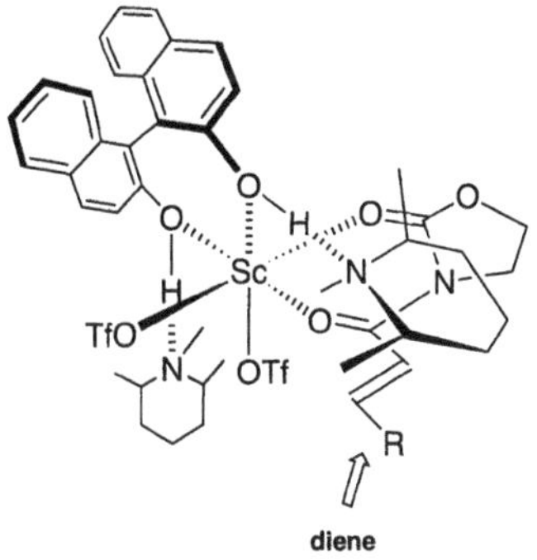

Scheme 13

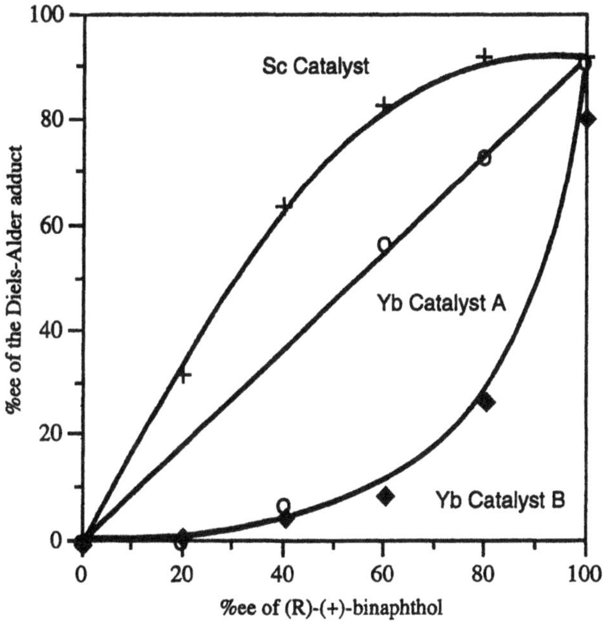

Yb Catalyst A: $Yb(OTf)_3$ + (*R*)-(+)-binaphtol + TMP+30
Yb Catalyst B: $Yb(OTf)_3$ + (*R*)-(+)-binaphtol + TMP+PAA
Sc Catalyst: $Sc(OTf)_3$ + (*R*)-(+)-binaphtol + TMP

Fig. 1. Correlation between the ee of the Diels-Alder adduct and the ee of (*R*)-(+)-binaphthol

scandium-catalyzed reactions can be rationalized by assuming an intermediate octahedral Sc(III)-dienophile complex (Scheme 13). The axial chirality of (*R*)-(+)-binaphthol is transferred to the amine, the *re* face of the acyl-1,3-oxazolidin-

2-one is effectively shielded by the amine part, and a diene approaches the dienophile from the *si* face to afford the adduct in a high enantioselectivity.

It was also suggested that aggregation of the catalysts influenced the selectivities in the Diels-Alder reactions, and the reaction of 3-(2-butenoyl)-1,3-oxazolidin-2-one with cyclopentadiene using (*R*)-(+)-binaphthol in lower enantiomeric excesses was examined [84]. The results are shown in Fig. 1. Very interestingly, a positive nonlinear effect was observed in the chiral Sc catalyst. In the chiral Yb catalysts, on the other hand, the effect was dependent on the additives. The extent of asymmetric induction in catalyst A did not deviate from the enantiomeric excesses of (*R*)-(+)-binaphthol in the range 60–100% ee [85], while a negative nonlinear effect was observed in catalyst B. These results can be ascribed to a difference in aggregation between the Sc catalyst, Yb catalyst A, and Yb catalyst B.

8 Asymmetric Aza Diels-Alder Reactions

While asymmetric reactions using chiral Lewis acids have been demonstrated to achieve several highly enantioselective carbon-carbon bond-forming processes using catalytic amounts of chiral sources [86], chiral Lewis acid-catalyzed asymmetric reactions of nitrogen-containing substrates are rare, probably because most chiral Lewis acids would be trapped by the basic nitrogen atoms to block the catalytic cycle. For example, aza Diels-Alder reactions are one of the most basic and versatile reactions for the synthesis of nitrogen-containing heterocyclic compounds [58, 87]. Although asymmetric versions using chiral auxiliaries or a stoichiometric amount of a chiral Lewis acid have been reported [88], examples using a catalytic amount of a chiral source are unprecedented.

In the previous section, lanthanide triflates were shown to be excellent catalysts for achiral aza Diels-Alder reactions. While stoichiometric amounts of Lewis acids are required in many cases, a small amount of the triflate effectively catalyzes the reactions. On the other hand, chiral lanthanide Lewis acids have been developed to realize highly enantioselective Diels-Alder reactions of 2-oxazolidin-1-one with dienes [89]. The reaction of *N*-benzylideneaniline with cyclopentadiene was first performed under the influence of 20 mol% of a chiral ytterbium Lewis acid prepared from ytterbium triflate ($Yb(OTf)_3$), (*R*)-(+)-1,1'-bi-naphthol (BINOL), and trimethylpiperidine (TMP). The reaction proceeded smoothly at room temperature to afford the desired tetrahydroquinoline derivative in a 53% yield, although no chiral induction was observed. At this stage, it was indicated that bidentate coordination between a substrate and a chiral Lewis acid would be necessary for reasonable chiral induction. *N*-Benzylidene-2-hydroxyaniline (**31a**) was then prepared, and the reaction with cyclopentadiene (**32a**) was examined. It was found that the reaction proceeded smoothly to afford the corresponding 8-hydroxyquinoline derivative (**33a**) [90] in a high yield. The enantiomeric excess of the *cis* adduct in the first trial was only 6%; however, the selectivity increased when diazabicyclo-[5,4,0]-undec-7-ene (DBU) was used in-

HO, N, Ph, H **31a** + **32a** → Chiral Yb Complex[a] (20 mol%) Additive, CH_2Cl_2, MS4A → H, H, Ph, N H, OH **33a**

Table 29. Effect of additive

Additive[b]/mol%	Temp/°C	Yield/%	*cis/trans*	ee/% (*cis*)
—	0	71	98 / 2	62
—	-15 to 0	48	99 / 1	68
MID (20)	-15 to 0	21	98 / 2	91
DTBP (20)	0	49	95 / 5	31
DTBP (100)	0	67	99 / 1	61
DMP (100)	0	14	98 / 2	56
DTBMP (100)	-15	82	>99 / 1	70
DTBP (100)	-15	92	>99 / 1	71

[a] Prepared from $Yb(OTf)_3$, (*R*)-(+)-BINOL, and DBU.
[b] MID: 1-Methylimidazole. DTBP: 2,6-di-*t*-butylpyridine. DMP: 2,6-dimethylpyridine. DTBMP: 2,6-di-*t*-butyl-4-methylpyridine.

stead of TMP (Table 29). It was also indicated that the phenolic hydrogen of **31a** would interact with DBU, which should interact with the hydrogen of (*R*)-(+)-BINOL [91], to decrease the selectivity. Additives which interact with the phenolic hydrogen of **31a** were then examined. When 20 mol% of *N*-methylimidazole (MID) was used, 91% ee of the *cis* adduct was obtained; however, the chemical yield was low. Other additives were screened and it was found that the desired tetrahydroquinoline derivative was obtained in a 92% yield with high selectivities (*cis*/*trans*>99/1, 71% ee), when 2,6-di-*t*-butyl-4-methylpyridine (DTBMP) was used.

Other substrates were tested, and the results are summarized in Table 30 [92]. Vinyl ethers (**32b–32d**) also worked well to afford the corresponding tetrahydroquinoline derivatives (**33b–33e**) in good to high yields with good to excellent diastereo- and enantioselectivities (entries 1–9). Use of 10 mol% of the chiral catalyst also gave the adduct in high yields and selectivities (entries 2, 6). As for additives, 2,6-di-*t*-butylpyridine (DTBP) gave the best result in the reaction of imine **31a** with ethyl vinyl ether (**32b**), while higher selectivities were obtained when DTBMP or 2,6-diphenylpyridine (DPP) was used in the reaction of imine **31a** with **32b.** This could be explained by the slight difference in the asymmetric environment created by $Yb(OTf)_3$, (*R*)-(+)-BINOL, DBU, and the additive (see below). While use of butyl vinyl ether (**32c**) decreased the selectivities (entry 7), dihydrofurane (**32d**) reacted smoothly to achieve high levels of selectivity (entries 8, 9). It was found that the imine (**31c**) prepared from cyclohexanecar-

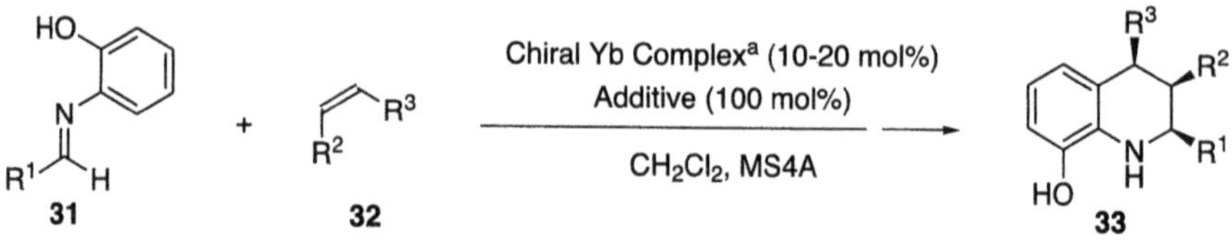

HO, N, R^1, H **31** + R^3, R^2 **32** — Chiral Yb Complex[a] (10-20 mol%), Additive (100 mol%), CH_2Cl_2, MS4A → R^3, R^2, R^1, N H, HO **33**

Table 30. Asymmetric synthesis of tetrahydroquinoline derivatives

Entry	R^1	Alkene	Additive[a]	Amount of Catalyst /mol%	Temp /°C	Product	Yield /%	*cis/trans*	Ee/% (*cis*)
1	Ph **(31a)**	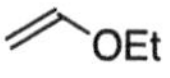 OEt **(32b)**	DTBP	20	-45	**33b**	58	94 / 6	61
2	Ph **(31a)**	**32b**	DTBP	10	-45	**33b**	52	94 / 6	77
3	α-Naph **(31b)**	**32b**	DTBP	20	-15	**33c**	69	>99/ 1	86
4	α-Naph **(31b)**	**32b**	DPP[b]	20	-15	**33c**	65	99/ 1	91
5	α-Naph **(31b)**	**32b**	DTBMP	20	-15	**33c**	74	>99/ 1	91
6	α-Naph **(31b)**	**32b**	DTBMP	10	-15	**33c**	62	98/ 2	82
7	α-Naph **(31b)**	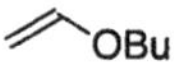 OBu **(32c)**	DTBMP	20	-15	**33d**	80	66/34	70
8	α-Naph **(31b)**	 O **(32d)**	DTBMP	20	-15	**33e**	90	91 / 9	78
9	α-Naph **(31b)**	**32d**	DPP[b]	20	-15	**33e**	67	93 / 7	86
10	α-Naph **(31b)**	 **(32a)**	DTBMP	20	-15	**33f**	69	>99/ 1	68
11[c]	c-C_6H_{11} **(31c)**	**32a**	DTBMP	20	-15	**33g**	58	>99/ 1	73

[a] See Table 29.
[b] 2,6-Diphenylpyridine.
[c] $Sc(OTf)_3$ was used.

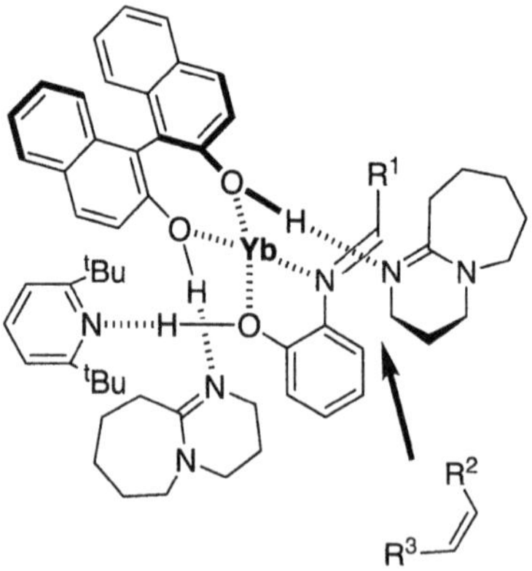

O, H, Yb, O, H, N, ^{t}Bu, N, H, O, ^{t}Bu, N, N, R^1, N, N, R^2, R^3

Scheme 14

boxaldehyde and 2-hydroxyaniline was unstable and difficult to purify. The asymmetric aza Diels-Alder reaction was successfully carried out using the three component coupling procedure (successively adding the aldehyde, the amine, and cyclopentadiene) in the presence of $Sc(OTf)_3$ (instead of $Yb(OTf)_3$), (*R*)-(+)-BINOL, DBU, and DTBMP.

Assumed transition state of this reaction is shown in Scheme 14. $Yb(OTf)_3$, (*R*)-(+)-BINOL, and DBU form a complex with two hydrogen bonds, and the axial chirality of (*R*)-(+)-BINOL is transferred via the hydrogen bonds to the amine parts. The additive would interact with the phenolic hydrogen of the imine, which is fixed by bidentate coordination to Yb(III). Since the top face of the imine is shielded by the amine, the dienophiles approach from the bottom face to achieve high levels of selectivity.

Thus, catalytic asymmetric aza Diels-Alder reactions of imines with alkenes have been developed using a chiral lanthanide Lewis acid, to afford 8-hydroxyquinoline derivatives in high yields with high diastereo- and enantioselectivities. Characteristic points of this reaction are as follows.

i. Asymmetric aza Diels-Alder reactions between achiral azadienes and dienophiles have been achieved using a catalytic amount of a chiral source.
ii. The unique reaction pathway in which the chiral Lewis acid activates not dienophiles but dienes, is revealed. In most asymmetric Diels-Alder reactions reported using chiral Lewis acids, the Lewis acids activate dienophiles [86, 93].
iii. A unique lanthanide complex including an azadiene and an additive, which is quite different from the conventional chiral Lewis acids, has been developed.

9 Catalytic Enantioselective 1,3-Dipolar Cycloadditions

1,3-Dipolar cycloadditions between nitrones and alkenes are most useful and convenient for the preparation of isoxazolidine derivatives, which are readily converted to 1,3-amino alcohol equivalents under mild reducing conditions [94]. In spite of the importance of chiral amino alcohol units for the synthesis of biologically important alkaloids, amino acids, β-lactams, and amino sugars, etc. [95], catalytic enantioselective 1,3-dipolar cycloadditions remain relatively unexplored [95, 96]. Catalytic enantioselective 1,3-dipolar cycloadditions of nitrones with alkenes using a chiral lanthanide catalyst were investigated [97, 98]. First, the 1,3-dipolar cycloaddition reaction of *N*-benzylidenebenzylamine *N*-oxide with 3-(2-butenoyl)-1,3-oxazolidin-2-one was performed in the presence of a chiral Yb(III) catalyst (20 mol%) prepared from $Yb(OTf)_3$, (*S*)-1,1'-binaphthol ((*S*)-BINOL), and triethylamine (Et_3N). The reaction proceeded smoothly at room temperature to afford the corresponding isoxazolidine derivative in a 65% yield with high *endo*/*exo* selectivity (99/1), and a moderate ee of the *endo* adduct was observed (Table 31). The enantiomeric excess was improved to 78% when *cis*-1,2,6-trimethylpiperidine (TMP) was used instead of Et_3N. Further-

Table 31. Effect of amines

Amine	Yield/%	*endo* /*exo*	ee/% [b]
Et_3N	65	99/1	63
iPr_2NEt	73	>99/1	62
cis -1,2,6-TMP [c]	73	99/1	78
(*R*)-MPEA [d]	92	>99/1	71
(*S*)-MPEA	80	97/3	35
(*R*)-MNEA [e]	92	99/1	96
(*S*)-MNEA	87	99/1	62

[a] Chiral Yb(III)=Yb$(OTf)_3$+(*S*)-BINOL+amine.
[b] Ee of the *endo* adducts.
[c] *cis*-1,2,6-Trimethylpiperidine.

more, it was found that *use of chiral amines influenced the selectivity dramatically, and combination of the chirality of BINOL and the amine was crucial for the selectivity.* Namely, 71% ee of the *endo* adduct was obtained in the model reaction using a catalyst prepared by the combination of (*S*)-BINOL and *N*-methyl-bis[(*R*)-1-phenylethyl]amine ((*R*)-MPEA), while only 35% ee was observed by the combination of (*S*)-BINOL and (*S*)-MPEA. Moreover, it was found that 96% ee of the *endo* adduct was obtained with an excellent yield (92%) and diastereoselectivity (*endo*/*exo*=99/1) by the combination of (*S*)-BINOL and a newly-prepared chiral amine, *N*-methyl-bis[(*R*)-1-(1-naphthyl)ethyl]amine ((*R*)-MNEA) [99]. The chiral Yb(III) catalyst thus prepared has two independent chiralities (hetero-chiral Yb(III) catalyst, see below) and it was found that the sense of the chiral induction in these reactions was mainly determined by BINOL and that the chiral amine increased or decreased the induction relatively.

Several examples of the 1,3-dipolar cycloadditions between nitrones and 3-(2-alkenoyl)-1,3-oxazolidin-2-ones using the novel hetero-chiral Yb(III) catalyst are shown in Table 32. In most cases, the desired isoxazolidine derivatives were obtained in excellent yields with excellent diastereo- and enantioselectivities. It is noted that high levels of selectivities were attained at room temperature. Nitrones derived from aromatic and heterocyclic aldehydes gave satisfactory re-

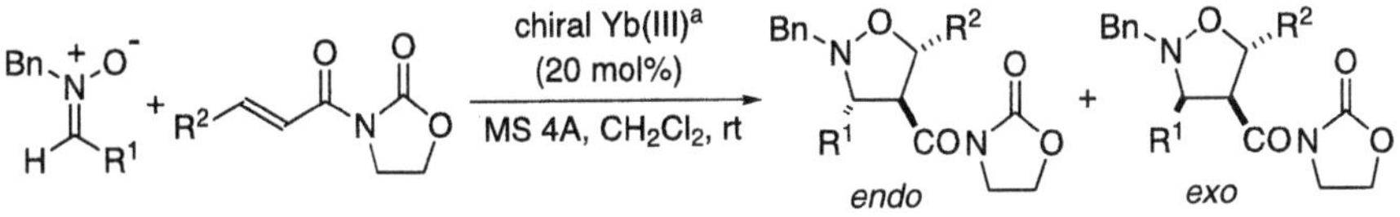

chiral Yb(III)[a] (20 mol%)
MS 4A, CH_2Cl_2, rt
endo
exo

Table 32. Catalytic enantioselective 1,3-dipolar cycloadditions

R^1	R^2	Yield/%	*endo/exo*	ee/%[b]
Ph	CH_3	92	99/1	96
p-Cl-Ph	CH_3	93	99/1	92
p-MeO-Ph	CH_3	82	95/5	90
2-furyl	CH_3	89	95/5	89
1-naphthyl	CH_3	88	98/2	85
Ph	H	91	>99/1	79
Ph	C_3H_7	89	98/2	93
C_2H_5	CH_3	88	53/47	96

[a] Chiral Yb(III)=Yb(OTf)$_3$+(*S*)-BINOL+(*R*)-MNEA.
[b] Ee of the *endo* adduts.

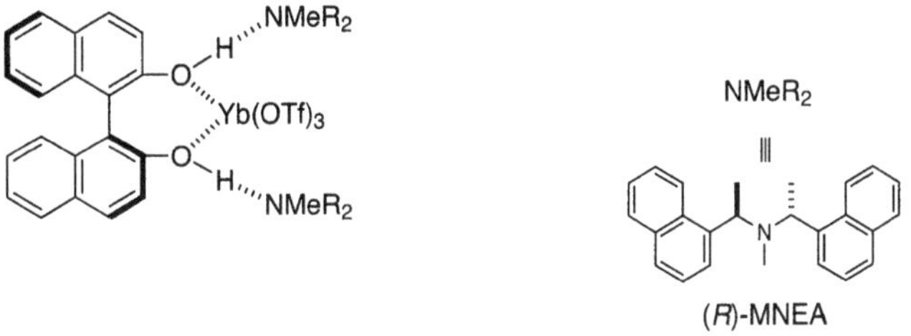

NMeR$_2$
Yb(OTf)$_3$
NMeR$_2$
(*R*)-MNEA

Scheme 15

sults, and even in the reaction using the nitrone derived from an aliphatic aldehyde, the cycloaddition proceeded smoothly to give the *endo* adduct in an excellent enantiomeric excess, albeit low *endo/exo* selectivity was observed. Moreover, it was found that alkenes which could be employed in the present 1,3-dipolar cycloaddition were not limited to 3-(2-alkenoyl)-1,3-oxazolidin-2-one derivatives. When *N*-phenylmaleimide was used as a dipolarophile, the desired isoxazolidine derivative was obtained in a 70% yield with *endo/exo*>99/1, and the enantiomeric excess of the *endo* adduct was 70% ee under the standard reaction conditions [100, 101].

As for the structure of the hetero-chiral Yb(III) catalyst, the following structure was supported (Scheme 15). Actually, the existence of hydrogen bonds between the phenolic hydrogens of (*S*)-BINOL and the nitrogens of (*R*)-MNEA was confirmed by the IR spectra of the catalyst [79, 102].

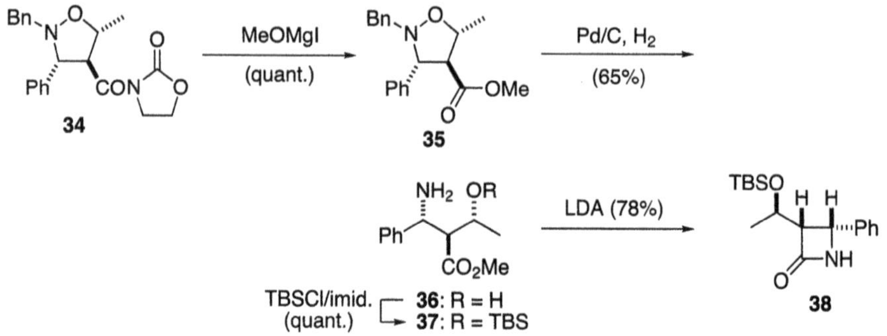

Scheme 16

3-Hydroxyethyl-β-lactam derivative was synthesized using the present reactions (Scheme 16). Isoxazolidine derivative **34**, prepared via the catalytic enantioselective 1,3-dipolar cycloaddition, was treated with methoxymagnesium iodide [103] to give methyl ester **35**. Reductive N-O bond cleavage and deprotection of the *N*-benzyl part of **35** was performed in the same pot using Pd/C under hydrogen atmosphere (10 kg/cm^2) [104] to afford amino ester **36**. After the resulting alcohol moiety was protected as its *t*-butyldimethylsilyl (TBS) ether, cyclization of **37** proceeded smoothly using lithiumdiisopropylamide (LDA) [105] to afford the corresponding β-lactam (**38**) [106] in a good yield.

10 Conclusions

Lanthanide triflates are new types of Lewis acids different from typical Lewis acids such as $AlCl_3$, BF_3, $SnCl_4$, etc. While most Lewis acids are decomposed or deactivated in the presence of water, lanthanide triflates are stable and works as Lewis acids in water solutions. Many nitrogen-containing compounds such as imines and hydrazones are also successfully activated by using a small amount of $Ln(OTf)_3$. Lanthanide triflates are also excellent Lewis acid catalysts in organic solvents. A catalytic amount of $Ln(OTf)_3$ is enough to complete reactions in most cases. In addition, $Ln(OTf)_3$ can be recovered after reactions are completed and can be reused.

Recently, polymer-supported lanthanide catalysts have beem of great interest, and these topics are discussed elsewhere. Use of lanthanide catalysts in solid-phase organic synthesis is now well-recognized [107]. There have also been many transformations other than carbon-carbon bond-forming reactions in organic synthesis using lanthanide triflates as catalysts, and all these will be reviewed in the near future.

Acknowledgments. Our work in this area was partially supported by CREST, Japan Science and Technology Corporation (JST), and a Grant-in-Aid for Scientific Research from the Ministry of Education, Science, Sports, and Culture, Japan. The author thanks and expresses his deep gratitude to his coworkers whose names appear in the references.

11
References and Notes

1. Schinzer D (ed) (1989) Selectivities in Lewis acid promoted reactions. Kluwer Academic, Dordrecht
2. Molander GA (1992) Chem Rev 92:29
3. Baes CF Jr, Mesmer RE (1976) The hydrolysis of cations. Wiley, New York, p 129
4. Thom KF (1971) US Pat 3,615,169; CA (1972) 76:5436a
5. (a) Forsberg JH, Spaziano VT, Balasubramanian TM, Liu GK, Kinsley SA, Duckworth CA, Poteruca JJ, Brown PS, Miller JL (1987) J Org Chem 52:1017; (b) Collins S, Hong Y (1987) Tetrahedron Lett 28:4391; (c) Almasio MC, Arnaud-Neu F, Schwing-Weill M-J (1983) Helv Chim Acta 66:1296; (d) Harrowfield JM, Kepert DL, Patrick JM, White AH (1983) Aust J Chem 36:483
6. (a) Mukaiyama T, Narasaka K, Banno T (1973) Chem Lett 1011; (b) Mukaiyama T, Banno K, Narasaka K (1974) J Am Chem Soc 96:7503
7. Mukaiyama T (1982) Org React 28:203
8. Kobayashi S, Murakami M, Mukaiyama T (1985) Chem Lett 1535
9. (a) Kawai M, Onaka M, Izumi Y (1986) Chem Lett 1581; (b) Kawai M, Onaka M, Izumi Y (1988) Bull Chem Soc Jpn 61:1237
10. (a) Noyori R, Yokoyama K, Sakata J, Kuwajima I, Nakamura E, Shimizu M (1977) J Am Chem Soc 99:1265; (b) Nakamura E, Shimizu M, Kuwajima I, Sakata J, Yokoyama K, Noyori R (1983) J Org Chem 48:932
11. Lanthanide(III) chlorides or some organolanthanide compounds catalyzed aldol reactions of ketene silyl acetals with aldehydes were reported: (a) Takai K, Heathcock CH (1985) J Org Chem 50:3247; (b) Vougioukas AE, Kagan HB (1987) Tetrahedron Lett 28:5513; (c) Gong L, Streitwieser A (1990) J Org Chem 55:6235, (d) Mikami K, Terada M, Nakai T (1993) J Chem Soc, Chem Commun 343 and references cited therein
12. (a) Lubineau A (1986) J Org Chem 51:2142; (b) Lubineau A, Meyer E (1988) Tetrahedron 44:6065
13. For example: (a) Hajos ZG, Parrish DR (1973) J Org Chem 38:3244; (b) Stork G, Isobe M (1975) J Am Chem Soc 97:4745; (c) Lucast DH, Wemple J (1976) Synthesis 724; (d) Ono N, Miyake H, Fujii M, Kaji A (1983) Tetrahedron Lett 24:3477; (e) Tsuji J, Nisar M, Minami I (1986) Tetrahedron Lett 27:2483; (f) Larsen SD, Grieco PA, Fobare WF (1986) J Am Chem Soc 108:3512
14. Snider BB and Yamamoto K respectively developed formaldehyde-organoaluminum complex as formaldehyde source in several reactions: (a) Snider BB, Rodini DJ, Kirk TC, Cordova R (1982) J Am Chem Soc 104:555; (b) Snider BB (1989) In: Schinzer D (ed) Selectivities in Lewis acid promoted reactions. Kluwer Academic Publishers, London, pp 147–167; (c) Maruoka K, Conception AB, Hirayama N, Yamamoto H (1990) J Am Chem Soc 112:7422; (d) Maruoka K, Concepcion AB, Murase N, Oishi M, Yamamoto H (1993) J Am Chem Soc 115:3943
15. Cf. TMSOTf-mediated aldol-type reaction of silyl enol ethers with dialkoxymethanes was also reported: Murata S, Suzuki M, Noyori R (1980) Tetrahedron Lett 21:2527
16. Kobayashi S (1991) Chem Lett 2187
17. Gaut H, Skoda J (1946) Bull Soc Chim Fr 13:308

18. Kobayashi S, Hachiya I (1992) Tetrahedron Lett 33:1625
19. Kobayashi S, Hachiya I (1994) J Org Chem 59:3590.
20. Kobayashi S, Hachiya I, Yamanoi Y (1994) Bull Chem Soc Jpn 67:2342
21. Haggin J (1994) Chem Eng News Apr 18:22
22. Kobayashi S, Hachiya I, Takahori T (1993) Synthesis 371
23. Kobayashi S, Hachiya I, Ishitani H, Araki M (1993) Synlett 472
24. $Y(OTf)_3$ and $Yb(OTf)_3$ catalyzed the aldol reaction of 2 with benzaldehyde at room temperature in dichloromethane to afford the aldol adduct in 58% (*syn*/*anti*=32/68) and 81% yields (*syn*/*anti*=29/71), respectively
25. It was reported that TMSOTf selectively activated acetals rather than aldehydes in the aldol-type reaction of silyl enolates: (a) Noyori R, Murata S, Suzuki M (1981) Tetrahedron 37:3899; (b) Mukai C, Hashizume S, Nagami K, Hanaoka M (1990) Chem Pharm Bull 38:1509. (c) Murata S, Suzuki M, Noyori R (1980) J Am Chem Soc 102:3248. Selective activation of acetals or aldehydes under certain non-basic conditions are now under investigation in our group; cf. (d) Mukaiyama T, Ohno T, Han JS, Kobayashi S (1991) Chem Lett 949
26. Kobayashi S, Wakabayashi T, Nagayama S, Oyamada H (1997) Tetrahedron Lett 38:4559
27. Although several organic reactions in micelles were reported, there was no report on Lewis acid catalysis in micelles, to the best of our knowledge. In addition, judging from the amount of the surfactant used in the present case, the aldol reaction would not proceed not only in micelle. Precise reaction mechanism is now under investigations: (a) Fendler JH, Fendler EJ (1975) Catalysis in micellar and macromolecular systems. Academic Press, London; (b) Holland PM, Rubingh DN (eds) Mixed surfactant systems. ACS, Washington, DC; (c) Cramer CJ, Truhlar DG (eds) (1994) Structure and reactivity in aqueous solution. ACS, Washington, DC; (d) Sabatini DA, Knox RC, Harwell JH (eds) (1995) Surfactant-enhanced subsurface remediation. ACS, Washington, DC
28. Hydrolysis of silyl enol ether 1 took place when lower yields were observed as in Table 8
29. (a) Yates D, Eaton PE (1960) J Am Chem Soc 82:4436; (b) Hollis TK, Robinson NP, Bosnich B (1992) J Am Chem Soc 114:5464; Review: (c) Carruthers W (1990) Cycloaddition reactions in organic synthesis. Pergamon, Oxford
30. Kobayashi S, Hachiya I, Takahori T, Araki M, Ishitani H (1992) Tetrahedron Lett 6815
31. Kobayashi S, Hachiya I, Araki M, Ishitani H (1993) Tetrahedron Lett 3755
32. Some Diels-Alder reactions in water without catalyst were reported. For example: (a) Rideout DC, Breslow R (1980) J Am Chem Soc 102:7816; (b) Grieco PA, Garner P, He Z (1983) Tetrahedron Lett 24:1897
33. Review: Yamamoto Y, Asao N (1993) Chem Rev 93:2207
34. Hachiya I, Kobayashi S (1993) J Org Chem 58:6958
35. For the reactions of carbonyl compounds with tetraallyltin: (a) Peet WG, Tam W (1983) J Chem Soc, Chem Commun 853; (b) Daude G, Pereyre M (1980) J Organometal Chem 190:43; (c) Harpp DN, Gingras M (1988) J Am Chem Soc 110:7737. See also (d) Fukuzawa S, Sato K, Fujinami T, Sakai S (1983) J Chem Soc, Chem Commun 853. Quite recently Yamamoto et al. has reported allylation reactions of aldehydes with tetraallyltin in the presence of hydrochloric acid: (e) Yanagisawa A, Inoue H, Morodome M, Yamamoto H (1993) J Am Chem Soc 115:10,356
36. Lewis acid-promoted allylation reactions of carbonyl compounds with allyltrialkyltin were reported: (a) Yamamoto Y, Yatagai H, Naruta Y, Maruyama K (1980) J Am Chem Soc 102:7107; (b) Pereyre M, Quintard J-P, Rahm A (1987) Tin in organic synthesis. Butterworths, London, p 216
37. (a) Schmid W, Whitesides GM (1991) J Am Chem Soc 113:6674; (b) Kim E, Gordon DM, Schmid W, Whitesides GM (1993) J Org Chem 58:5500

38. Kleinman EF (1991) In: Trost BM (ed) Comprehensive organic synthesis, vol 2. Pergamon, Oxford, p 893
39. Ojima I, Inaba S-I, Yoshida K (1977) Tetrahedron Lett 3643
40. (a) Ikeda K, Achiwa K, Sekiya M (1983) Tetrahedron Lett 24:4707; (b) Mukaiyama T, Kashiwagi K, Matsui S (1989) Chem Lett 1397; (c) Mukaiyama T, Akamatsu H, Han JS (1990) Chem Lett 889; (d) Onaka M, Ohno R, Yanagiya N, Izumi Y (1993) Synlett 141. For a stoichiometric use: (e) Dubois J-E, Axiotis G (1984) Tetrahedron Lett 25:2143; (f) Colvin EW, McGarry DG (1985) J Chem Soc, Chem Commun 539; (g) Shimada S, Saigo K, Abe M, Sudo A, Hasegawa M (1992) Chem Lett 1445. For an enantioselective version: (h) Hattori K, Yamamoto H (1994) Tetrahedron 50:2785
41. Kobayashi S, Araki M, Ishitani H, Nagayama S, Hachiya I (1995) Synlett 233
42. Kobayashi S, Araki M, Yasuda M (1995) Tetrahedron Lett 36:5773
43. Lucchini V, Prato M, Scorrano G, Tecilla P (1988) J Org Chem 53:2251
44. Guanti G, Narisano E, Banfi L (1987) Tetrahedron Lett 28:4331
45. Selected examples using hydrazones as electrophiles or equivalents are as follows. As far as we know, there have been no examples of Lewis acid-mediated reactions of hydrazones with nucleophiles: (a) Kodata I, Park J-Y, Yamamoto Y (1996) J Chem Soc, Chem Commun 841; (b) Enders D, Ward D, Adam J, Raabe G (1996) Angew Chem, Int Ed Engl 35:981; (c) Denmark SE, Weber T, Piotrowski DW (1987) J Am Chem Soc 109:2224; (d) Enders D, Schubert H, Nubling C (1986) Angew Chem, Int Ed Engl 25:1109; (e) Claremon DA, Lumma PK, Phillips BT (1986) J Am Chem Soc 108:8265; (f) Takahashi H, Tomita K, Otomasu H (1979) J Chem Soc, Chem Commun 668
46. In pioneering work, rhodium-catalyzed enantioselective hydrogenation of acylhydrazones derived from ketones was reported: Burk MJ, Feaster JE (1992) J Am Chem Soc 114:6266
47. Oyamada H, Kobayashi S (1998) Synlett 249
48. Kobayashi S, Furuta T, Sugita K, Oyamada H (1998) Synlett (in press)
49. It has been shown that lanthanide triflates are excellent catalysts for the catalytic activation of imine and related compounds, while most Lewis acids are decomposed or deactivated in the presence of basic nitrogen atoms. For example, (a) Kobayashi S, Busujima T, Nagayama S (1998) J Chem Soc, Chem Commun 19; (b) Kobayashi S, Nagayama S (1997) J Am Chem Soc 119:10,049; (c) Kobayashi S, Akiyama R, Kawamura M, Ishitani H (1997) Chem Lett 1039; (d) Kobayashi S, Ishitani H, Ueno M (1997) Synlett 115
50. $Yb(OTf)_3$ was effective for the activation of the acylhydrazone, although the yield was ca. 10% lower than that using $Sc(OTf)_3$ in a preliminary experiment
51. It is also known that the imines derived from α,β-unsaturated aldehydes are often difficult to prepare due to their instability
52. Alexakis A, Lensen N, Mangeney P (1991) Synlett 625. Cf. Seebach D, Wykypiel W (1979) Synthesis 423. See also [45]
53. The structure was confirmed by X-ray analysis of the corresponding *N*-benzoyl derivative
54. Cf. Overman LE, Rogers BN, Tellew JE, Trenkle WC (1997) J Am Chem Soc 119:7159
55. Kobayashi S, Ishitani H (1995) J Chem Soc, Chem Commun 1379
56. Other lanthanide triflates can also be used. In the reaction of phenylglyoxal•monohydrate, *p*-chloroaniline, and 2-methoxypropene, 90% ($Sm(OTf)_3$), 94% ($Tm(OTf)_3$), and 91% ($Sc(OTf)_3$) yields were obtained
57. Grieco et al. reported in situ generation and trapping of ammonium salts under Mannich-like conditions: (a) Larsen SD, Grieco PA (1985) J Am Chem Soc 107:1768; (b) Grieco PA, Parker DT (1988) J Org Chem 53:3325 and references cited therein
58. (a) Weinreb SM (1991) In: Trost BM (ed) Comprehensive organic synthesis, vol 5. Pergamon, Oxford, p 401; (b) Boger DL, Weinreb SM (1987) Hetero Diels-Alder methodology in organic synthesis. Academic Press, San Diego, chaps 2 and 9

59. Danishefsky S, Kitahara T (1974) J Am Chem Soc 96:7807
60. (a) Boger DL (1983) Tetrahedron 39:2869; (b) Grieco PA, Bahsas A (1988) Tetrahedron Lett 29:5855 and references cited therein
61. (a) Kobayashi S, Ishitani H, Nagayama S (1995) Chem Lett 423; (b) Kobayashi S, Ishitani H, Nagayama S (1995) Synthesis 1195
62. As for the reactions of vinyl ethers, see (a) Joh T, Hagihara N (1967) Tetrahedron Lett 4199; (b) Povarov LS (1967) Russian Chem Rev 36:656; (c) Worth DF, Perricine SC, Elslager EF (1970) J Heterocyclic Chem 7:1353; (d) Kametani T, Takeda H, Suzuki Y, Kasai H, Honda T (1986) Heterocycles 24:3385; (e) Cheng YS, Ho E, Mariano PS, Ammon HL (1985) J Org Chem 56:5678 and references cited therein. As for the reactions of vinyl sulfides, see Narasaka K, Shibata T (1993) Heterocycles 35:1039
63. Fujiwara et al. reported a similar type of reaction: Makioka Y, Shindo T, Taniguchi Y, Takaki K, Fujiwara Y (1995) Synthesis 801
64. Sisko J, Weinreb SM (1989) Tetrahedron Lett 30:3037
65. When typical Lewis acids such as $BF_3.OEt_2$ and $ZnCl_2$ (100 mol%) were used instead of the lanthanide triflates under the same reaction conditions, lower yields were observed (23 and 12%, respectively)
66. Nomura Y, Kimura M, Takeuchi Y, Tomoda S (1978) Chem Lett 267
67. (a) Hashimoto S, Komeshita N, Koga K (1979) J Chem Soc, Chem Commun 437; (b) Narasaka K, Iwasawa N, Inoue M, Yamada T, Nakashima M, Sugimori J (1989) J Am Chem Soc 111:5340; (c) Chapuis C, Jurczak J (1987) Helv Chim Acta 70:436; (d) Maruoka K, Itoh T, Shirasaki T, Yamamoto H (1988) J Am Chem Soc 110:310; (e) Corey EJ, Imwinkelried R, Pikul S, Xiang YB (1989) J Am Chem Soc 111:5493; (f) Kaufmann D, Boese R (1990) Angew Chem, Int Ed Engl 29:545; (g) Corey EJ, Imai N, Zhang HY (1991) J Am Chem Soc 113:728; (h) Sartor D, Saffrich J, Helmchen G, Richard CJ, Lambert H (1991) Tetrahedron Asym 2:639; (i) Mikami K, Terada M, Motoyama Y, Nakai T (1991) Tetrahedron Asym 2:643; (j) Evans DA, Miller SJ, Lectka T (1993) J Am Chem Soc 115:6460; (k) Ishihara K, Gao Q, Yamamoto H (1993) J Am Chem Soc 115:10,412; (l) Maruoka K, Murase N, Yamamoto H (1993) J Org Chem 58:2938; (m) Seerden JG, Scheeren HW (1993) Tetrahedron Lett 34:2669. Review: (n) Narasaka K (1991) Synthesis 1; (o) Kagan HB, Riant O (1992) Chem Rev 92:1007
68. (a) Bednarski M, Maring C, Danishefsky S (1983) Tetrahedron Lett 24:3451. See also (b) Quimpere M, Jankowski K (1987) J Chem Soc, Chem Commun 676. Quite recently, Shibasaki et al. reported catalytic asymmetric nitro aldol reactions using a chiral lanthanum complex as a base: (c) Sasai H, Suzuki T, Itoh N, Tanaka K, Date T, Okumura K, Shibasaki M (1993) J Am Chem Soc 115:10,372 and references cited therein
69. Cotton FA, Wilkinson G (1988) Advanced inorganic chemistry, 5th edn. Wiley, New York, p 973
70. While $Yb(OTf)_3$ or (*R*)-(+)-binaphthol dissolved only sluggishly in dichloromethane, the mixture of $Yb(OTf)_3$, (*R*)-(+)-binaphthol, and an amine became an almost clear solution
71. Narasaka et al. first reported an excellent chiral titanium catalyst for this reaction: Narasaka K, Iwasawa N, Inoue M, Yamada T, Nakashima M, Sugimori J (1989) J Am Chem Soc 111:5340
72. Kobayashi S, Hachiya I, Ishitani H, Araki M (1993) Tetrahedron Lett 34:4535
73. Kobayashi S, Ishitani H (1994) J Am Chem Soc 116:4083
74. Stinson SC (1993) Chem Eng News Sept 27:38
75. 3-Acyl-1,3-oxazolidin-2-ones are weakly coordinating ligands, whereas PAA is stronger coordinating. From the experiments, site A seems to be more easily available for coordination than site B
76. Kobayashi S, Araki M, Hachiya I (1994) J Org Chem 59:3758
77. Narasaka K, Tanaka H, Kanai F (1991) Bull Chem Soc Jpn 64:387

78. Review: Hart, FA (1987) In: Wilkinson G (ed) Comprehensive coordination chemistry, vol 3. Pergamon, New York, p 1059
79. Fritsch J, Zundel G (1981) J Phys Chem 85:556
80. (a) Maruoka K, Yamamoto H (1989) J Am Chem Soc 111:789; (b) Bao J, Wulff WD, Rheingold AL (1993) J Am Chem Soc 115:3814
81. Hattori K, Yamamoto H (1992) J Org Chem 57:3264
82. (a) Reetz MT, Kyung S-H, Bolm C, Zierke T (1986) Chem Ind 824; (b) Mikami K, Terada M, Nakai T (1990) J Am Chem Soc 112:3949
83. Since the amine part can be freely chosen, the design of the catalyst was easier than other catalysts based on (*R*)-(+)-binaphthol. Although some "modified" binaphthols were reported to be effective as chiral sources, their preparations often require long steps: Bao J, Wulff WD, Rheingold AL (1993) J Am Chem Soc 115:3814. See also [68a]
84. (a) Oguni N, Matsuda Y, Kaneko T (1988) J Am Chem Soc 110:7877; (b) Kitamura M, Okada S, Suga S, Noyori R (1989) J Am Chem Soc 111:4028; (c) Puchot C, Samuel O, Dunach E, Zhao S, Agami C, Kagan H (1986) J Am Chem Soc 108:2353
85. Lower enantiomeric excesses of the product were observed when (*R*)-(+)-binaphthol in less than 60% ee was used. This may be ascribed to turnover of the catalyst
86. (a) Maruoka K, Yamamoto H (1993) In: Ojima I (ed) Catalytic asymmetric synthesis. VCH, Weinheim, p 413; (b) Narasaka K (1991) Synthesis 1
87. (a) Kametani T, Kasai H (1989) Studies in Natural Product Chem 3:385; (b) Grigos VI, Povarov LS, Mikhailov BM (1965) Izv Akad Nauk SSSR, Ser Khim 2163; CA (1966) 64:9680
88. (a) Waldmann H (1994) Synthesis 535; (b) Borrione E, Prato M, Scorrano G, Stiranello M (1989) J Chem Soc, Perkin Trans 1 2245; (c) Ishihara K, Miyata M, Hattori K, Tada T, Yamamoto H (1994) J Am Chem Soc 116:10,520
89. Kobayashi S, Ishitani H, Hachiya I, Araki M (1994) Tetrahedron 50:11,623
90. Some interesting biological activities have been reported in 8-hydroxyquinoline derivatives. For example: (a) Rauckman BS, Tidwell MY, Johnson JV, Roth B (1989) J Med Chem 32:1927; (b) Johnson JV, Rauckman BS, Baccanari DP, Roth B (1989) J Med Chem 32:1942; (c) Ife RJ, Brown TH, Keeling DJ, Leach CA, Meeson ML, Parsons ME, Reavill DR, Theobald CJ, Wiggall KJ (1992) J Med Chem 35:3413; (d) Sarges R, Gallagher A, Chambers TJ, Yeh L-A (1993) J Med Chem 36:2828; (e) Mongin F, Fourquez J-M, Rault S, Levacher V, Godard A, Trecourt F, Queguiner G (1995) Tetrahedron Lett 36:8415
91. Kobayashi S, Ishitani H, Araki M, Hachiya I (1994) Tetrahedron Lett 35:6325
92. Ishitani H, Kobayashi S (1996) Tetrahedron Lett 37:7357
93. Inverse electron-demand asymmetric Diels-Alder reactions of 2-pyrone derivatives were reported: (a) Posner GH, Carry J-C, Lee JK, Bull DS, Dai H (1994) Tetrahedron Lett 35:1321; (b) Markó IE, Evans GR (1994) Tetrahedron Lett 35:2771
94. (a) Tufariello JJ (1984) In: Padwa A (ed) 1,3-Dipolar cycloaddition chemistry, vol 2. Wiley, Chichester, p 83; (b) Torssell KBG (1988) Nitrile oxides, nitrones and nitronates in organic synthesis. VCH, Weinheim
95. Review: Frederickson M (1997) Tetrahedron 53:403
96. Pioneering works in this field: (a) Seerden J-PG, Scholte OP, Reimer AWA, Scheeren HW (1994) Tetrahedron Lett 35:4419; (b) Gothelf KV, Jørgensen KA (1994) J Org Chem 59:5687; (c) Seerden J-PG, Kuypers MMM, Scheeren HW (1995) Tetrahedron Asym 6:1441; (d) Gothelf KV, Thomsen I, Jørgensen KA (1996) J Am Chem Soc 118:59; (e) Seebach D, Marti RE, Hintermann T (1996) Helv Chim Acta 79:1710; (f) Hori K, Kodama H, Ohta T, Furukawa I (1996) Tetrahedron Lett 37:5947; (g) Jensen KB, Gothelf KV, Hazell RG, Jørgensen KA (1997) J Org Chem 62:2471 and references cited therein
97. It was recently found that lanthanide triflates are excellent catalysts in achiral 1,3-dipolar cycloadditions between nitrones and alkenes and also in three-component coupling reactions of aldehydes, hydroxylamines, and alkenes: Kobayashi S, Akiyama R,

Kawamura M, Ishitani H (1997) Chem Lett 1039. Cf. Minakata S, Ezoe T, Ilhyong R, Komatsu M, Ohshiro Y (1997) The 72nd Annual Meeting of the Chemical Society of Japan, Tokyo, 2F3 37. See also [98]

98. Quite recently, Jøgensen et al. reported similar asymmetric 1,3-dipolar cycloadditions using $Yb(OTf)_3$-PyBOX; however, enantiomeric excesses obtained were up to 73%: Sanchez-Blanco AI, Gothelf KV, Jøgensen KA (1997) Tetrahedron Lett 38:7923
99. (*R*)-MNEA was prepared from (*R*)-1-(1-naphthyl)ethylamine
100. It is believed that bidentate coordination (ex. Yb(III)-3-(2-alkenoyl)-1,3-oxazolidin-2-one) is necessary to obtain high selectivities in many chiral lanthanide-catalyzed reactions [101]. These results are very interesting and promising because it has been shown that even monodentate coordination can achieve good selectivities by using the hetero-chiral Yb(III) catalyst
101. [72, 73, 76, 89, 91]. See also (a) Marko IE, Evans GR (1994) Tetrahedron Lett 35:2771; (b) Burgess K, Lim H-J, Porte AM, Sulikowski GA (1996) Angew Chem, Int Ed Engl 35:220
102. A bond pair (953 and 987 cm^{-1}), which indicated hydrogen bonds (the OH•••N and O^-•••H^+N equilibrium), was observed in the area from 930 to 1000 cm^{-1} in the IR spectra of the hetero-chiral Yb(III) catalyst [103]. Direct coordination of the amine to Yb(III) is doubtful in light of the fact that the 1,3-dipolar cycloaddition proceeded very slowly when $Yb(OTf)_3$ and (*R*)-MNEA were first combined and then (*S*)-BINOL was added
103. Evans DA, Morrissey MM, Dorow RL (1985) J Am Chem Soc 107:4346
104. Kametani T, Huang SP, Nakayama A, Honda T (1982) J Org Chem 47:2328
105. Sekiya M, Ikeda K, Terao Y (1981) Chem Pharm Bull 29:1747
106. As for carbapenems and penems: (a) Palomo C (1990) In: Lukacs G, Ohno M (eds) Re cent progress in the chemical synthesis of antibiotics. Springer, Berlin Heidelberg New York, p 565; (b) Perrone E, Franceschi G (1990) In: Lukacs G, Ohno M (eds) Recent progress in the chemical synthesis of antibiotics. Springer, Berlin Heidelberg New York, p 613
107. Kobayashi S (1998) Chem Soc Rev (in press)

Lanthanide- and Group 3 Metallocene Catalysis in Small Molecule Synthesis

Gary A. Molander* and Eric D. Dowdy

Department of Chemistry and Biochemistry, University of Colorado at Boulder, Boulder, CO 80309-0215 USA
* *e-mail:* gary.molander@colorado.edu

Although the polymerization prowess of organolanthanide complexes has been known for some time, efforts to apply these catalysts to small molecule synthesis have only recently begun. The selectivity of these metallocenes is predominantly steric in nature, and they are compatible with a wide variety of organic functional groups. A review of their use in olefin hydrogenation, hydrosilylation, and polyene cyclization with emphasis on chemoselectivity and diastereoselectivity is presented here. The various ways in which the catalysts and reagents can be tuned to produce the desired products is also discussed.

Keywords: Metallocenes, Catalysis, Cyclization, Hydrogenation, Silylation

1 **Introduction** 120

2 **General Features of the Catalytic Systems and the Olefin Insertion Reaction** 121

3 **Catalytic Hydrogenation Reactions** 123

4 **Catalytic Hydrosilylation Reactions** 127

4.1 Silylation of Alkenes 128
4.2 Silylation of Alkynes 130

5 **Catalytic Cyclization Reactions** 133

5.1 Termination by β-Hydride Elimination 134
5.2 Termination by Hydrogenation 135
5.3 Termination by Silylation 136
5.3.1 Cyclization/Silylation of Terminal Dienes and Trienes 137
5.3.2 Cyclization/Silylation of Hindered Dienes and Trienes. 145
5.3.3 Cyclization/Silylation of Enynes and Dienynes 147

6 **Conclusions** 151

7 **References** 152

1 Introduction

Carbon-carbon bond-forming reactions constitute the heart and soul of synthetic organic chemistry. Nowhere are these reactions more prolific than in commercial polymerization reactions, where worldwide production of polyethylene and polypropylene is carried out on enormous scale. The polymerization of α-olefins employing Ziegler-Natta catalysts [1], organolanthanides and group 3 organometallics [2], palladium and nickel cationic complexes [3], and related catalysts [4] have revolutionized the controlled synthesis of polymers derived from terminal alkenes. Group 3 organometallics and organolanthanides in particular are among the most active known catalysts in olefin polymerization chemistry [5]. However, in spite of the spectacular ability of these complexes to generate new carbon-carbon bonds in polymerization reactions, relatively little effort has been made in applying these catalytic systems to small molecule synthesis through cyclization reactions of dienes, enynes, and related substrates

cat. "Cp^*_2LnH" — $LnCp^*_2$ — FG (1)

To be sure, the extrapolation of reactivity from ethylene and propylene to highly functionalized organic molecules is daunting, and many questions arise concerning the suitability of the aforementioned catalysts within this context. For example, only recently have available catalysts been able to polymerize anything but monosubstituted alkenes. Even allylic substitution of such alkenes has presented difficulties for the sterically sensitive organometallic complexes [6]. To be broadly applicable, catalysts would have to be developed that would allow insertion of more highly substituted alkenes. Functional group compatibility was also of concern. Most polar functional groups (even ethers) were reported to either react with the organometallics [7] or irreversibly bind so as to inhibit catalysis [8]. Again, to be considered as part of a general synthetic method it was imperative that organometallic catalysts be developed that could tolerate a wide range of common organic functionality. Many of the complexes, particularly the organolanthanides and group 3 organometallics, are extraordinarily air-sensitive and require glove-box or Schlenk-line techniques for their handling. The synthesis of reasonably air-stable complexes would facilitate the introduction of these catalysts to a broader range of applications. Finally, several issues of selectivity would have to be addressed. First, chemoselectivity in the insertion of a single alkene or alkyne in a polyunsaturated system would be required. A high degree of regioselectivity in this insertion is also essential. Diastereoselectivity in reactions of chiral substrates and prochiral unsaturated systems would necessitate examination. Finally, chiral, nonracemic complexes would have to be developed in order to meet the challenges of modern synthetic organic chemistry.

This contribution is not intended to be a comprehensive review of the application of group 3 metal and lanthanide metallocenes to selective organic synthesis, but rather seeks to highlight some of the general characteristics of these catalysts and to provide a critical discussion of the reactions to which they have been applied. Emphasis has been placed upon the means by which some of the potential problems delineated above have been addressed and upon the application of these catalytic species to novel methods in synthetic organic chemistry.

2
General Features of the Catalytic Systems and the Olefin Insertion Reaction

The key step in nearly all of the catalytic processes to be discussed is olefin insertion into a metal hydride [Eq. (2)] or organometallic species [Eq. (3)]. These hydrometallation and carbometallation processes also form the basis for the polymerization of alkenes. Olefin insertions generally occur with the same regioselectivity as hydroboration reactions [9], with the bulky metal and associated ligands residing at the least hindered site of the two carbon reactive unit.

$$\text{"}Cp^*_2LnH\text{"} + CH_2{=}CHR \longrightarrow Cp^*_2LnCH_2CH_2R \tag{2}$$

$$Cp^*_2LnR + CH_2{=}CHR' \longrightarrow Cp^*_2LnCH_2CHRR' \tag{3}$$

In most instances these reactions are extremely exothermic and effectively irreversible under reasonable reaction conditions [10]. Even *tert*-butyl-substituted organolanthanides (prepared by other methods) exhibit significant kinetic stability [11].

The organolanthanide- and group 3 organometallic catalysts are highly electrophilic species. Although there is a significant electronic driving force in the olefin insertion process, for the most part steric factors predominate. Much of this can be attributed to the effects of coordinative unsaturation at the metal center. High reactivity is associated with free coordination sites and terminal (non-bridging) ligands [12]. The requirement for free coordination sites dictates that noncoordinating solvents be utilized for the catalytic reactions. Thus common ether solvents lead to low catalytic turnover rates [10] and can even deactivate catalysts via ether cleavage reactions [7, 13]. Consequently, hydrocarbon solvents are used exclusively in the catalytic reactions.

Although it would seem logical to utilize the most sterically unhindered ligands about the metal to achieve maximum reactivity with hindered alkenes, in fact there is a delicate balance that must be achieved between the "openness" of the metal center and the tendency for the organometallic hydrides to undergo a deactivating ligand redistribution [14] or to dimerize, forming hydride-bridged dimers [Eq. (4)] [10].

$$2\,Cp_2LnH \rightleftharpoons Cp_2Ln(\mu\text{-}H)_2LnCp_2 \tag{4}$$

These dimers are less reactive or unreactive in catalytic reactions of interest, and thus some substitution on the cyclopentadienyl ligands is necessary to prevent dimer formation through steric hindrance to association. Pentamethylcyclopentadienyl (Cp*) ligands are useful for this, and lead to catalytic systems that are highly reactive and yet exquisitely selective in the insertion of monosubstituted alkenes. One such precatalyst system is the yttrium complex $Cp^*_2YMe{\cdot}THF$. Although Lewis bases normally depress catalytic activity because they compete for empty coordination sites on the catalyst, in this precatalyst the single THF of solvation appears to catalyze the hydrometallation process [Eq. (5)] [15] in the same manner that Lewis bases catalyze the hydroboration of olefins with 9-BBN [16]:

$$Cp^*_2Y\langle^{H}_{H}\rangle YCp^*_2 \quad + \quad 2\ THF \longrightarrow 2\ Cp^*_2YH{\cdot}THF \tag{5}$$

A related complex lacking a THF of association, $Cp^*_2YCH(TMS)_2$, is orders of magnitude less reactive than $Cp^*_2YMe{\cdot}THF$ in reactions with monosubstituted alkenes. Addition of one molar equivalent of THF per equivalent of $Cp^*_2YCH(TMS)_2$ restores catalytic activity to the level of the methyl complex [17]. Apparently, this THF also depresses the rate of σ-bond metathesis (see below) relative to olefin insertion [10], with important ramifications for cyclization/termination processes to be discussed later. Curiously, this THF effect appears unique to the Cp^*_2YR system.

Unfortunately, the yttrium hydride complex Cp^*_2YH is very hindered about the reactive metal center and does not react readily with 1,1-disubstituted alkenes and more highly hindered olefins. Three general catalyst modifications have been utilized to overcome this difficulty. The first is the incorporation of lanthanide metals with larger ionic radii. Because of the lanthanide contraction, early lanthanides have a larger ionic radius than the late lanthanides. Placement of one of the larger metal ions into the ligand system has the effect of opening it up, allowing access to more hindered alkenes. This strategy has its limits, and to create even more accessible metal centers the ligand system itself must be changed. Incorporation of a single bulky substituent onto the cyclopentadienyl units has been utilized to open the reactive metal center. Substituents such as *tert*-butyl groups and trimethylsilyl groups appear to prevent extensive hydride dimer formation, but at the same time leave the metal center relatively open for interaction with olefins. An example of such a precatalyst is displayed in Fig. 1. Although the methyl precatalyst is a dimer, the active hydride catalyst is most likely a monomer, and is capable of rapid olefin insertion of most 1,1-disubstituted alkenes [18]. Another interesting feature of the catalyst is its unusual stability. Members of this class of precatalysts have been weighed in the air and utilized in hydrosilylation reactions performed with normal procedures for the benchtop handling of air-sensitive materials [9, 19].

The third strategy that has been utilized to provide more open access to the metal is to utilize "hinged" cyclopentadienyl ligands. A one atom bridge between two cyclopentadienyl units serves to increase the angle between them, and again

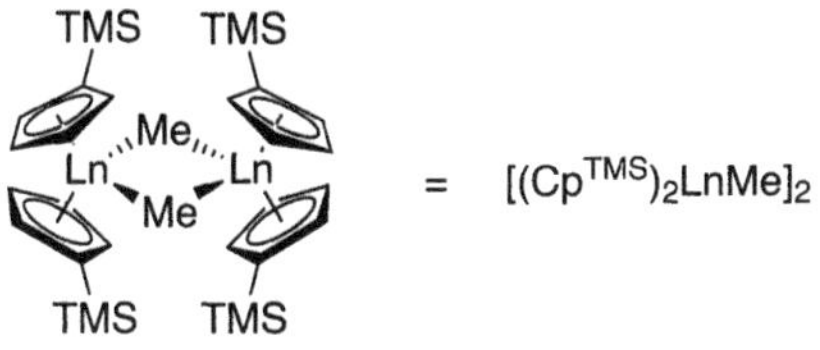

TMS TMS
Me
Ln Ln
Me
TMS TMS
= $[(Cp^{TMS})_2LnMe]_2$

Ln = Sm, Y, Lu

Fig. 1. Unhindered catalyst systems for reaction with hindered alkenes

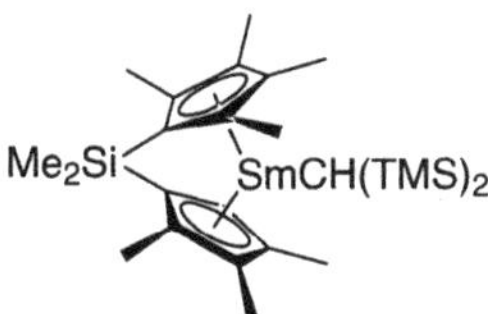

Fig. 2. Hinged catalyst system for reaction with hindered alkenes

provides more facile approach of substrate to the metal. Figure 2 depicts one such precatalyst [20].

With a wide range of metals and ligands available [2a, 21], organolanthanide and group 3 organometallic catalysts are readily "tuned" to provide the desired reactivity and selectivity patterns in reactions of interest.

3
Catalytic Hydrogenation Reactions

Organolanthanides and group 3 organometallics are extraordinarily reactive hydrogenation catalysts [22]. They are also relatively challenging to prepare and handle. This, combined with their lack of tolerance to reactive functional groups, makes them less attractive than many transition metal-based catalysts in standard hydrogenation reactions. It is instructive, however, to examine the catalytic cycle of hydrogenation reactions because it serves as a useful model for the other catalytic reactions to be discussed subsequently. Additionally, useful information concerning selectivity has been derived from studies of the hydrogenation reaction.

Transition metal-based hydrogenation reactions most often operate on a catalytic cycle that involves oxidative addition, olefin insertion, and reductive elimination. The mechanistic basis for organolanthanide hydrogenation is quite different, and involves olefin insertion and σ-bond metathesis (Fig. 3).

A σ-bond metathesis initiates the process, converting an organometallic precatalyst into the catalytic hydride species through a four-centered exchange of

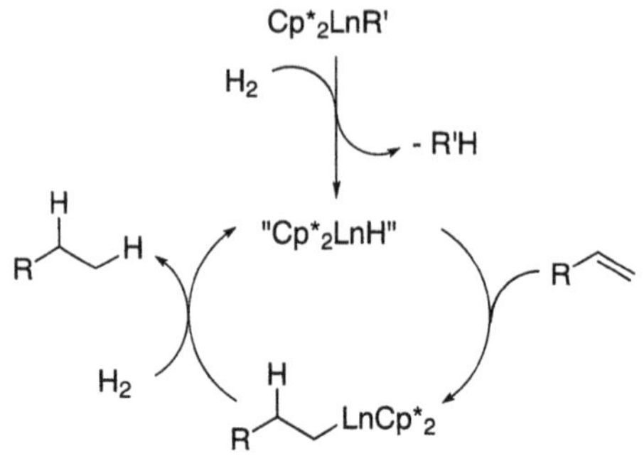

Cp*$_2$LnR' → (+ H_2, − R'H) → "Cp*$_2$LnH"; + R-CH=CH$_2$ → R-CH(H)-CH$_2$-LnCp*$_2$; + H_2 → R-CH(H)-CH$_2$-H

Fig. 3. Catalytic cycle for hydrogenation

ligands [Eq. (6)] [23]. Olefin insertion is the product-determining step of the process, and is generally irreversible under optimal conditions for hydrogenation [10]. A σ-bond metathesis reaction constitutes the final step of the process, wherein the organometallic reacts with hydrogen to release the hydrocarbon and regenerate the active catalyst.

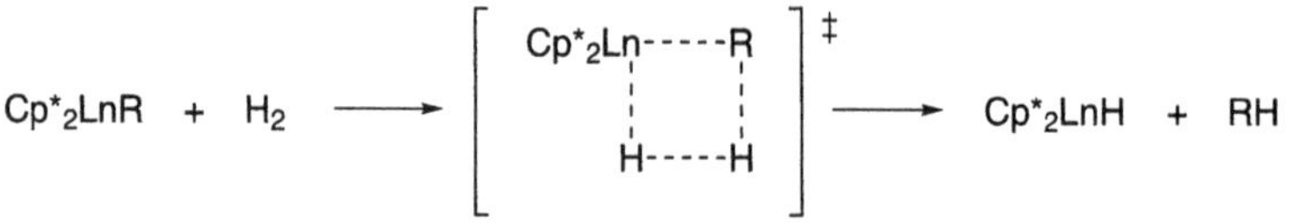

$$Cp^*_2LnR + H_2 \longrightarrow \left[\begin{matrix} Cp^*_2Ln \cdots R \\ \vdots \quad\quad \vdots \\ H \cdots H \end{matrix} \right]^{\ddagger} \longrightarrow Cp^*_2LnH + RH \tag{6}$$

Detailed kinetic studies for organolanthanide-catalyzed hydrogenations have been performed [22]. Although there are some exceptions, for reactive olefins the rate of the reaction is proportional to the product of the catalyst concentration and hydrogen concentration, indicating a rapid olefin insertion and rate limiting σ-bond metathesis. With the exception of cyclohexene, the relative rate constants for hydrogenation of monosubstituted alkenes and a variety of 1,2-disubstituted alkenes are remarkably similar. Nevertheless, synthetically useful selectivities can be achieved with a highly discriminating catalyst (Eqs. 7–9) [24]. For most diene substrates wherein the two alkenes exhibit different reactivities, hydrogen uptake need not be monitored. Even with a highly reactive, strained alkene such as is found in the bicyclo[2.2.1] system, the monosubstituted alkene is so much more reactive that a single equivalent of hydrogen is utilized even in the presence of a vast excess of hydrogen (Eq. 7). Perhaps even more impressive is the selectivity displayed in Eq. (8), wherein substitution at one allylic position of a diene system is sufficient to permit virtually complete selectivity. Most common alcohol protecting groups are tolerated (Eq. 9), but allylic acetate and allylic halide functionalities inhibit catalytic turnover.

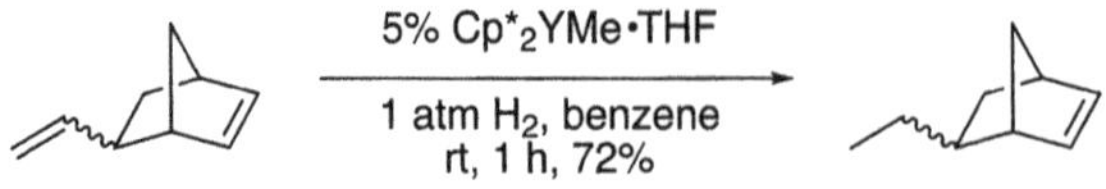

5% Cp*$_2$YMe·THF; 1 atm H_2, benzene; rt, 1 h, 72% (7)

OTr — 5% Cp*$_2$YMe·THF; 1 atm H_2, benzene, rt, 1 h, 96% → OTr (8)

Me Me R — 2% Cp*$_2$YMe·THF; 1 atm H_2, benzene, rt, 1 h; R = OTBS, OMe, OBn; yield = 74-99%; R = OAc, Cl; yield = 0% → Me Me R (9)

More hindered, less reactive alkenes such as 1,1-disubstituted alkenes require a more open catalyst [25]. For the hydrogenation reaction, the simple alternative of utilizing a samarium- or ytterbium-based catalyst permits these substrates to react effectively, often with high diastereoselectivities (Eqs. 10–12). A variety of alkyl and aryl substituents are tolerated, including a tertiary amine (Eq. 11). Allylic ether substitution, however, inhibits the reaction, even under more vigorous conditions. After initial insertion of the olefin, the Lewis acidic metal center can form a stable five-membered ring chelate with the oxygen, preventing further reaction. In general, the diastereoselectivity of the olefin insertion diminishes when the existing stereogenic center occupies a position that causes less interaction with the bulky organometallic reagent (Eq. 12). The stereochemistry of the products is that predicted by the approach of the bulky organometallic to the less hindered face of the exo-methylene unit.

10% Cp*$_2$SmCH(TMS)$_2$; 1 atm H_2, cyclopentane, rt, 84% (10)

R — 3% Cp*$_2$SmCH(TMS)$_2$; 1 atm H_2, cyclopentane; R = Ph, $(CH_2)_3NMe_2$, 50 °C, 76-96%, >10 : 1 ds; R = OMe, 70 °C, 0% → R (11)

t-Bu — 3% Cp*$_2$YbCH(TMS)$_2$; 1 atm H_2, cyclopentane, -20 °C, 73%, 3.3 : 1 ds → *t*-Bu (12)

Enantioselective catalytic hydrogenation reactions have been reported and appear to offer synthetically useful results, although for a very limited set of alkenes [10, 26]. Thus deuteration of styrene and hydrogenation of substituted styrenes, molecules with essentially no functional group available to serve as a stereochemical control element, can be accomplished with modest to high asymmetric induction using a series of elegantly designed chiral, nonracemic catalysts (Eq. 13).

0.7% cat.

1 atm. H_2
heptane, rt, <1 h
95%, 71% ee

cat. = $(TMS)_2CHSm$ $SiMe_2$ R* R* =

(13)

Although a model was proposed in which olefin insertion occurred to place the metal on the terminus of the alkene ("1,2-addition") [10, 26], based upon subsequent mechanistic and synthetic studies of the hydrosilylation reaction of styrenes (see below), this model would appear to be incorrect [27]. Thus an irreversible, stereochemically determinant "2,1-insertion" probably initiates the reaction, with subsequent σ-bond metathesis completing the process. Most remarkable is the fact that, if correct, this model demands that the olefin insertion takes place to orient the highly hindered metal center at a tertiary carbon center, and that apparently little, if any, β-hydride elimination occurs from the resultant organometallic.

A "frontal trajectory" has been suggested to explain the sense of asymmetric induction in these reactions (Fig. 4), and seems most valid based on steric effects. However, a "lateral trajectory" cannot be ruled out based upon the evidence available to date [10, 26].

Alkynes also undergo selective hydrogenation to generate *cis*-alkenes [22, 28]. The process has not been developed, however, and at any rate it is unlikely to compete with more established methods.

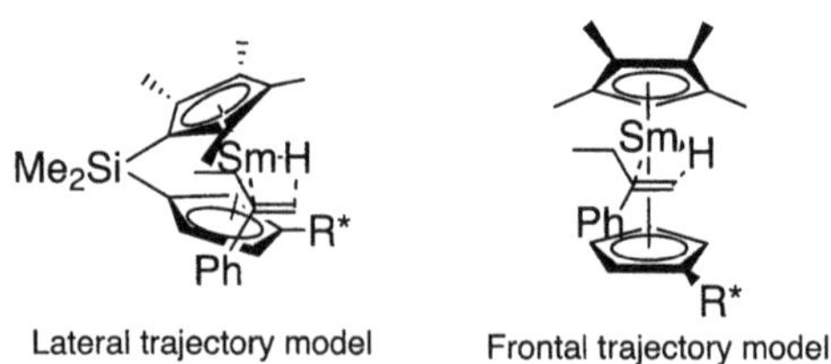

Fig. 4. Models for asymmetric induction in hydrogenation of styrene derivatives

4
Catalytic Hydrosilylation Reactions

Metal catalyzed hydrosilylation reactions provide the most efficient and economical route to organosilanes [29]. Organosilanes can be readily oxidized to the corresponding alcohols utilizing a Tamao procedure or related reactions [30]. The overall process thus constitutes the synthetic equivalent of a catalytic hydroboration/oxidation sequence [31]. One advantage of the silylation/oxidation protocol is the stability of the intermediate silane which allows the unmasking of the alcohol to be performed at a synthetically convenient time. In spite of the potential of hydrosilylation reactions in selective organic synthesis, relatively little effort has been made to develop procedures for the selective hydrosilylation of polyfunctional alkenes and alkynes, especially when compared to the analogous hydroboration reaction.

Organolanthanide and group 3 organometallic catalysts provide an alternative to the more traditional platinum-based catalysts for the selective hydrosilylation of alkenes and alkynes. Mechanistically, the transformation is analogous to the catalytic hydrogenation reaction detailed previously [22]. When silane is utilized in place of hydrogen, the σ-bond metathesis occurs to place the silane moiety on the alkyl unit, and the organometallic hydride is again regenerated (Fig. 5). In the overall process a rapid, exothermic, and essentially irreversible olefin insertion is followed by the slower, rate-determining σ-bond metathesis. Because the group 3 metallocenes and organolanthanides are highly effective catalysts for dehydrogenative polysilylation [32], the desired process demands that olefin insertion and σ-bond metathesis occur much more rapidly than the generation of polysilanes.

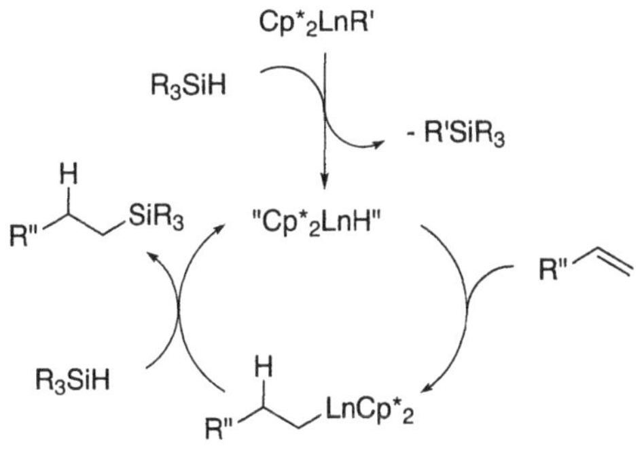

Cp^*_2LnR'
R_3SiH
$- R'SiR_3$
$"Cp^*_2LnH"$
R''
SiR_3
R_3SiH
$LnCp^*_2$

Fig. 5. Catalytic cycle for hydrosilylation

4.1 Silylation of Alkenes

The earliest reported examples of group 3 or organolanthanide catalyzed hydrosilylation reactions emphasized reactions with simple alkenes (e.g., 1-octene) (Eq. 14) [33]. The lutetium- and neodymium-based catalysts utilized for these studies typically required 2 d at high temperatures (80–90 °C) to react and provided modest to good yields of the desired terminal (linear) organosilanes. Styrene provided mixtures of linear and branched organosilanes [Eq. (15)].

$$\text{CH}_2{=}\text{CH}{-}n\text{-}C_6H_{13} \xrightarrow[\text{MePhSiH}_2,\ 90\ ^\circ\text{C, 2 d; 75\%, 28 : 1}]{5\%\ Cp_2LuC_6H_4CH_3\cdot THF} \text{PhMeHSi}{-}CH_2CH_2{-}n\text{-}C_6H_{13} + CH_3CH(SiHMePh){-}n\text{-}C_6H_{13} \quad (14)$$

$$\text{CH}_2{=}\text{CH}{-}\text{Ph} \xrightarrow[\text{PhSiH}_3,\ 80\ ^\circ\text{C, 2 d; 88\%, 4.5 : 1}]{4\%\ Cp^*_2NdCH(TMS)_2} CH_3CH(SiH_2Ph){-}\text{Ph} + \text{PhH}_2\text{Si}{-}CH_2CH_2{-}\text{Ph} \quad (15)$$

The regioselectivity of olefin insertion varies with the complex used in the reaction [27, 34]. In the hydrosilylation of a monosubstituted olefin, the use of complexes with larger metals and more open ligands provide increased yields of the product derived from reversed ("2,1") insertion (Eq. 16). These results reveal that a variety of complexes give excellent selectivity for terminal insertion, but the conditions to elevate the amount of "2,1" insertion remain elusive.

$$\text{CH}_2{=}\text{CH}{-}n\text{-}C_8H_{17} \xrightarrow[\text{PhSiH}_3,\ \text{rt, 1-24 h}]{5\%\ \text{catalyst}} \text{PhH}_2\text{Si}{-}CH_2CH_2{-}n\text{-}C_8H_{17} + CH_3CH(SiH_2Ph){-}n\text{-}C_8H_{17} \quad (16)$$

Catalyst	Yield (% isolated)	ds
$Cp^*_2LuMe\cdot THF$	98	100 : 0
$Cp^*_2YbCH(TMS)_2$	91	100 : 0
$Cp^*_2YMe\cdot THF$	84	100 : 0
$Cp^*_2SmCH(TMS)_2$	90	11 : 1
$Cp^*_2NdCH(TMS)_2$	85	3.2 : 1
$Cp^*_2LaCH(TMS)_2$	90	1.9 : 1
$Me_2SiCp''_2YCH(TMS)_2$	84	31 : 1
$Me_2SiCp''_2SmCH(TMS)_2$	98	1 : 2
$Me_2SiCp''_2NdCH(TMS)_2$	89	1 : 2

The development of organoyttrium catalysts provided a major breakthrough in terms of efficiency and selectivity in the synthesis of organosilanes [35]. These catalysts reacted with monosubstituted alkenes and reactive disubstituted alkenes within hours at room temperature, providing high yields of the desired organosilanes (Eq. 17). It was also demonstrated that common organic functional groups (halides, ethers, and acetals) could be tolerated under the reaction conditions (Eq. 18).

3% $Cp^*_2YCH(TMS)_2$
$PhSiH_3$, benzene
rt, 3.5 h, 85%
SiH_2Ph
H
(17)

3% $Cp^*_2YCH(TMS)_2$
$PhSiH_3$, benzene
rt, 83-96%
X = Cl, OBn, OTBDMS, OTHP
X
X
SiH_2Ph
(18)

In the same study, remarkable steric selectivity was demonstrated in the hydrosilylation of dienes with varying substitution patterns [35]. Virtually complete selectivity was observed for the reaction of monosubstituted olefins in the presence of disubstituted alkenes (Eqs. 19, 20) and for the silylation of a 1,1-disubstituted olefin in preference to a trisubstituted double bond (Eq. 21).

3% $Cp^*_2YCH(TMS)_2$
$PhSiH_3$, benzene, rt
92 h, 91%
SiH_2Ph
(19)

3% $Cp^*_2YCH(TMS)_2$
$PhSiH_3$, toluene, rt
18 h, 97%
SiH_2Ph
(20)

3% $Cp^*_2YCH(TMS)_2$
$PhSiH_3$, benzene, rt
96 h, 61%, 2:1 ds
SiH_2Ph
(21)

The yttrium catalysts are less effective for more sterically hindered olefins, but the flexibility afforded by being able to alter both the metal and the ligand system provides a means to adjust reactivity in a manner that allows hydrosilylation of more highly substituted alkenes. The simple modification of increasing the ionic radius of the metal permits the hydrosilylation of 1,1-disubstituted alkenes (Eq. 22) [25]. This effect dominates over slight ligand modifications, as a complex with more hindered ligands (C_5Me_4i-Pr) (Eq. 23) shows similar reactivity to the C_5Me_5-derived yttrium complex (Eq. 21) [36]. A silicon-hinged catalyst further increases turnover frequency over nonbridged systems by a factor of eight (Eq. 24) [27]. Unfortunately, there is a trade-off with the more open catalysts. Although increased reactivity with more highly substituted alkenes is observed, monosubstituted alkenes react with poor regioselectivity.

5% $Cp^*_2SmCH(TMS)_2$
$PhSiH_3$, cyclohexane
70 °C, 12 h
63-92%, 9 : 1 ds
R = Me, *i*-Bu, Bn, $(CH_2)_3NMe_2$
(22)

5% $(C_5Me_4i\text{-}Pr)_2SmCH(TMS)_2$
$PhSiH_3$, cyclohexane
rt, 96 h, 82%
(23)

cat. $Me_2SiCp'_2SmCH(TMS)_2$
$PhSiH_3$, benzene-d_6
60 °C, 98%
Cp' = C_5Me_4
(24)

Styrene derivatives react with "2,1-" regioselectivity [27]. This reversal of selectivity varies considerably with the metal ionic radius and the ligand array present, with larger metals and bridged ligands giving higher ratios of the "2,1" product. As with the catalytic hydrogenation of styrene derivatives discussed previously, the olefin insertion reaction defines the regiochemistry and stereochemistry of the final product. Thus the olefin insertion is essentially irreversible under the reaction conditions, and the σ-bond metathesis presumably occurs with retention of configuration to provide the observed products with remarkably high ee's considering the overall nature of the transformation (Eq. 25).

0.5% cat.
$PhSiH_3$, benzene-d_6
rt, 98%, 68% ee
(25)

cat. = $(TMS)_2CHSm$ ... $SiMe_2$ ... R*   R* =

4.2 Silylation of Alkynes

Although terminal alkynes are metallated with organolanthanide hydrides and therefore cannot be hydrometallated [37], internal alkynes do undergo effective hydrosilylation [38]. As expected, *cis*-addition of the organometallic hydride to the alkyne is observed (Eq. 26). Branching at one of the propargyl positions is necessary for high regioselectivity in unsymmetrical alkynes. A variety of branched substituents are suitable for use (Eqs. 27–31) widening the possibilities of subsequent synthetic steps. Placing a tertiary group on the alkyne (Eq. 30) slows the reaction, allowing the competitive dehydrogenative polymerization of silane to lower the yield of the desired product.

n-C_6H_{13} — 5% $(C_5Me_4i\text{-}Pr)_2YCH(TMS)_2$, $PhSiH_3$, 50 °C, 12 h, 78%, 4 : 1 → PhH_2Si, H, n-C_6H_{13} + H, SiH_2Ph, n-C_6H_{13} (26)

5% $Cp^*{}_2YMe{\cdot}THF$, $PhSiH_3$, cyclohexane, 50 °C, 24 h, 80% → SiH_2Ph (27)

n-C_5H_{11} — 5% $Cp^*{}_2YMe{\cdot}THF$, $PhSiH_3$, 50 °C, 24 h, 85% → H, n-C_5H_{11}, SiH_2Ph (28)

TBDMSO, n-$C_{10}H_{21}$ — 5% $Cp^*{}_2YMe{\cdot}THF$, $PhSiH_3$, cyclohexane, 90 °C, 24 h, 89% → TBDMSO, n-$C_{10}H_{21}$, SiH_2Ph (29)

t-Bu—≡—n-$C_{10}H_{21}$ — 5% $Cp^*{}_2YMe{\cdot}THF$, $PhSiH_3$, 55 °C, 48 h, 28% → t-Bu, H, n-$C_{10}H_{21}$, SiH_2Ph (30)

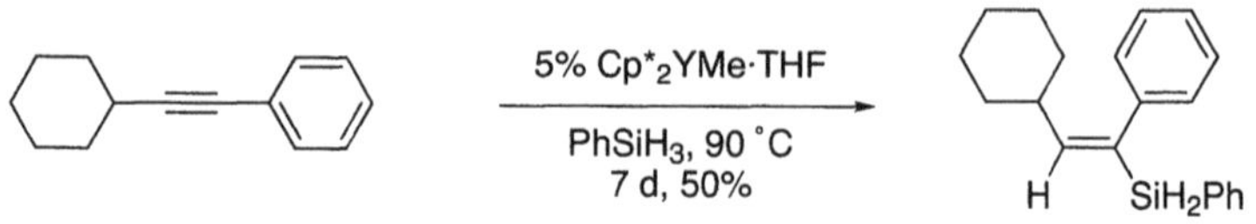

5% $Cp^*_2YMe{\cdot}THF$
$PhSiH_3$, 90 °C
7 d, 50%
H SiH_2Ph
(31)

Equation (31) illustrates an interesting facet of this regioselectivity [38]. Although both substituents are branched, the phenyl group is effectively smaller than the cyclohexyl moiety, allowing good selectivity. The steric preference is likely buttressed by the same electronic directing effects of aryl groups observed for styrene derivatives [27].

The critical and in fact rate-limiting step of alkyne hydrosilylation is the σ-bond metathesis [38]. Both the increased strength of a C_{sp2}-metal bond and the geometric requirements for the σ-bond metathesis combine to require the reactions to be heated overnight to achieve high yields of the desired alkenylsilanes. The hydrosilylation of alkynes is tolerant of a wide variety of functional groups (Eq. 32) including halides, amines, and several alcohol protecting groups [38]. A collection of substrates that were unreactive at temperatures up to 90 °C is pictured in Fig. 6. The strongly Lewis acidic complexes probably cause catalyst-deactivating ionization of the propargylic acetals.

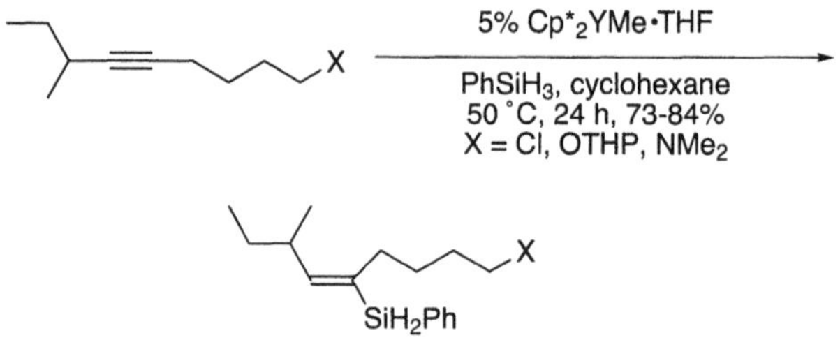

X
5% $Cp^*_2YMe{\cdot}THF$
$PhSiH_3$, cyclohexane
50 °C, 24 h, 73-84%
X = Cl, OTHP, NMe_2
(32)
X
SiH_2Ph

The chemoselectivity of the catalyst for alkynes over alkenes is of interest. Excellent discrimination is achieved in substrates containing an alkyne paired with a hindered olefin (Eqs. 27, 33) [38]. When offered a monosubstituted olefin (Eq. 34) the catalyst is less selective, producing mixtures of alkyl- and vinylsilanes. As previously noted for the hydrogenation of dienes, the addition of a group allylic to the alkene sterically shields the double bond and can electroni-

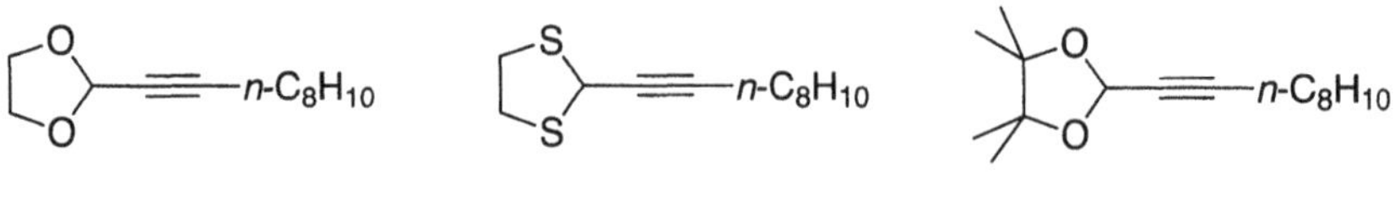

O O n-C_8H_{10} S S n-C_8H_{10} O O n-C_8H_{10}

Fig. 6. Alkynes that are unreactive to catalytic hydrosilylation conditions

cally deactivate it as well (Eq. 35). This allows virtually complete selectivity for alkyne insertion.

5% $Cp^*_2YMe \cdot THF$; $PhSiH_3$, 50 °C, 24 h, 90% (33)

5% $Cp^*_2YMe \cdot THF$; 0.5 equiv. $PhSiH_3$, 50 °C, 12 h (34)

5% $Cp^*_2YMe \cdot THF$; $PhSiH_3$, rt, 21 h, 93% (35)

5
Catalytic Cyclization Reactions

The propensity for organolanthanides and group 3 organometallics to undergo olefin insertion reactions leading to the polymerization of α-olefins provides the possibility of cyclizing dienes and other polyunsaturated substrates. A reasonable catalytic cycle for such a transformation is depicted in Fig. 7. There are several requirements for successful cyclization. In an unsymmetrical diene, selective reaction at a single alkene is necessary to avoid a mixture of regioisomeric products. As noted previously, this requirement has been met with organoyttrium catalysts (Eq. 8) [24]. If the reactions are carried out under a hydrogen atmosphere, intramolecular olefin insertion must occur more rapidly than σ-bond metathesis of the newly formed organometallic with hydrogen. Diastereoselectivity is established in the cyclization, and should be predictable based upon a simple chair transition structure for the cyclization. Finally, in the absence of hydrogen the catalytic cycle can be completed by β-hydride elimination to afford the exomethylene-substituted cycloalkane.

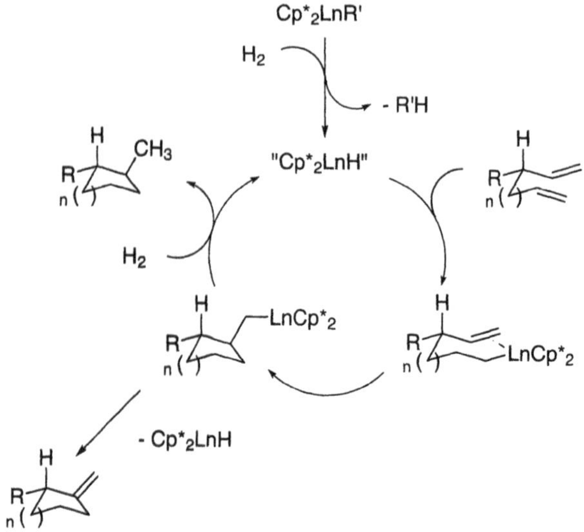

Fig. 7. Catalytic cycle for the cyclization of dienes

5.1 Termination by β-Hydride Elimination

Carbocycles and heterocycles ranging from five-membered to nine-membered have been synthesized utilizing organoscandium catalysts (Eqs. 36–39) [8]. These reactions are carried out in the absence of hydrogen or other "chain terminators". Consequently, the organometallic generated after initial olefin insertion is more persistent, and is therefore provided the time to cyclize to ring sizes that are normally inaccessible. Diastereoselectivity issues were not addressed in this study, nor was the tolerance of the catalysts for a wide range of functional groups. From the reported studies it is clear that discrimination of the catalysts for a monosubstituted alkene in the presence of an allylically substituted monosubstituted alkene is not high (Eq. 39).

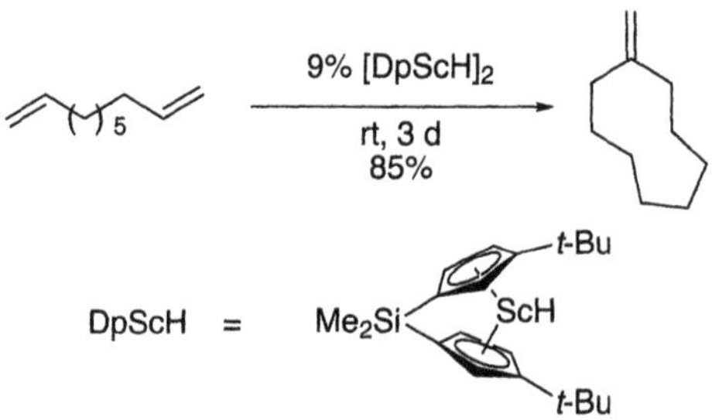

9% $[DpScH]_2$, rt, 3 d, 85%

DpScH = Me_2Si(C₅H₃-t-Bu)₂ScH

(36)

5% $[DpScH]_2$, 80 °C, 10 min, quant (37)

4% $Me_2SiCp'_2Sc(H)(PMe_3)$, 80 °C, 10 min, 93% (38)

6% $[DpScH]_2$, 80 °C, 10 min, 99%, 2 : 1 (39)

5.2 Termination by Hydrogenation

Organoyttrium catalysts have been utilized to effect the cyclization of dienes under reductive conditions [39]. Excellent selectivity is achieved in these reactions between two monosubstituted alkenes leading to a single regioisomeric product (Eq. 40), and the diastereoselectivity is consistent with the simple chair transition structure model (Fig. 7). Both acetals and thioacetals are tolerated (Eq. 41), whereas nitriles, esters, and sulfones preclude product formation (Eq. 42).

OTr — 5% $Cp^*_2YMe{\cdot}THF$, 1 atm H_2, benzene, rt, <1 h, 99%, 21 : 1 ds → OTr (40)

5% $Cp^*_2YMe{\cdot}THF$, 1 atm H_2, benzene, rt, <1 h, X = O, 85%, X = S, 97% (41)

R, R — 5% $Cp^*_2YMe{\cdot}THF$, 1 atm H_2, benzene, R = CN, CO_2Me, SO_2Ph → no reaction (42)

A reduced yield of the cyclized product accompanied by reduced uncyclized material is obtained in the reaction of 1,2-divinylbenzene (Eq. 43) [39]. This could be caused by the rigid aryl group skewing the geometry of the transition state, making cyclization more difficult. Additionally, it seems likely that a substantial amount of inverse ("2,1") addition occurs. The failure of the secondary benzylic organometallic thus formed to cyclize would lead to formation of the uncyclized hydrogenated product.

5% $Cp^*_2YMe\cdot THF$, 1 atm H_2, benzene, rt, 1 h → 53% + 26% (43)

Fully reduced uncyclized material is also obtained in the hydrogenation of diallyldimethylsilane (Eq. 44) [39]. This is undoubtedly because the relatively long Si-C bonds perturb the cyclization transition state.

Me_2Si(allyl)$_2$ — 5% $Cp^*_2YMe\cdot THF$, 1 atm H_2, benzene, rt, 1 h, 64% → Me_2Si(propyl)$_2$ (44)

More highly hindered (e.g., 1,1-disubstituted) alkenes cannot be accommodated by the Cp^*_2YH catalysts. However, more sterically open catalysts permit rapid cyclization under extraordinarily mild conditions (Eq. 45) [10].

0.5% cat., 1 atm H_2, neat, rt, 2 h (45)

cat. = TMS, $(TMS)_2CHY$, $SiMe_2$, R*   R* =

5.3 Termination by Silylation

Although the cyclization reactions described above represent a potentially powerful means for carbon-carbon bond formation, under reductive (hydrogenolysis) conditions the method utilizes a highly functionalized diene substrate and leaves an essentially unfunctionalized product in its wake. The utilization of silylation as a chain terminating event provides a way to place functionality back into the molecule after the cyclization (Fig. 8).

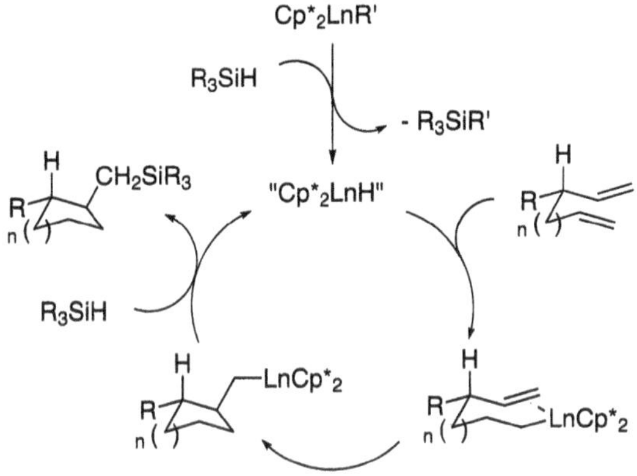
Cp*2LnR'
R3SiH
- R3SiR'
"Cp*2LnH"
CH2SiR3
R3SiH
LnCp*2
LnCp*2

Fig. 8. Catalytic cycle for the cyclization/silylation of dienes

5.3.1
Cyclization/Silylation of Terminal Dienes and Trienes

Early studies centered on the utilization of organolutetiums (Eq. 46) [32a], organoneodymiums [33c], and organosamariums [27] for the cyclization of 1,5-hexadiene and homologs. The early studies of the cyclization/silylation process included only unsubstituted dienes, leaving questions of regioselection, diastereoselection, and functional group toleration unanswered. A more thorough study of this chemistry that focused on the application to small molecule synthesis was performed utilizing organoyttrium complexes [40]. The organoyttrium-catalyzed process employed on monosubstituted dienes appears to be quite general for the synthesis of both five- and six-membered rings. For the synthesis of five-membered rings, phenylsilane is a convenient "chain terminator." It provides high yields of cyclized/silylated products with no stereocenters introduced as a result of the incorporation of the silicon atom. High diastereoselectivities are achieved in many instances (Eq. 47).

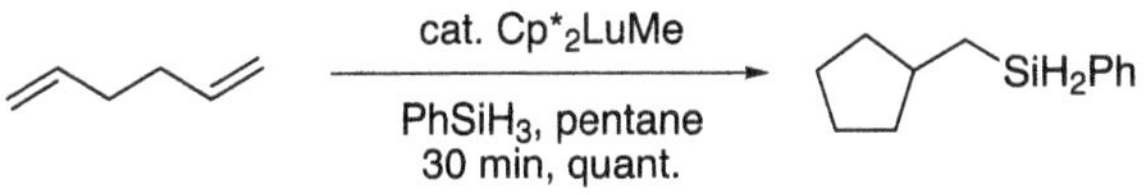
cat. Cp*2LuMe
PhSiH3, pentane
30 min, quant.
SiH2Ph

(46)

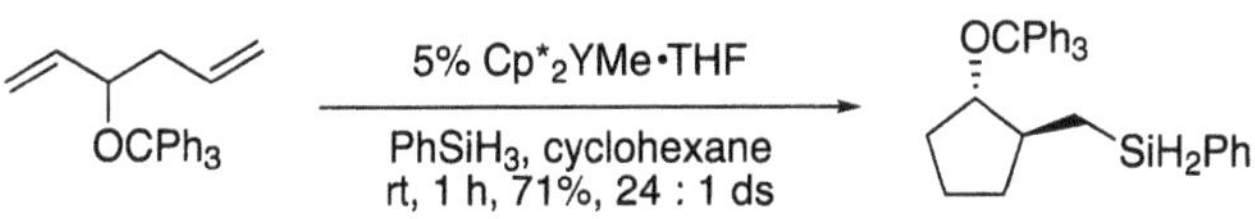
OCPh3
5% Cp*2YMe•THF
PhSiH3, cyclohexane
rt, 1 h, 71%, 24 : 1 ds
OCPh3
SiH2Ph

(47)

Noteworthy in all of these metallocene-catalyzed reactions is the fact that they proceed with complete "atom economy" [41], i.e., all of the atoms from the substrates and reagents are incorporated into the desired product and there are no byproducts produced. Many of these reactions are so clean, in fact, that pouring the reaction mixture through a short bed of Florisil to remove the catalyst, followed by evaporation of the solvent and bulb-to-bulb distillation, leads to essentially quantitative yields of analytically pure product.

The synthetic equivalency of the silane and an alcohol can be easily demonstrated by subjecting the crude silane product to any of a variety of available oxidizing conditions (Eq. 48) [30, 40].

5% $Cp^*_2YMe{\cdot}THF$
$PhSiH_3$
cyclohexane, rt, 1 h

Ph
SiH_2Ph

1. $HBF_4{\cdot}OEt_2$
$CHCl_3$, 0 °C, 1 h
2. KF, $KHCO_3$, H_2O_2
THF, MeOH
heat, 18 h

Ph
OH

74% overall yield
9 : 1 ds

(48)

Cyclohexane formation is entropically less favorable than cyclopentane generation, and treatment of 1,6-dienes under the conditions listed in Eq. (47) leads to the production of uncyclized, disilylated products (Eq. 49) [35] or silicon bridged dimers [40]. To avoid these problems, phenylmethylsilane can be employed as the chain terminator. Utilizing this more hindered silane slows the σ-bond metathesis sufficiently to prevent dimerization (Eq. 50). The trapping step can be retarded even further by the use of diphenylsilane (Eq. 51). Thus, not only can the metal and the ligand array be manipulated to bring about the desired result in the catalytic process, but the properties of the silane reagent itself can also be adjusted to meet the demands of the synthesis at hand.

3% $Cp^*_2YCH(TMS)_2$
$PhSiH_3$, benzene, rt
26 h, 96%

PhH_2Si SiH_2Ph (49)

OTBDMS

5% $Cp^*_2YMe{\cdot}THF$
$MePhSiH_2$, cyclohexane
rt, 2 h, 99%, 1.5 : 1 ds

OTBDMS
SiHMePh

(50)

t-Bu
5% $Cp^*_2YMe{\cdot}THF$
$MePhSiH_2$, cyclohexane
rt, 2 h, 97%
SiHMePh
5% $Cp^*_2YMe{\cdot}THF$
Ph_2SiH_2, cyclohexane
rt, 20 h, 65%
$SiHPh_2$
(51)

The functional group compatibility of this process is similar to that described previously [40]. Protected alcohols (Eq. 50), tertiary amines (Eq. 51), and protected ketones (Eq. 52) are all inert to the reaction conditions.

5% $Cp^*_2YMe{\cdot}THF$
$PhMeSiH_2$
cyclohexane, rt, 1 h
99%
SiHMePh
(52)

In the reaction of triallylamine (Eq. 53) [40], after the first ring forming event the organometallic is trapped by σ–bond metathesis with the silane instead of undergoing an entropically unfavorable intramolecular olefin insertion to yield a bridged bicyclic structure. The remaining double bond then competes effectively for the catalyst, making the isolation of monocyclic silylated product bearing a free allyl group impossible.

5% $Cp^*_2YMe{\cdot}THF$
$MePhSiH_2$, cyclohexane
rt, 1 h, 90%
SiHMePh
SiHMePh
(53)

Monosubstituted diene systems have been employed for the synthesis of bicyclic systems as well as monocyclics [17]. The simplest way to accomplish this is to construct a second ring onto an existing structure (Eq. 54). This cyclization process is initiated at the alkene lacking allylic substitution. The formation of the six-membered ring necessitates the use of methylphenylsilane as the silylating reagent. Because the silane itself comprises a new stereocenter, it must be removed by oxidation to assess accurately the diastereoselectivity of cyclization. Fluxionality of the five-membered ring results in a mixture of diastereomers at the silylmethyl-substituted stereocenter.

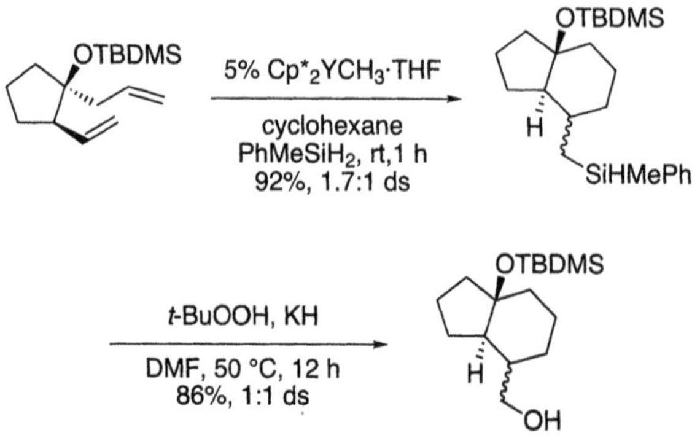

OTBDMS
5% $Cp^*_2YCH_3 \cdot THF$
cyclohexane
$PhMeSiH_2$, rt,1 h
92%, 1.7:1 ds
OTBDMS
H
SiHMePh

(54)

t-BuOOH, KH
DMF, 50 °C, 12 h
86%, 1:1 ds
OTBDMS
H
OH

Because the stereochemistry at the ring juncture is passed undisturbed from the substrate, by generating substrates with a fixed relationship between the olefin bearing substituents either ring fusion may be produced at will (Eq. 55) [17]. After oxidation of the silane product and deprotection to the corresponding diol, the stereochemistry of the single isomer was determined by X-ray crystal analysis. In this case the high diastereoselectivity observed may be explained by the transition structures shown in Fig. 9. The bulky organometallic prefers to be away from the existing ring, leading to the observed *cis*-decalin product.

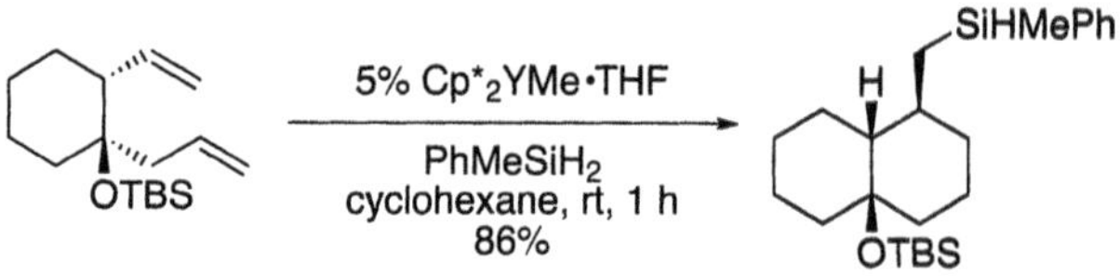

OTBS
5% $Cp^*_2YMe \cdot THF$
$PhMeSiH_2$
cyclohexane, rt, 1 h
86%
SiHMePh
H
OTBS

(55)

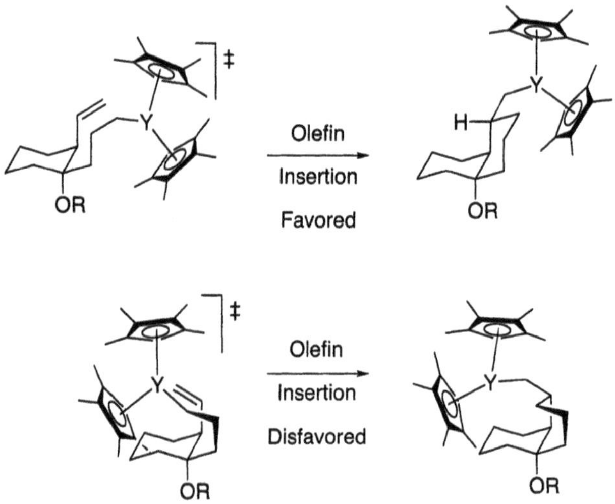

‡
Y
OR
Olefin
Insertion
Favored
H
Y
OR

‡
Y
OR
Olefin
Insertion
Disfavored
Y
OR

Fig. 9. Steric factors governing *cis*-decalin formation

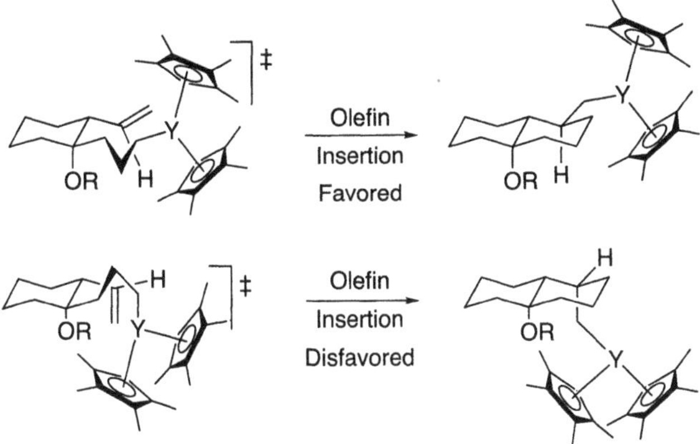

Fig. 10. Steric factors governing *trans*-decalin formation

A *trans*-decalin system can be produced by using a substrate with the opposite diastereomeric relationship (Eq. 56 and Fig. 10) [17]. Of interest is the heightened stereoselectivity observed when the relatively rigid cyclohexane unit is substituted for the fluxional cyclopentane moiety (Eq. 54). This rigidity accentuates the steric interactions governing the stereochemical course of the reaction.

OTMS — 5% $Cp^*_2YMe \cdot THF$, $PhMeSiH_2$, cyclohexane, rt, 1 h, 89% → H, SiHMePh, OTMS (56)

If the substituents are arranged to form a *trans*-decalin with complementary substitution, no cyclization occurs (Eq. 57) [17]. The slower six-membered ring formation at the hindered alkene is interrupted by silylation. This allows impressive chemoselectivity in polyene substrates. When presented with a substrate possessing three monosubstituted olefins with varying steric environments, the catalyst can select the least hindered alkene for initial insertion and will only cyclize onto the less hindered of the two remaining double bonds (Eq. 58). A mechanistic outline and depiction of the steric environments is given in Fig. 11.

THPO — 5% $Cp^*_2YCH_3 \cdot THF$, cyclohexane, $PhMeSiH_2$, rt, 1 h, 98% → THPO, SiHMePh (57)

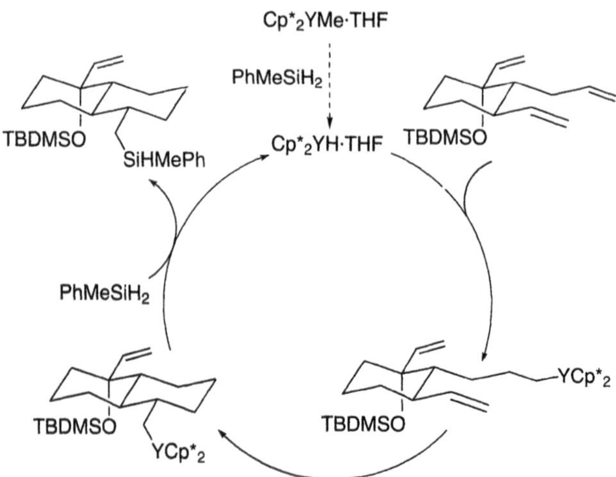

Fig. 11. Catalytic cycle for the selective reaction of a triene

TBSO → TBSO, H, SiHMePh

5% $Cp^*_2YMe{\cdot}THF$, $PhMeSiH_2$, cyclohexane, rt, 1 h, 86% (58)

The failure to insert quaternary vinyl groups in a cyclization process is limited to 1,6-diene systems [17]. When a similar competition for selective insertion is attempted on a 1,5-diene, cyclization occurs because of the inherent entropic advantage of five-membered ring formation (Eq. 59). The close approach of the bulky organometallic to the axial alkoxy substituent in the chair-like transition structure (Fig. 12b) causes the reaction to proceed through the less hindered boat formation (Fig. 12a).

OTMS → TMSO, SiHMePh, H

5% $Cp^*_2YMe{\cdot}THF$, $PhMeSiH_2$, cyclohexane, rt, 6 h, 80% (59)

In addition to their ability to assemble bicyclic structures on a monocyclic scaffold, the organoyttrium catalysts can also convert trienes to bicyclics in a sequential process (Eqs. 60, 61) [17]. Both five- and six-membered rings can be constructed at will. In these cases the stereochemistry at the bridgeheads is a result of the chair-like transition structures operative during the intramolecular olefin insertions (Schemes 1, 2). There are two notable features of this reaction. The first is that after the initial olefin insertion a 5-exo cyclization at the allyli-

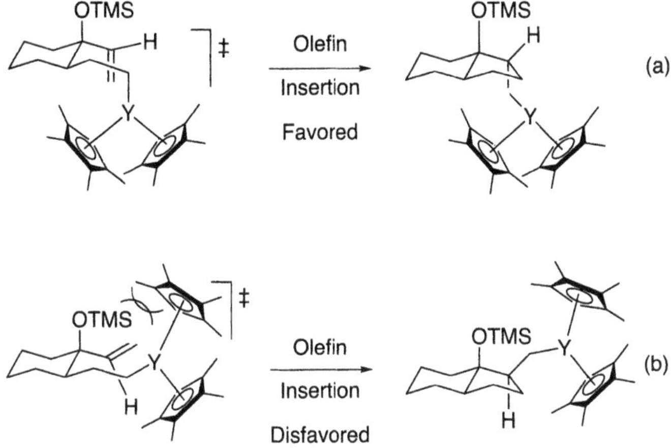

Fig. 12. Steric factors governing bicyclic product formation

cally substituted alkene is chosen over a 6-exo cyclization at an unhindered alkene. The second is the formation of the *trans*-bicyclo[3.3.0] ring system. Currently, the number of methods available to prepare this highly strained system is quite limited [42]. The formation of single product isomers attests to the high chemo- and stereoselectivity of each individual reaction step.

TMSO — 5% $Cp^*_2YMe{\cdot}THF$, $PhSiH_3$, cyclohexane, rt, 1 h → *t*-BuOOH, KH, DMF, 50 °C, 12 h → OH, H, OH 73% (60)

5% $Cp^*_2YMe{\cdot}THF$, $PhMeSiH_2$, cyclohexane, rt, 1 h, 68% → H, H, SiHMePh (61)

The chemoselectivity of these catalysts allows the selective construction of bicyclic systems bearing additional substitution (Eqs. 62, 63) [17]. By varying the stereochemistry of the substituents, the three-dimensional structure of the product can be altered as desired.

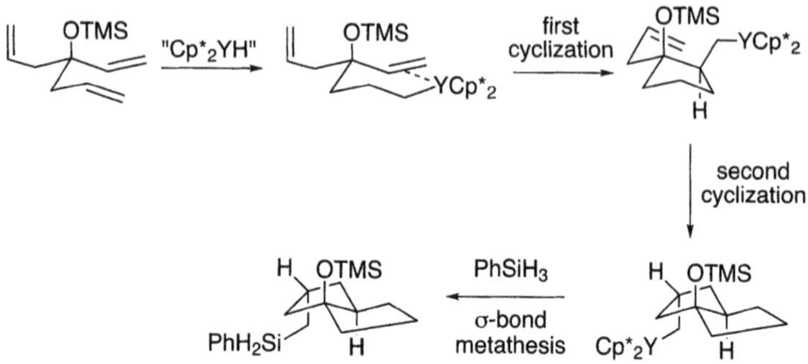

OTMS "Cp*2YH" OTMS YCp*2 first cyclization OTMS YCp*2 H second cyclization H OTMS Cp*2Y H PhSiH3 σ-bond metathesis H OTMS PhH2Si H

Scheme 1

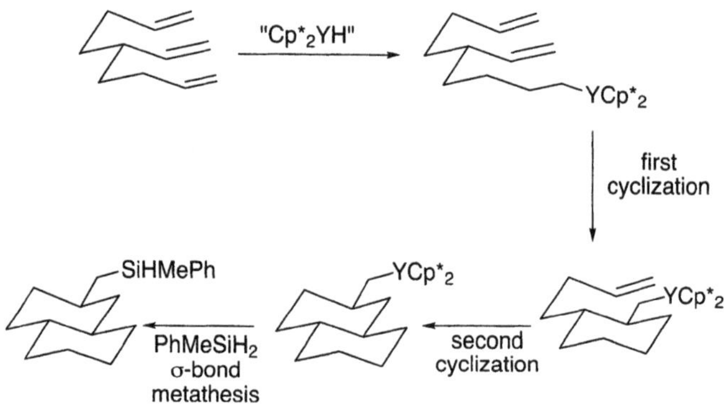

"Cp*2YH" YCp*2 first cyclization YCp*2 second cyclization YCp*2 PhMeSiH2 σ-bond metathesis SiHMePh

Scheme 2

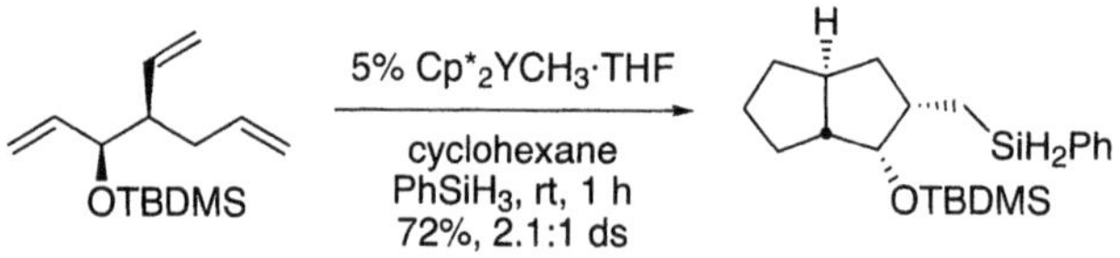

OTBDMS 5% Cp*2YCH3·THF cyclohexane PhSiH3, rt, 1 h 72%, 2.1:1 ds H SiH2Ph OTBDMS (62)

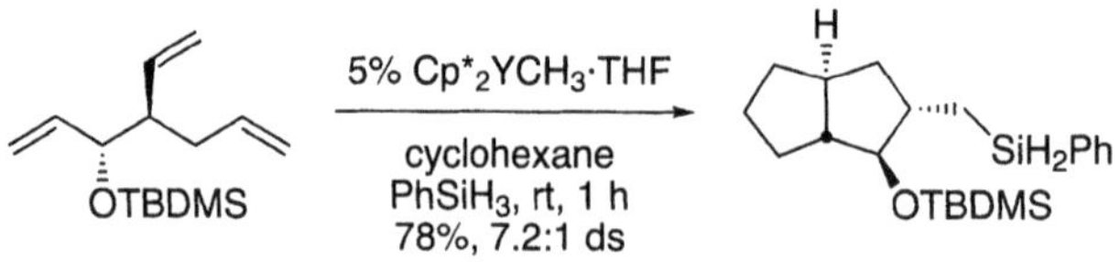

OTBDMS 5% Cp*2YCH3·THF cyclohexane PhSiH3, rt, 1 h 78%, 7.2:1 ds H SiH2Ph OTBDMS (63)

In spite of their recognized Lewis acidity and the propensity to complex with Lewis bases (particularly in an intramolecular chelate) [8], the organoyttrium complexes can be utilized for the synthesis of nitrogen heterocycles. The protocol has been employed in a concise synthesis of (±)-epilupinine (Scheme 3) [43].

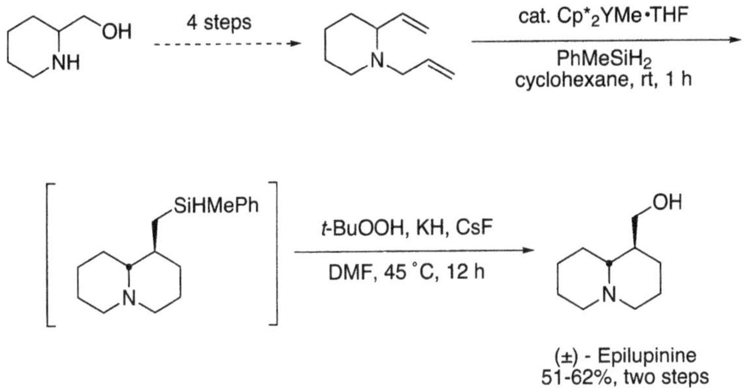

4 steps

cat. $Cp^*_2YMe \cdot THF$

$PhMeSiH_2$
cyclohexane, rt, 1 h

SiHMePh

t-BuOOH, KH, CsF

DMF, 45 °C, 12 h

(±) - Epilupinine
51-62%, two steps

Scheme 3

5.3.2
Cyclization/Silylation of Hindered Dienes and Trienes

As mentioned previously, more highly hindered alkene systems cannot be accommodated by the relatively closed organoyttrium catalysts, and thus more open ligand arrays have been enlisted to allow incorporation of 1,1-disubstituted alkenes. A first generation catalyst was developed that worked quite well in these systems, albeit at somewhat elevated temperatures and protracted reaction times [44]. The ligand possessed a silicon hinge that served to open the clamshell comprised of the cyclopentadienyl rings, and the rings themselves possessed a single trimethylsilyl group to prevent hydride dimer formation. Monosubstituted alkenes reacted in preference to 1,1-disubstituted alkenes (Eq. 64). Remarkably, both alkenes can be disubstituted, generating a quaternary center with complete stereochemical control (Eq. 65). Trienes reacted in a sequential process, in one case leading to a spirocyclic system (Eq. 66).

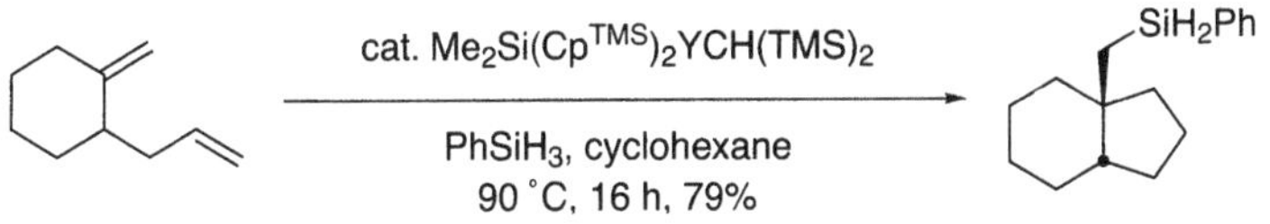

cat. $Me_2Si(Cp^{TMS})_2YCH(TMS)_2$

$PhSiH_3$, cyclohexane
90 °C, 16 h, 79%

SiH_2Ph

(64)

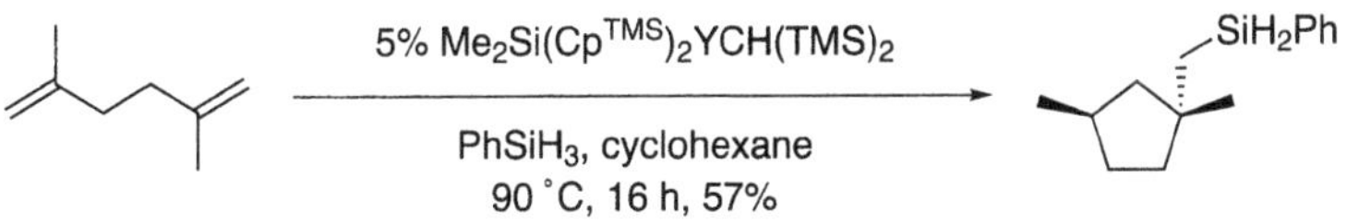

5% $Me_2Si(Cp^{TMS})_2YCH(TMS)_2$

$PhSiH_3$, cyclohexane
90 °C, 16 h, 57%

SiH_2Ph

(65)

5% $Me_2Si(Cp^{TMS})_2YCH(TMS)_2$
$PhSiH_3$, cyclohexane
90 °C, 24 h, 64%
SiH_2Ph
(66)

Recognizing the need to carry out these reactions under milder conditions, a second generation catalyst was tested in similar systems [44]. This new system, lacking the silicon hinge, proved to be extraordinarily reactive, in fact orders of magnitude more reactive than any other neutral catalyst tested to date in similar substrates (compare Eqs. 66, 67). As described previously, the precatalyst has modest air stability, and yet the catalyst itself is remarkably reactive. Catalyst loadings as low as 0.5 mol% have been employed in reasonably large scale reactions (10–20 mmol), and the catalyst also supports the presence of the standard functional groups for this general class of catalysts (Eq. 68). One problem with the catalyst is that it apparently becomes deactivated over time because of hydride dimer formation. Ideal reaction conditions for slow-reacting substrates thus involve addition of smaller portions of the catalyst at fixed intervals to maintain an active concentration of the catalyst.

5% $[(Cp^{TMS})_2YMe]_2$
$PhSiH_3$, cyclohexane
rt, 1 h, 88%
SiH_2Ph
(67)

OTBDPS
5% $[(Cp^{TMS})_2YMe]_2$
$PhSiH_3$, cyclohexane
rt,16 h, 74%
OTBDPS
SiH_2Ph
Me
(68)

The expanded capabilities of this catalyst allowed the construction of additional ring systems [44]. A transannular olefin insertion formed a bicyclo[3.3.1]nonane structure (Eq. 69). Multiple insertion steps can selectively transform a monocyclic substrate into a propellane derivative (Eq. 70). The catalytic cycle begins at the monosubstituted olefin, followed by the insertion of the remaining double bonds to form five membered rings.

5% $[(Cp^{TMS})_2YMe]_2$
$PhSiH_3$, cyclohexane
rt, 20 h, 50%
SiH_2Ph
(69)

5% $[(Cp^{TMS})_2YMe]_2$
$PhSiH_3$, cyclohexane
rt, 1 h

SiH_2Ph
75%
+
SiH_2Ph
14%

(70)

This catalyst is surprisingly intolerant of branching on the hindered olefin (Eq. 71) [44]. The addition of two methylene units between the alkene and the branch point is required to provide good yields of cyclized product. Another limitation is the inability of the complex to insert an endocyclic olefin (Eq. 72).

5% $[(Cp^{TMS})_2LnMe]_2$
$PhSiH_3$, cyclohexane
rt, 1 h

n = 0, Ln = Lu
n = 2, Ln = Y

SiH_2Ph
0%
84%
+
PhH_2Si
84%
0%

(71)

5% $[(Cp^{TMS})_2LuMe]_2$
$PhSiH_3$, cyclohexane
rt,1 h, 93%

SiH_2Ph

(72)

5.3.3
Cyclization/Silylation of Enynes and Dienynes

Enynes have been found to be excellent substrates for the cyclization/silylation sequential reaction [45]. The electron rich alkyne undergoes preferential reaction, thus requiring propargyl branching to direct the regiochemistry of the initial insertion. Although sp^2 hybridized, the organometallic originally formed is converted smoothly via the cyclization/silylation process to afford the desired products. A variety of functional groups can be tolerated at key positions in these substrates (Eqs. 73, 74). The diastereoselection varies reliably with the size of the substituent, as would be expected from the chair-like transition structure for insertion. Six-membered rings can also be generated in high yields and with modest diastereoselection when phenylmethylsilane is employed (Eq. 75). In this case varying the silylating reagent causes a drastic change in the outcome of the reaction (compare with Eq. 35).

OCPh$_3$ 5% Cp*$_2$YMe·THF, PhSiH$_3$, cyclohexane, rt, 2 h, 88%, 35 : 1 ds → PhH$_2$Si, OCPh$_3$ (73)

R 5% Cp*$_2$YMe·THF, PhSiH$_3$, cyclohexane, rt, 2h → R, SiH$_2$Ph + R, SiH$_2$Ph (74)

R	% Yield	ds
OTBDMS	93	6.5 : 1
OTIPS	93	12 : 1
OTr	84	24 : 1
Et	88	>50 : 1
CH_2OMe (100 °C)	80	20 : 1
$CH_2OTBDMS$	76	>50 : 1
—N (isoindolinyl)	91	40 : 1

TBDMSO 12% Cp*$_2$YMe·THF, MePhSiH$_2$, cyclohexane, 50 °C, 8 h, quant., 4 : 1 ds → PhMeHSi, OTBDMS (75)

The cyclization protocol can be carried a step further by judiciously placing another olefin in the substrate [38]. The incorporation of an additional allylic group only results in low yields of bicyclic products (Eq. 76). The reduced yield is undoubtedly because of the poor chemoselectivity between the alkyne and the allylic olefin. Alkyl substitution on the allyl group gives complete selectivity for initial alkyne insertion, but the hindered olefin prevents bicyclic product formation (Eq. 77). Geminal dimethyl substitution at the allylic position, however, provides selectivity in the initial insertion without stopping the second intramolecular insertion from taking place (Eq. 78). This dienyne cyclization allows another entry into the strained *trans*-bicyclo[3.3.0]octane system, and contains an additional handle for further functionalization.

5% $Cp^*_2YMe{\cdot}THF$; benzene-d_6, $PhSiH_3$, rt, 2.5 h, 33% → H, SiH_2Ph (76)

5% $Cp^*_2YMe{\cdot}THF$; benzene-d_6, $PhSiH_3$, rt, 24 h, 84% → SiH_2Ph (77)

5% $Cp^*_2YMe{\cdot}THF$; benzene-d_6, $PhSiH_3$, rt, 18 h, 76%, 3.9 : 1 ds → H, SiH_2Ph (78)

A single substituent allylic to the monosubstituted olefin is sterically large enough to make the initial insertion selective (Eqs. 79, 80) [38]. Either substrate diastereomer can be prepared, leading to different steric interactions in the transition states (Figs. 13, 14), and different product stereochemistry patterns.

O, O, OTr; 5% $Cp^*_2YMe{\cdot}THF$; benzene-d_6, $PhSiH_3$, rt, 90%, >40 : 1 ds → OTr, SiH_2Ph, H, O, O (79)

conformation leading to the formation of the major product

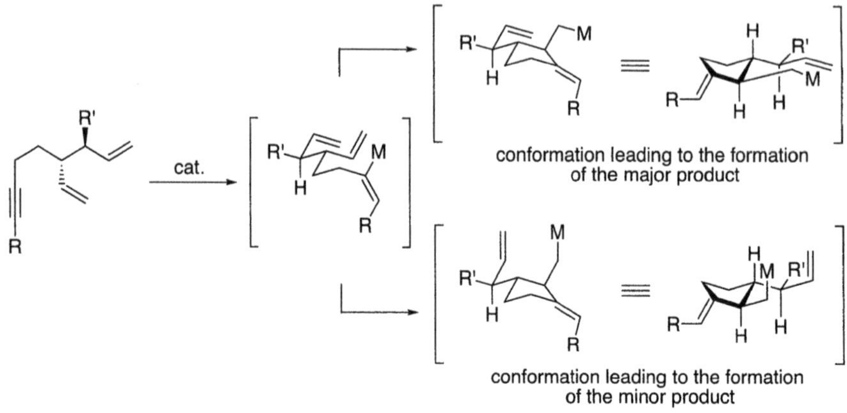

conformation leading to the formation of the minor product

Fig. 13. Steric factors governing dienyne bicyclization

conformation leading to the formation of the minor product

conformation leading to the formation of the major product

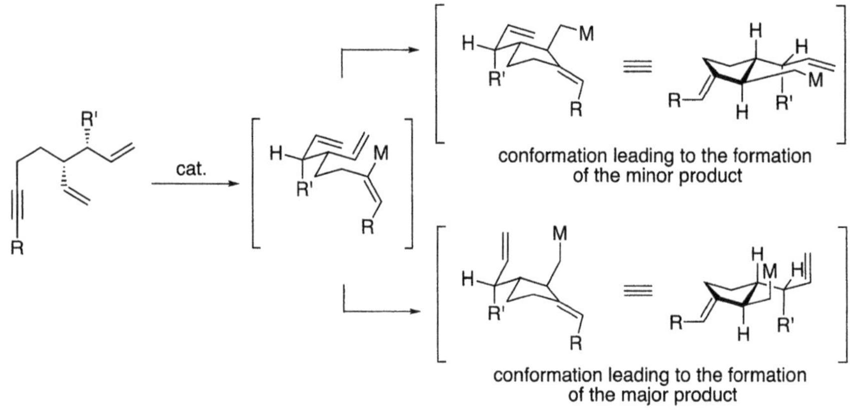

Fig. 14. Steric factors governing dienyne bicyclization

5% $Cp^*_2YMe{\cdot}THF$

benzene-d_6, $PhSiH_3$

rt, 89%, 30 : 1 ds

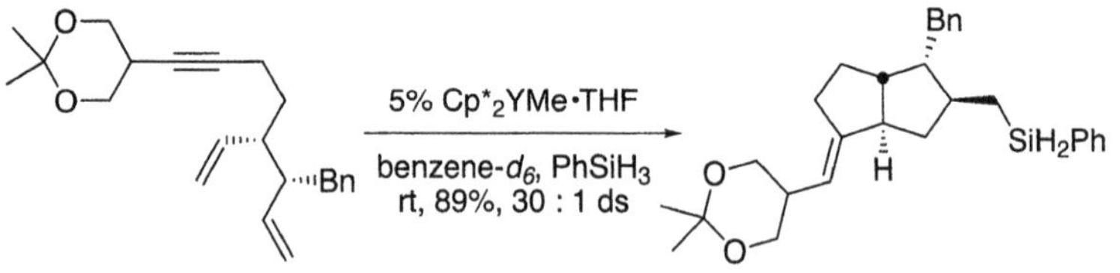

(80)

If the alkyl chain is shortened by one carbon, a substrate is generated that has the possibility for bicyclo[2.2.1]heptane generation (Eq. 81) [38]. After the initial alkyne insertion, the catalyst must choose between cyclobutane and cyclopentane formation. The five-membered ring is formed because of the lower strain involved, but the remaining olefin is not inserted because the lowest energy conformation of the organometallic intermediate places the olefin out of reach of the carbon-metal bond (Eq. 82).

OTBDMS; 5% $Cp^*_2YMe{\cdot}THF$; benzene-d_6, $PhSiH_3$; rt, 2h, 70%; SiH_2Ph; OTBDMS; SiH_2Ph (81)

H; OTBDMS; R; YCp^*_2; ‡; H; H; H; R; OTBDMS; YCp^*_2; ‡ (82)

6
Conclusions

The lanthanide- and group 3 metallocenes exhibit a rich chemistry that can be exploited for the selective synthesis of small molecules. At least one class of catalysts that are easily synthesized in a single pot exhibit reasonable stability in the air. Consequently, they should be readily accessible to practicing synthetic organic chemists. Catalysts have been developed that permit reactions of monosubstituted and 1,1-disubstituted alkenes, as well as internal alkynes. Practical solutions for the inability of more highly substituted alkenes to insert have yet to be reduced to practice. Terminal alkynes are rapidly metallated by these catalysts, and are unlikely to be adaptable to many of the processes outlined in this contribution. A variety of functional groups are tolerated by the catalysts (halides, acetals, thioacetals, ethers, and amines), and thus highly functionalized substrates of interest in complex molecule synthesis should be amenable to utilization in selected processes. A high degree of chemoselectivity can be achieved in polyunsaturated systems, and regiochemistry in the olefin insertion reactions can also be controlled. Both of these conspire to provide an effective means to control the direction of cyclization in unsymmetrical systems. Excellent diastereoselectivity can often be achieved, taking advantage of both the inherent selectivity of the catalysts themselves as well as the highly ordered transition structures involved in the intramolecular processes they promote. Finally, elegant asymmetric catalysts have been synthesized and utilized in selective reactions. Further developments in this arena are certain to produce a class of readily available catalysts that provide rapid and efficient entry to a wide range of complex structures in enantiopure form.

Acknowledgments
Our work in this area was supported by a research grant from the National Institutes of Health (GM48580). The authors thank a very talented group of coworkers whose names appear in the references below for their dedicated effort to this chemistry.

7 References

1. (a) Gavens PD, Bottrill M, Kelland JW, McMeeking J (1982) Ziegler-Natta catalysis. In: Wilkinson G, Stone, FGA, Abel EW (eds) Comprehensive organometallic chemistry, vol 3. Pergamon, Oxford, chap 22.5; (b) Ballard DGH (1973) Adv Catal 23:263; (c) Ballard DGH, Burnham DR, Twose DL (1976) J Catal 44:116; (d) Sinn H, Kaminsky W (1980) Adv Organomet Chem 18:99; (e) Chien JCW (ed) (1979) Coordination polymerization. Academic Press, New York; (f) Boor J (1979) Ziegler-Natta catalysts and polymerizations. Academic Press, New York; (g) Keim W, Behr A, Röper M (1982) Alkene and alkyne oligomerization, cooligomerization and telomerization reactions. In: Wilkinson G, Stone FGA, Abel EW (eds) Comprehensive organometallic chemistry, vol 8. Pergamon, Oxford, chap 52; (h) Gates BC, Katzer JR, Schuit GCA (1979) Chemistry of catalytic processes. McGraw-Hill, New York; (i) Parshall G (1980) Homogeneous catalysis. Wiley-Interscience, New York; (j)Pino P, Mülhaupt R (1980) Angew Chem, Int Ed Engl 19:857; (k) Coates GW, Waymouth RM (1995) Transition metals in polymer synthesis: Ziegler-Natta reaction. In: Abel EW, Stone FGA, Wilkinson G (eds) Comprehensive organometallic chemistry II, vol 12. Pergamon, Oxford, chap 12.1; (l) Zucchini U, Dall'Occo T, Resconi L (1993) Indian J Technol 31:247; (m) Simonazzi T, Gianni U (1994) Gazz Chim Ital 124:533; (n) Brintzinger HH, Fischer D, Waymouth RM (1995) Angew Chem, Int Ed Eng 34:1143; (o) Keii T, Soga K (eds) (1986) Catalytic polymerization of olefins. Kodenska, Tokyo; (p) Reichert KH (1983) In: Quirk RP (ed) Transition metal catalyzed polymerizations: alkenes and dienes. Harwood Academic, New York
2. (a) Marks TJ, Ernst RD (1982) Scandium, yttrium and the lanthanides and actinides. In: Wilkinson G, Stone FGA, Abel EW (eds) Comprehensive organometallic chemistry, vol 3. Pergamon, Oxford, chap 21; (b) Watson PL, Parshall GW (1985) Acc Chem Res 18:51; (c) Edelmann FT (1995) Scandium, yttrium, and the lanthanide and actinide elements, excluding their zero oxidation state complexes. In: Abel EW, Stone FGA, Wilkinson G (eds) Comprehensive organometallic chemistry II, vol 4. Pergamon Press, Oxford, chap 2; (d) Edelmann FT (1996) Top Curr Chem 179:247
3. (a) Johnson LK, Killian CM, Brookhart M (1995) J Am Chem Soc 117: 6414; (b) Brookhart M, Wagner MI (1996) J Am Chem Soc 118:7219; (c) Feldman J, McLain SJ, Parthasarathy A, Marshall WJ, Calabrese JC, Arthur SD (1997) Organometallics 16:1514; (d) Killian CM, Johnson LK, Brookhart M (1997) Organometallics 16: 2005
4. (a) Emrich R, Heinemann O, Jolly PW, Krüger C, Verhovnik GPJ (1997) Organometallics 16:1511; (b) Scollard JD, McConville DH, Rettig SJ (1997) Organometallics 16:1810
5. Jeske G, Lauke H, Mauermann H, Swepston PN, Schumann H, Marks TJ (1985) J Am Chem Soc 107:8091
6. (a) Borricello A, Busico V, Sudemijer O (1996) Macromolecular Rapid Commun 17:589; (b) Sinclair KB, Wilson RB (1994) Chem Ind 857
7. (a) Watson PL (1983) J Chem Soc, Chem Commun 276; (b) Evans WJ, Chamberlain LR, Ulibarri TA, Ziller JW (1988) J Am Chem Soc 110:6423
8. Piers WE, Shapiro PJ, Bunel EE, Bercaw JE (1990) Synlett 74
9. Brown HC (1975) Organic syntheses via boranes. Wiley, New York
10. Haar CM, Stern CL, Marks TJ (1996) Organometallics 15:1765

11. (a) Schumann H, Genthe W, Bruncks N (1981) Angew Chem, Int Ed Engl 20:119; (b) Evans WJ, Wayda AL, Hunter WE, Atwood JL (1981) J Chem Soc, Chem Commun 292
12. Evans WJ (1987) Polyhedron 6:803
13. Deelman BJ, Booij M, Meetsma A, Teuben JH, Kooijman H, Spek AL (1995) Organometallics 14:2306
14. Stern D, Sabat M, Marks TJ (1990) J Am Chem Soc 112:9558
15. den Haan KH, Wielstra Y, Teuben JH (1987) Organometallics 6:2053
16. Wang KK, Brown HC (1982) J Am Chem Soc 104:7148
17. (a) Nichols PJ (1997) PhD thesis. University of Colorado at Boulder; (b) Molander GA, Nichols PJ, Noll BC (1998) J Org Chem 63:2292
18. Markus Keitsch (unpublished results)
19. Shriver DF, Drezdzon MA (1986) The manipulation of air-sensitive compounds, 2nd edn. Wiley-Interscience, New York
20. Jeske G, Schock LE, Swepston PN, Schumann H, Marks TJ (1985) J Am Chem Soc 107:8103
21. (a) Evans WJ (1982) In: Hartley FR, Patai S (eds) The chemistry of the metal-carbon bond. Wiley, Chichester; (b) Schumann H, Genthe W (1984) In: Gschneidner KA Jr, Eyring L (eds) Handbook on the chemistry and physics of rare earths. Elsevier, Amsterdam; (c) Schumann H (1984) Angew Chem, Int Ed Engl 23:474; (d) Evans WJ (1985) Adv Organomet Chem 24:131; (e) Schumann H, Meese-Marktscheffel JA, Esser L (1995) Chem Rev 95:865; (f) Schaverien CJ (1994) Adv Organomet Chem 36:283
22. Jeske G, Lauke H, Mauermann H, Schumann H, Marks TJ (1985) J Am Chem Soc 107:8111
23. (a) Zinnen HA, Pluth JJ, Evans WJ (1980) J Chem Soc, Chem Commun 810; (b) Brothers PJ (1981) Prog Inorg Chem 28:1; (c) Evans WJ, Meadows JH, Wayda AL, Hunter WE, Atwood JL (1982) J Am Chem Soc 104:2008; (d) Schumann H, Genthe WJ (1981) J Organomet Chem 213:C7; (e) Marks TJ (1987) Inorg Chim Acta 140:1; (f) Thompson ME, Baxter SM, Bulls AR, Burger BJ, Nolan MC, Santarsiero BD, Schaefer WP, Bercaw JE (1987) J Am Chem Soc 109:203
24. Molander GA, Hoberg JO (1992) J Org Chem 57:3266
25. Molander GA, Winterfeld J (1996) J Organomet Chem 524:275
26. Giardello MA, Conticello VP, Brard L, Gagné MR, Marks TJ (1994) J Am Chem Soc 116:10,241
27. Fu PF, Brard L, Li Y, Marks TJ (1995) J Am Chem Soc 117:7157
28. Evans WJ, Bloom I, Hunter WE, Atwood JL (1981) J Am Chem Soc 103:6507
29. (a) Speier JL (1979) Adv Organomet Chem 17:407; (b) Ojima I (1989) The hydrosilylation reaction. In: Patai S, Rappoport Z (eds) The chemistry of organic silicon compounds. Wiley, London; (c) Yamamoto K, Takemae M (1990) Synlett 259; (d) Takahashi T, Hasegawa M, Suzuki N, Saburi M, Rousset CJ, Fanwick PE, Negishi E (1991) J Am Chem Soc 113:8564; (e) Boudjouk P, Han BH, Jacobsen JR, Hauck BJ (1991) J Chem Soc, Chem Commun 1424; (f) Kesti MR, Abdulrahman M, Waymouth RM (1991) J Organomet Chem 417:C12; (g) Kesti MR, Waymouth RM (1992) Organometallics 11:1095
30. (a) Tamao K, Ishida N, Tanaka T, Kumada M (1983) Organometallics 2:1649; (b) Bergens SH, Nogeda P, Whelan J, Bosnich B (1992) J Am Chem Soc 114:2121; (c) Taber DF, Yet L, Bhamidipati RS (1995) Tetrahedron Lett 36:351; (d) Fleming I, Hennings R, Plaut H (1984) J Chem Soc, Chem Commun 29; (e) Fleming I, Sanderson PE (1987) Tetrahedron Lett 28:4229; (f) Fleming I, Winter SBD (1993) Tetrahedron Lett 34:7287; (g) Smitrovich JH, Woerpel KA (1996) J Org Chem 61:6044
31. Burgess K, Ohlmeyer MJ (1991) Chem Rev 91:1179
32. (a) Watson PL, Tebbe FN (1991) Chem Abstr 114:123,331p; (b) Forsyth CM, Nolan SP, Marks TJ (1991) Organometallics 10:1450; (c) Radu NS, Tilley TD, Rheingold AL (1992) J Am Chem Soc 114:8293; (d) Radu NS, Tilley TD (1995) J Am Chem Soc 117:5863

33. (a) Beletskaya IP, Voskoboinikov AZ, Parshina IN, Magomedov GKI (1990) Izv Akad Nauk SSR Ser Khim 693; (b) Sakakura T, Lautenschlager HJ, Tanaka M (1991) J Chem Soc, Chem Commun 40; (c) Onozawa S, Sakakura T, Tanaka M (1994) Tetrahedron Lett 35:8177
34. Molander GA, Dowdy ED, Noll BC (1998) Organometallics 17:3754
35. Molander GA, Julius M (1992) J Org Chem 57:6347
36. Schumann H, Keitsch MR, Winterfeld J, Mühle S, Molander GA (1998) J Organomet Chem 559:181
37. (a) Heeres HJ, Heeres A, Teuben JH (1990) Organometallics 9:1508; (b) Heeres HJ, Teuben JH (1991) Organometallics 10:1980
38. (a) Molander GA, Retsch WH (1995) Organometallics 14:4570; (b) Retsch WH (1997) PhD thesis, University of Colorado at Boulder
39. Molander GA, Hoberg JO (1992) J Am Chem Soc 114:3123
40. Molander GA, Nichols PJ (1995) J Am Chem Soc 117:4415
41. (a) Trost BM (1995) Angew Chem, Int Ed Engl 34:259; (b) Trost BM (1991) Science 254:1471
42. (a) Bailey WF, Khanolkar AD, Gavaskar KV (1992) J Am Chem Soc 114:8053; (b) Cooke MP, Gopal D (1994) Tetrahedron Lett 35:2837; (c) Davis JM, Whitby RJ, Jaxa-Chamiec A (1992) Tetrahedron Lett 33:5655; (d) Rousset CJ, Swanson DR, Lamaty F, Negishi E (1989) Tetrahedron Lett 30:5105; (e) Doi T, Yanagisawa A, Takahashi T, Yamamoto K (1996) Synlett 145; (f) Piers E, Kaller AM (1996) Tetrahedron Lett 37:5857
43. Molander GA, Nichols PJ (1996) J Org Chem 61:6040
44. Molander GA, Dowdy ED, Schumann H (1998) J Org Chem 63:3386
45. Molander GA, Retsch WH (1997) J Am Chem Soc 119:8817

Influence of Solvents or Additives on the Organic Chemistry Mediated by Diiodosamarium

Henri B. Kagan*, Jean-Louis Namy

Laboratoire de Synthèse Asymétrique, Institut de Chimie Moléculaire d'Orsay, Bat. 420, Université Paris Sud, 91405 Orsay, France
**e-mail:* kagan@icmo.u-psud.fr

The reactivity of diiodosamarium in solvents other than THF or in mixtures of solvents are discussed. The influence of additives able to coordinate to samarium are then considered (amides, amines or ethers). Proton donors sometimes drastically modify the selectivity of reactions induced by diiodosamarium; some metal salts [such as Fe(III) or Ni(II)] in catalytic amounts may also strongly accelerate or modify reactions induced by diiodosamarium. Thanks to the above tuning of the diiodosamarium reactivity, rich and diversified organic transformations have been performed, some examples of which are presented.

Keywords: Samarium diiodide, Electron transfer, Organosamarium, Radicals, Selectivity

1	**Introduction**	156
2	**Influence of Solvents**	156
2.1	Tetrahydrofuran	156
2.2	Tetrahydropyran	157
2.3	Ethers Other Than THF or THP	160
2.4	Nitriles	160
2.5	Benzene	162
3	**Influence of Additives**	163
3.1	Electron-Rich Additives	163
3.1.1	Hexamethylphosphoramide	163
3.1.2	N,N'-Dimethylpropyleneurea	175
3.1.3	Other Nitrogen Ligands	177
3.1.4	Miscellaneous	179
3.2	Proton Sources	179
3.2.1	Water	179
3.2.2	Alcohols	181
3.2.2.1	Reduction of Carbonyl and C=N Groups	181
3.2.2.2	Pinacol Formation	182
3.2.2.3	Carbonyl-Ene Couplings	182

3.2.2.4 Cleavage of Carbon–Heteroatom Bonds 185
3.2.2.5 Conclusions 187
3.2.3 Acids 188
3.3 Metal Salts 188
3.3.1 Iron Salts 188
3.3.2 Nickel Salts 189
3.3.3 Cobalt Salts 191
3.3.4 Copper Salts 191
3.3.5 Chromium Salts 191
3.3.6 Palladium Complexes 192
3.3.7 Lithium Salts 193

4 Conclusions 194

5 References 194

1 Introduction

Diiodosamarium was prepared for the first time by Matignon and Cazes in 1906 by heating triiodosamarium at high temperature under an atmosphere of hydrogen [1]. For a long time, subsequent preparations used high temperature technology and solid state chemistry (for example, see [2]).

In 1977 we described the mild preparation of diiodosamarium at room temperature by reaction between samarium powder and 1,2-diiodoethane in THF [3]. The dark-green solution of SmI_2 in THF (0.1 M) was used as a convenient reducing agent. We reported in 1980 the basic organic transformations induced by this new reagent [4]. Since that time we, as well as many other groups, have discovered plenty of new reactions, often performed under smooth conditions and with a high selectivity, through a combination of radical and organometallic chemistry. Diiodosamarium is becoming a major reagent in organic chemistry, many reviews on its reactivity are available (for example, see [5–12]). In this article we shall discuss the various ways to tune the reducing properties of diiodosamarium by changing THF to another solvent or by introduction of various additives which can coordinate to samarium or act as catalysts (for a short review, see [13]).

2 Influence of Solvents

2.1 Tetrahydrofuran (THF)

The chemistry of diiodosamarium has been developed in THF mainly because the mild preparation of SmI_2 from samarium powder and 1,2-diiodosamarium is performed in THF to give stable solutions (under inert atmosphere) of the reagent [3,4,14]. During the period 1977–1986, the basic organic transformations induced by SmI_2 involved reactions in THF. In 1987 Inanaga et al. described the acceleration of SmI_2-mediated reactions by addition of some HMPA (usually 5%) in THF [15,16]. This effect is discussed in Sect. 3. It involves the coordinating properties of HMPA towards Sm(II) and Sm(III) ions. Another way to alter the redox properties of the Sm(II)/Sm(III) couple is to replace THF by another solvent. One method is to evaporate THF and to add the desired solvent. This approach has been used to prepare SmI_2 complexed by various nitriles (vide infra). The difficulty comes from the necessity to fully remove the last THF molecules which remain bound to SmI_2, and which may influence its reactivity [17]. The best strategy, therefore, is to try to generate SmI_2 directly in the desired solvent. This has been achieved successfully in nitriles, in tetrahydropyran (THP), in tetraglyme and even in benzene. There are several incentives for performing SmI_2-induced chemistry in various solvents. One can expect some of the following consequences:

1. Acceleration of some reactions,
2. An improved selectivity,
3. New reactions, and
4. Elimination of by-products generated from competitive reactions with THF (H-abstraction, ring opening, etc.).

2.2 Tetrahydropyran (THP)

In 1992, Ito et al. reported the possibility of replacing THF by tetrahydropyran (THP) in some reactions induced by SmI_2 [19,20]. These authors wanted to prepare carbanions α to an amino nitrogen, through the strategy depicted in Scheme 1. The reactions were performed in the mixture THP/HMPA, since the

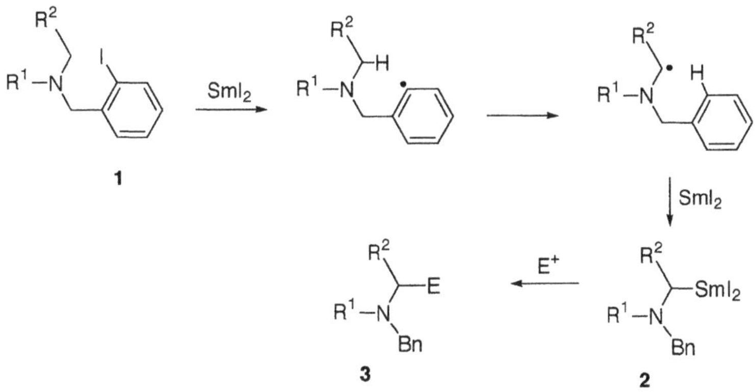

R2
R1–N
I
SmI2
1
SmI2
E+
E
Bn
3
SmI2
2

Scheme 1

yields were slightly better than in THF which acts as an intermolecular hydrogen donor to the aryl radical. The appended orthoiodobenzyl in **1** thus generated organosamarium **2** which then produced *N*-benzyl amines **3** by reaction with various electrophiles. SmI_2 was directly prepared in THP. Undheim et al. applied this reaction to the alkylation of saturated heterocycles α to nitrogen [21].

In 1994 we set up independently a preparation of SmI_2 in THP from samarium metal and 1,2-diodoethane [22]. We wanted to explore the possibility of improving the yields of the acid chloride chemistry, where by-products arise by ring opening of THF. Indeed there was a complete absence of the above by-products. Interestingly, acid chlorides with a branched α-carbon react with SmI_2 to give stable THP solutions of acylsamariums **4** which can then react with an aldehyde or ketone to give the mixed α-ketols **5** (Scheme 2). This two-step process avoids the competitive pinacol formation, especially fast when an aldehyde is in the presence of SmI_2. Some examples are indicated in Scheme 2. The acylsamarium structure **4** is well supported by the formation of deuterated aldehyde **8**, by reduction of acid chloride **7** and by reaction with D_2O.

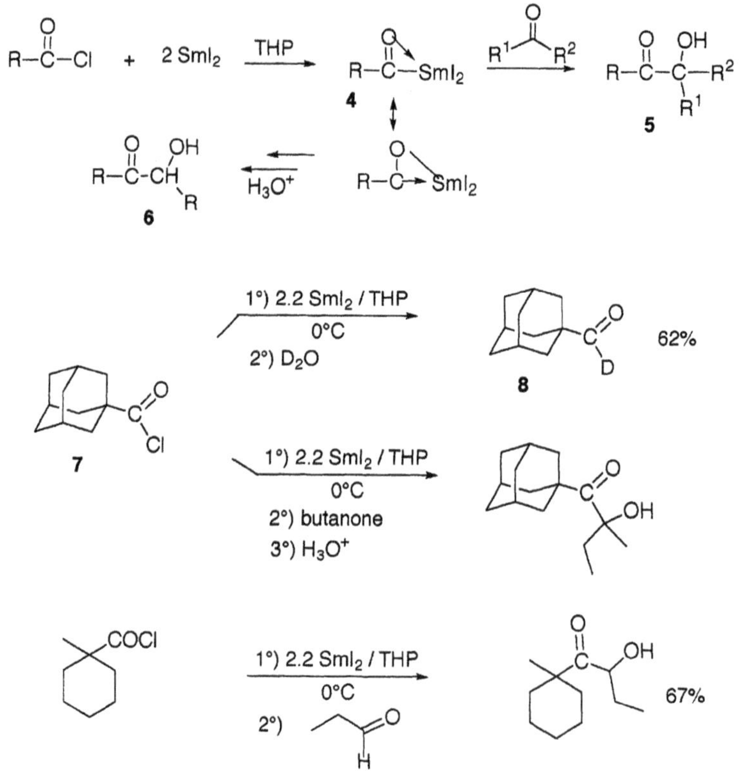

Scheme 2

Acid chlorides with an α-carbon with at least one hydrogen are prone to give self-coupling to symmetrical α-ketols **6**. However, the Barbier procedure in THP allows the preparation of mixed α-ketols **5** in good yields even when using aldehydes.

Recently we discovered that allylic and benzylic samarium compounds could be generated in THP [23]. For example, allylic iodide **9** treated at –15 °C with a THP solution of SmI_2 (2 equiv) followed by addition of butanone at the same temperature gave in 90% isolated yield homoallylic alcohol **10** devoid of the usual branched isomer (Scheme 3). Treatment of the allylic organosamarium with D_2O provided a fair yield of deuterated alkene. Allylsamarium itself is smoothly prepared at 0 °C and reacts with various types of substrates. Especially interesting is the high *endo* stereoselectivity of addition to camphor, the allylation of imine **11** and the addition on β-keto ester **12** which is easily enolizable. Diiodobenzylsamarium may be prepared at –15 °C from benzyl bromide and reacts with 2-octanone to give the tertiary alcohol in good yield.

In conclusion, THP may in many cases improve the efficiency of reactions initially performed in THF and stabilize organosamarium species which are generated and which may subsequently react with various electrophiles.

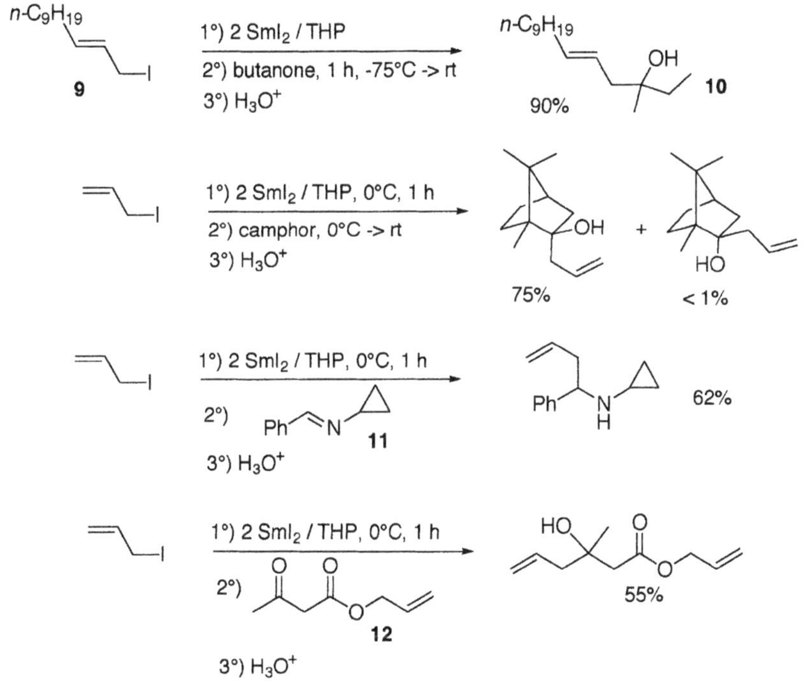

n-C_9H_{19} **9** — 1°) 2 SmI_2 / THP; 2°) butanone, 1 h, -75°C -> rt; 3°) H_3O^+ — n-C_9H_{19} OH **10** 90%

1°) 2 SmI_2 / THP, 0°C, 1 h; 2°) camphor, 0°C -> rt; 3°) H_3O^+ — OH 75% + HO < 1%

1°) 2 SmI_2 / THP, 0°C, 1 h; 2°) Ph N **11**; 3°) H_3O^+ — Ph N H 62%

1°) 2 SmI_2 / THP, 0°C, 1 h; 2°) O O O **12**; 3°) H_3O^+ — HO O O 55%

Scheme 3

2.3 Ethers Other Than THF or THP

There are almost no reports of preparation of SmI_2 in ethers other than THF or THP. Inanaga et al. prepared SmI_2 in 1,3-dioxolane and then added 10% acetonitrile to produce a clear solution. This solution was used in the masked formylation of aldehydes or ketones in the presence of iodobenzene [24]. Iodobenzene is transformed into the benzene radical which generates the 1,3-dioxolanyl radical. The latter is presumably reduced to the corresponding organosamarium which then adds on the carbonyl. This process is illustrated in Scheme 4 for cyclododecanone. It is essential to avoid the presence of THF which will compete with 1,3-dioxolane as a hydrogen donor to the benzene radical.

Tetraglyme (2,5,8,11,14-pentaoxapentadecane) has been used in the hydroxymethylation of aldehydes via addition of benzyl chloromethyl ether in the presence of SmI_2. Addition of tetraglyme suppresses the competitive pinacol formation, presumably by a complex formation with SmI_2 as evidenced by a purple color [25]. Tetraglyme has also been used as a cosolvent in an intramolecular Barbier reaction involving an iodoaldehyde and a THF solution of diiodosamarium [26].

2.4 Nitriles

Diiodosamarium in THF induces a reductive coupling of acid chlorides to form α-ketols **6** (Scheme 5) [27]. A mixed coupling between an acid chloride and a ketone may also give ketols **13**. We found that the by-product **14** was produced in many cases by the classical acylation of THF which involves a ring opening catalyzed by a Lewis acid [here the Sm(III) species]. Ruder discovered that SmI_2

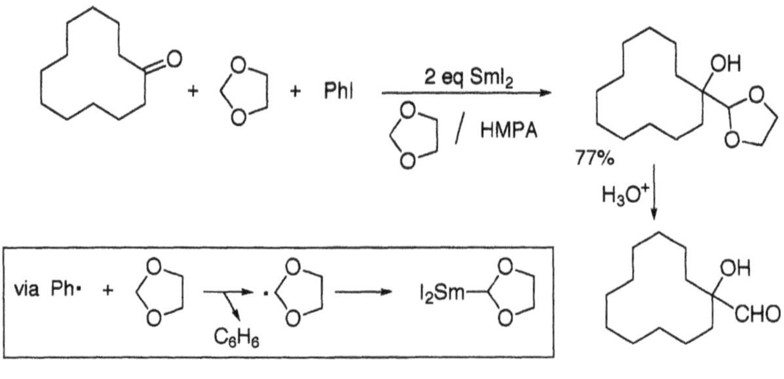

Scheme 4

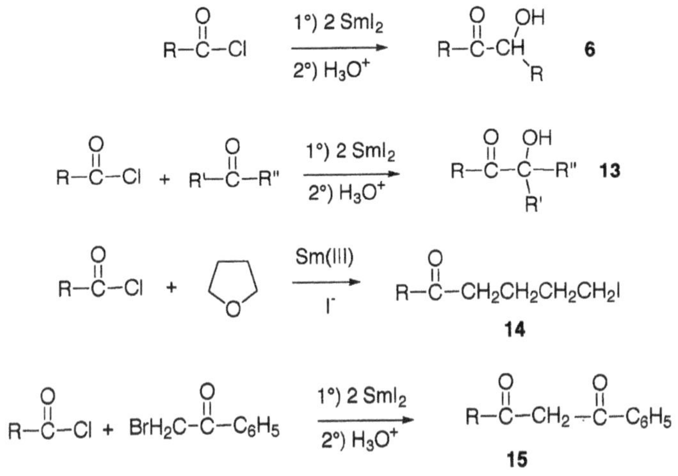

$$R{-}C({=}O){-}Cl \xrightarrow[2°)\ H_3O^+]{1°)\ 2\ SmI_2} R{-}C({=}O){-}CH(OH)R \quad \mathbf{6}$$

$$R{-}C({=}O){-}Cl + R'{-}C({=}O){-}R'' \xrightarrow[2°)\ H_3O^+]{1°)\ 2\ SmI_2} R{-}C({=}O){-}C(OH)(R'){-}R'' \quad \mathbf{13}$$

$$R{-}C({=}O){-}Cl + \text{THF} \xrightarrow[I^-]{Sm(III)} R{-}C({=}O){-}CH_2CH_2CH_2CH_2I \quad \mathbf{14}$$

$$R{-}C({=}O){-}Cl + BrH_2C{-}C({=}O){-}C_6H_5 \xrightarrow[2°)\ H_3O^+]{1°)\ 2\ SmI_2} R{-}C({=}O){-}CH_2{-}C({=}O){-}C_6H_5 \quad \mathbf{15}$$

Scheme 5

could be readily prepared in acetonitrile [28]. This author claimed that mixed α-ketols **13** were produced by a two-step procedure and in good yields in acetonitrile. We reinvestigated this work and found that it gave a mixture of products where the pinacol derived from the ketone was the major product [29]. Reactivity of SmI_2 in acetonitrile towards acid chlorides is roughly the same as in THF. Many by-products are formed involving some reactions with acetonitrile. In order to avoid the C-H acidity in the α of the nitrile group we investigated the use of pivalonitrile. It is possible to prepare SmI_2 (as a slurry) in pivalonitrile from samarium powder and 1,2-diiodoethane [29]. An X-ray crystal structure of SmI_2-$(NCCMe_3)_2$ was performed by Sen et al. and shows a distorted octahedron around samarium with a bent Sm-N=C structure [30]. The cross-coupling between acid chlorides and ketones gave α-ketols **2** in moderate yields. Barbier reactions are slower in acetonitrile than in THF; however, the regioselectivity of the reactions between allylic halides and ketones is significantly improved (in favor of the unbranched isomer). Surprisingly, HMPA does not enhance the reactivity of SmI_2 in nitriles (Barbier reactions or pinacolization of ketones). Reactions are accelerated by catalytic amounts of some transition-metal salts (see Sect. 3.3.6). Diiodosamarium may be prepared in propanenitrile or octanenitrile, but these solvents do not offer special advantages over pivalonitrile.

It is worth pointing out that acetonitrile was used as a cosolvent by Inanaga et al. in 1987 [4]. These authors later studied SmI_2-promoted aryl radical cyclization with olefins and obtained quite good yields in acetonitrile [31].

SmI_2 in acetonitrile has been used for the preparation in good yields of 1,3-diketones **15** [32]. The authors first added the α-halo ketone to the SmI_2 solution and subsequently added the desired acid chloride or anhydride. No comparisons were given between reactions run in THF or acetonitrile.

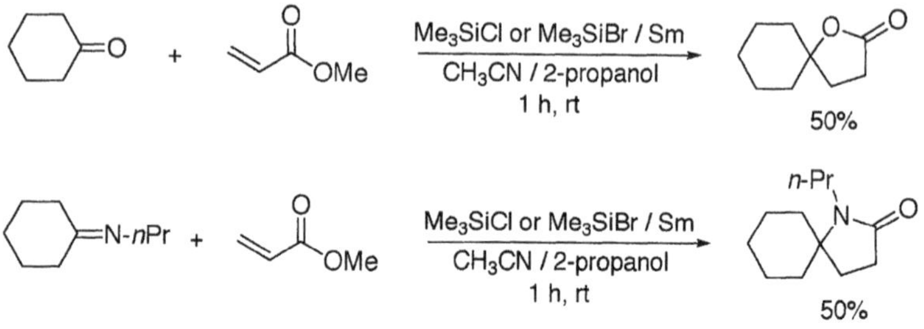

Scheme 6

Ishii et al. prepared an Sm(II) reagent in acetonitrile from NaI, $ClSiMe_3$ and samarium grain. A soluble species was produced with the deep-green color of SmI_2 but its structure was not established [33]. At –40 °C this Sm(II) reagent is able to dehalogenate α-chloro or α-bromo ketones or esters, if methanol is present as the proton source. Acetonitrile was superior to THF or DME for this reaction. The reagent gave a faster reaction compared to SmI_2/THF for the formation of pinacols of acetophenone [34]. Curiously, benzaldehyde did not produce hydrobenzoin unless HMPA was added. A lactone formation was the result of the coupling of methyl acrylate and a ketone or imine (Scheme 6). Reformatsky-type reactions between ethyl α-bromoacetate and ketones gave satisfactory yields of β-hydroxy esters, without showing any special improvement with respect to the SmI_2/THF system. However, it is interesting to note that some Reformatsky or Barbier reactions have been achieved in good yields with octanal, without perturbation by pinacol formation.

2.5 Benzene

It is not possible to prepare SmI_2 in benzene from samarium and 1,2-diiodoethane or iodine. However, it was discovered that addition of 10% HMPA as a cosolvent to benzene promoted the reaction of samarium and 1,2-diiodoethane [35]. The reagent reduced various 1,1-dibromoalkenes to the corresponding rearranged alkynes, presumably through the generation of alkylidenecarbenes (Scheme 7). The use of benzene instead of THF avoided the formation of reduced products obtained by hydrogen abstraction. This process was extended to the synthesis of cyclopentenes by C-H insertion on the intermediate carbene [36]. The reaction works especially well from 1,1-diiodoalkenes (Scheme 7). The same authors also used the benzene/HMPA solvent to achieve some SmI_2-induced Wittig rearrangements via 1,5-hydrogen transfer of vinyl radicals [37] (Scheme 7). Here again, it is crucial to avoid THF which will induce a competitive intermolecular hydrogen abstraction.

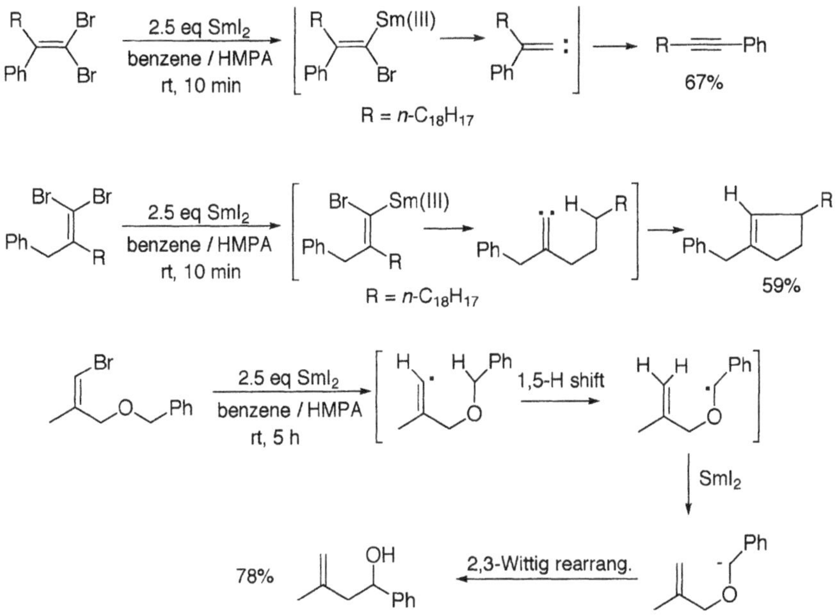

Scheme 7

3
Influence of Additives

3.1
Electron-Rich Additives

3.1.1
Hexamethylphosphoramide (HMPA)

The highly promoted effect of HMPA on reactions of SmI_2 was reported for the first time in 1986 by Inanaga et al. [38] for the reductive cross-coupling of carbonyl compounds with α,β-unsaturated esters (Scheme 8). HMPA was used as a cosolvent with THF (ca. 5%). Enhancement of the coupling rates and yields were quite remarkable; however, the diastereoselectivity was sometimes diminished in the presence of HMPA.

Independently, Fukuzawa and co-workers [39] reported intramolecular reactions leading to bicyclic γ-lactones from keto or aldo α,β-unsaturated esters (Scheme 9). The reaction was mediated by SmI_2 in the presence of HMPA (THF/HMPA=10/1) but the beneficial effect of HMPA was less obvious in that case.

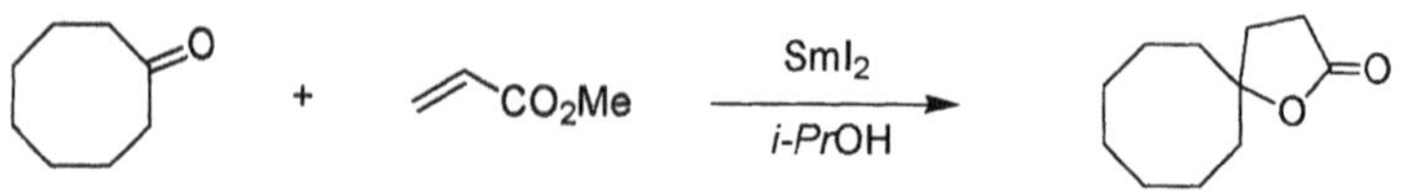

SmI_2
i-PrOH

none: 4 h; yield = 82%

HMPA: 1 min; yield = 95%

Scheme 8

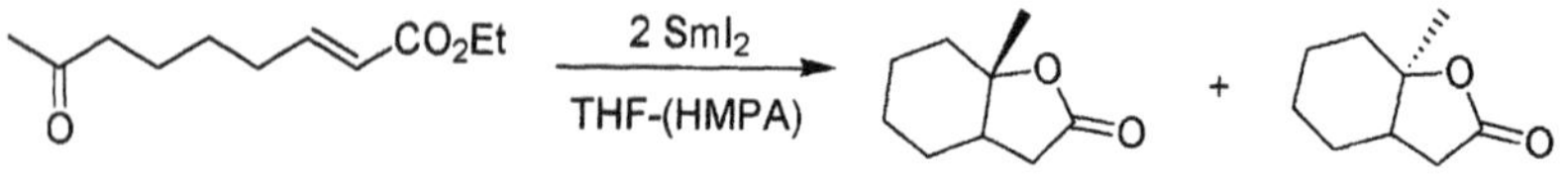

2 SmI_2
THF-(HMPA)

16 **17**

none: 4 h; 65°C; yield = 56%; **16**/**17** = 64/36

HMPA: 1 h; 65°C; yield = 47%; **16**/**17** = 79/21

Scheme 9

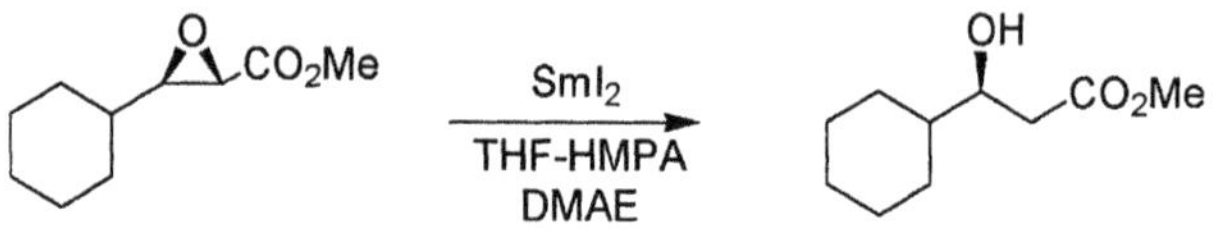

SmI_2
THF-HMPA
DMAE

Scheme 10

Inanaga et al. have developed the use of HMPA for the highly regioselective reduction of α,β-epoxy esters and δ,γ-epoxy α,β-unsaturated esters [40] (Scheme 10).

The best results were obtained when HMPA (5 equiv) and dimethylamino ethanol (DMAE, 2 equiv) were used as additives. The authors also reported the reductive cross-coupling of 1,3-dioxolane with carbonyl compounds to yield α-hydroxy derivatives [24]; the reaction was performed in the presence of HMPA (ca. 5% of the solvent) and was complete within 5 min at room temperature (Scheme 11).

A remarkable effect by HMPA was observed in the reduction of organic halides with SmI_2 in THF [41] (Scheme 12).

As mentioned in Sect. 2.3, Inanaga and co-workers have demonstrated that Barbier-type reactions of organic halides with carbonyl compounds are promoted by addition of HMPA [16]. They reported a mild convenient method for the direct synthesis of lactones from bromo esters and ketones or aldehydes by using a HMPA-promoted Barbier-type reaction with SmI_2 (Scheme 13). They also found that the SmI_2/THF-HMPA system was highly useful for the generation of

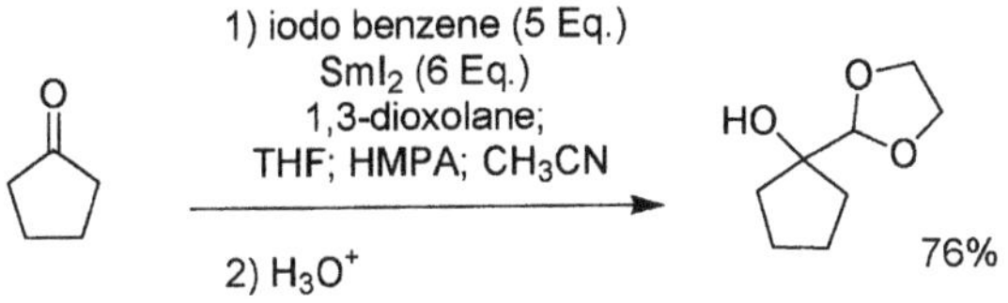

1) iodo benzene (5 Eq.)
SmI_2 (6 Eq.)
1,3-dioxolane;
THF; HMPA; CH_3CN
2) H_3O^+
HO
76%

Scheme 11

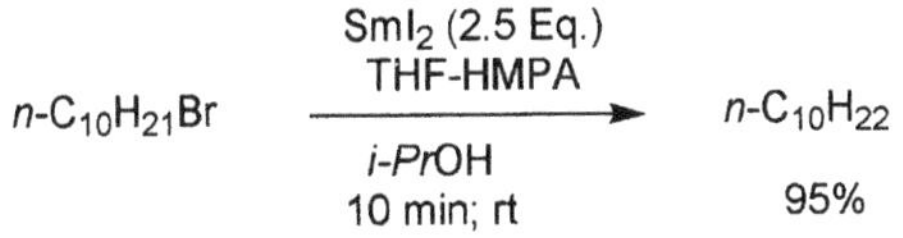

SmI_2 (2.5 Eq.)
THF-HMPA
i-PrOH
10 min; rt
n-$C_{10}H_{21}Br$ → *n*-$C_{10}H_{22}$
95%

Scheme 12

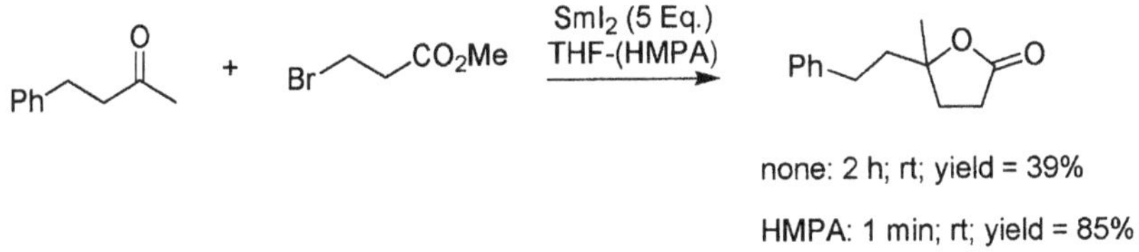

Ph + Br CO_2Me
SmI_2 (5 Eq.)
THF-(HMPA)
Ph
none: 2 h; rt; yield = 39%
HMPA: 1 min; rt; yield = 85%

Scheme 13

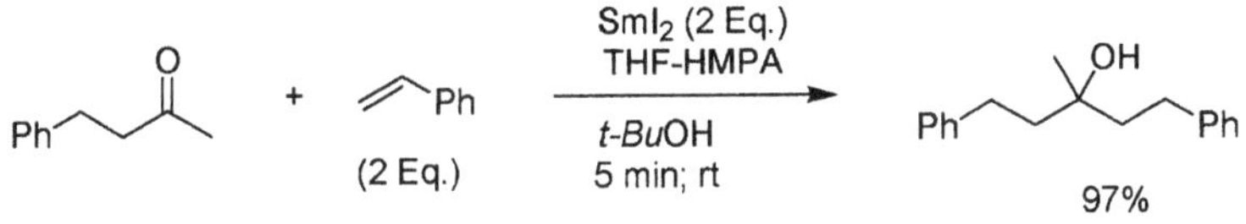

Ph + Ph (2 Eq.)
SmI_2 (2 Eq.)
THF-HMPA
t-BuOH
5 min; rt
Ph OH Ph
97%

Scheme 14

ketyls and their intramolecular addition to a variety of activated olefins [42] (Scheme 14). In addition, this system has been used to successfully reduce organo heteroatom oxides including triphenyl phosphine oxide [43].

Thus the most popular way to increase the rate of reactions of diiodosamarium has become the use of a mixture of THF–HMPA (90–95:10–5) as solvent. Although the coordination of HMPA to the Sm(II) ion was thought to be responsible for this effect, the role played by HMPA remained unclear until 1994.

In order to clarify this problem, Hou and Wakatsuki [44] prepared SmI_2/HMPA complexes by addition of HMPA to a blue solution of SmI_2 in THF. The resulting

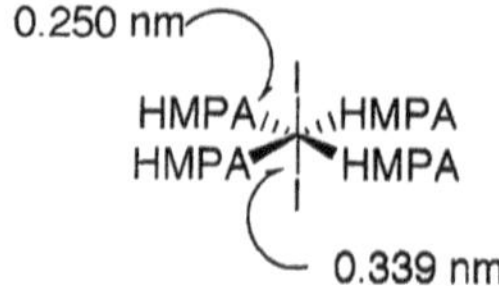

Scheme 15

purple solution was concentrated and toluene was added. After a few days at room temperature, black-purple blocks of $SmI_2(HMPA)_4$ were deposited in 90% yield, and they were structurally characterized. It appears that the Sm(II) ion sits on an inversion center and is bonded by two I^- anions and four HMPA ligands in a distorted octahedron. The central Sm(II) ion and the four HMPA ligands are exactly coplanar and the two iodide anions are mutually trans (Scheme 15).

In a recent work, electrochemical studies of the reducing power of SmI_2 in THF and the effect of HMPA cosolvent as a function of concentration have been reported [45]. This effect was studied by recording a linear sweep voltamogram for each cosolvent addition. The oxidation potential vs. the $Ag/AgNO_3$ reference electrode for 0 to 6 equiv of HMPA cosolvent was measured. The addition of 3 equiv of HMPA to SmI_2 had a drastic effect on the redox potential, increasing the oxidation potential from −1.33 to −1.95 V. The addition of 4 equiv of HMPA to SmI_2 increased the potential even further to −2.05. Further addition of HMPA showed no effect on the redox potential.

These observations are consistent with the structure determined by Hou et al.; the complex $SmI_2(HMPA)_4$ should be the reactive intermediate responsible for the unique reactivity of the SmI_2/THF–HMPA system.

Preparation, structural characterization and reactivity studies of SmI_3–HMPA complexes have also been performed [46]. The composition of the dried complex was found to be $SmI_3(HMPA)_4$. The reactivity of this complex was examined in comparison with that of SmI_3 in THF only. In contrast to the latter complex, the HMPA complex was not reduced to low-valent samarium species by *n*-butyl lithium or *sec*-butyl lithium. These results might be interpreted by considering that this Sm(III) complex is well stabilized by the coordination of HMPA which could be one of the driving forces for the facile electron transfer from SmI_2–HMPA complexes.

Molander et al. [47] suggested that other factors could be involved in the effective role of this additive: disaggregation of SmI_2, f-orbital perturbation due to the ligand field effect in the presence of the strong donor ligand HMPA raising the energy of the HOMO (electron-donating orbital). A combination of these effects might also be considered.

The effect of HMPA on the spectroscopic properties of SmI_2 has also been examined [48].

We wish to review here some studies that stress the specific role played by HMPA in samarium diiodide reactions.

Reduction of 1-iodo-5-hexene with SmI_2/THF–HMPA has been studied by Curran and co-workers [49] (Scheme 16). They found that reaction is slow without HMPA. A reasonable reaction time (<10 min) is attained with 2 equiv of HMPA per SmI_2. In the range from 1.3 to 4.4 equiv, the ratio of directly reduced product to the rearranged product increases dramatically (from 8:92 to 54:46). Increasing the amount of HMPA beyond 5 equiv results in a very slow decrease in the ratio. These results are consistent with the above results concerning the structure of the complex SmI_2–HMPA [44] and the electrochemical studies [45]. The easier the reduction of the 5-hexenyl radical to organosamarium species, the less the cyclopentyl-carbinyl radical is formed.

Curran et al. demonstrated that reduction of aryl halides to radicals is also strongly accelerated in the presence of HMPA, as well as reduction of alkyl radicals to alkyl samarium [50,51] (Scheme 17).

HMPA is also necessary for the in situ preparation of the organosamarium species from alkyl halides and SmI_2 in THF [52]. A similar effect was also observed in the case of reactions performed in tetrahydropyran [23].

Molander et al. studied the intramolecular coupling of unactivated olefinic ketones by a reductive ketyl-olefin radical cyclization, mediated by SmI_2 in the presence of HMPA [47] (Scheme 18).

HMPA was required to promote efficient ketyl-olefin cyclization in the desired manner. It seems that factors other than reduction potentials also play a

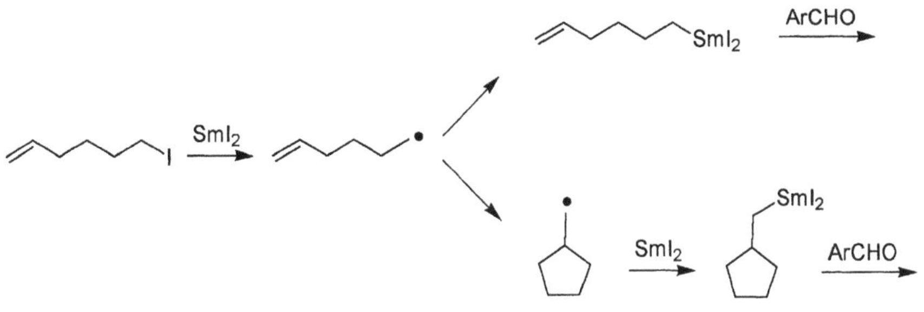

SmI2 ArCHO SmI2 SmI2 SmI2 ArCHO

Scheme 16

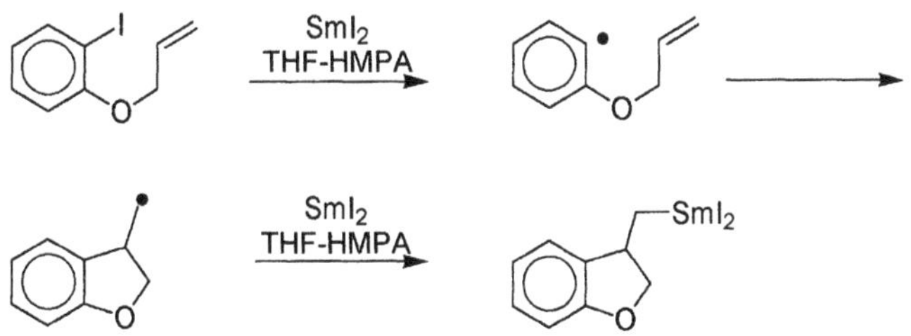

SmI2 THF-HMPA SmI2 THF-HMPA SmI2

Scheme 17

role in the efficiency of HMPA as an additive in SmI_2-promoted reactions: ketyl radical anions appear to have greater persistence when generated in the presence of HMPA than without additives. By progressively increasing the amount of HMPA, from 0 to 8 equiv, the percentage of **20** (generated by hydrogen abstraction from the solvent) decreased and the stereoselectivity of **19** increased. These enhanced diastereoselectivities were attributed to an added steric component in the cyclization transition states. As an application, SmI_2 in the presence of HMPA has been used to promote 8-*endo* cyclization of various keto olefins affording a variety of eight-membered ring products in fair to excellent yields [53]. This success was in part attributed to the formation of ketyls exhibiting remarkable persistence in the presence of HMPA perhaps because the HMPA excludes the proton source (*t*-BuOH) from the coordination sphere of the Sm(III) ion.

The successful intramolecular coupling between vinyl oxiranes and ketones mediated by SmI_2 in the presence of HMPA as a cosolvent has been reported [54] (Scheme 19). The use of HMPA as a cosolvent allowed the couplings to be performed at lower temperatures and improved the regioselectivity of the process as the 1,5-diols were then the only products isolated; the *E/Z* ratio was also affected.

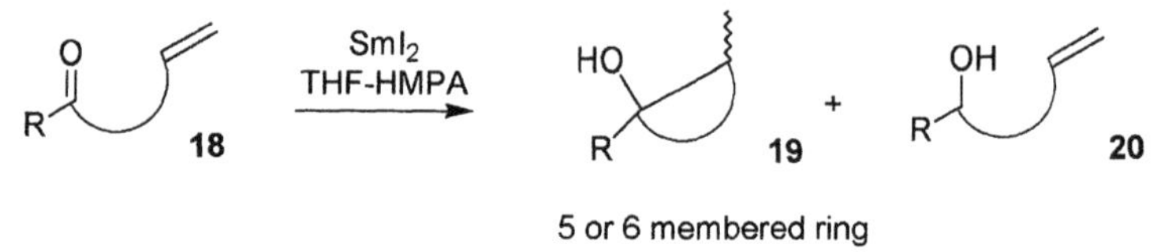

SmI_2 THF-HMPA

18 → **19** + **20**

5 or 6 membered ring

Scheme 18

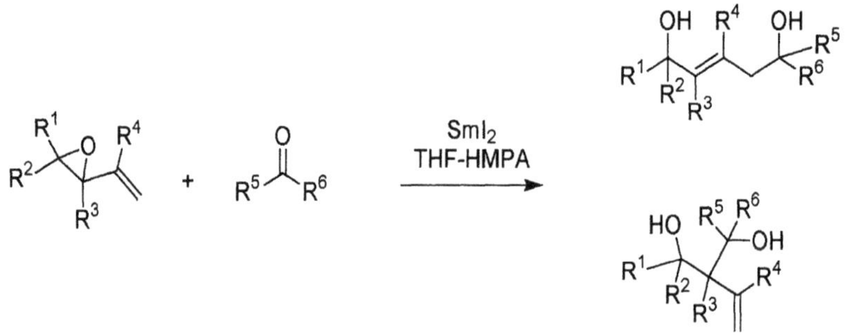

SmI_2 THF-HMPA

Scheme 19

Studies by Molander et al. have demonstrated that samarium diiodide efficiently promotes the intramolecular coupling of ketones with distal epoxy olefins in the presence of HMPA, in good yields and often with high diastereoselectivity (Scheme 20). When tetramethylguanidine was used instead of HMPA, the desired carbocycle was obtained in good yield but the diastereoselectivity was diminished [55]. Evidence has been presented that suggests that the reaction proceeds via an initial radical cyclization route although reaction via an allyl samarium species cannot be ruled out.

In order to achieve the total synthesis of (–)-anastrephin, a *D*-glucose-derived α,β-unsaturated ester, was subjected to an SmI_2-mediated intramolecular reductive coupling reaction with the SmI_2/THF–HMPA system. The reaction proceeded with moderate stereoselectivity leading to a diastereomeric mixture of hexahydro benzofuran-2(*3H*)-ones in which a *cis*-fused product was the major isomer [56] (Scheme 21). The presence of HMPA was essential for smooth coupling; significant reduction of the aldehyde to a hydroxymethyl group occurred without HMPA.

The use of HMPA can modify the course of a reaction. It is well known that aromatic aldehydes or aromatic ketones readily couple to pinacols **21** on treatment with SmI_2 in THF, but addition of 2.8 equiv of HMPA leads to the formation of product **22** that results from connecting a *para* carbon with the carbonyl of another molecule (Scheme 22).

It is proposed that HMPA molecules coordinate with samarium ions through their oxygen atom because the ketyls and ortho carbons are hindered by the HMPA ligands; coupling at the *para* position with a second molecule of benzaldehyde would be favored [57,58].

Scheme 20

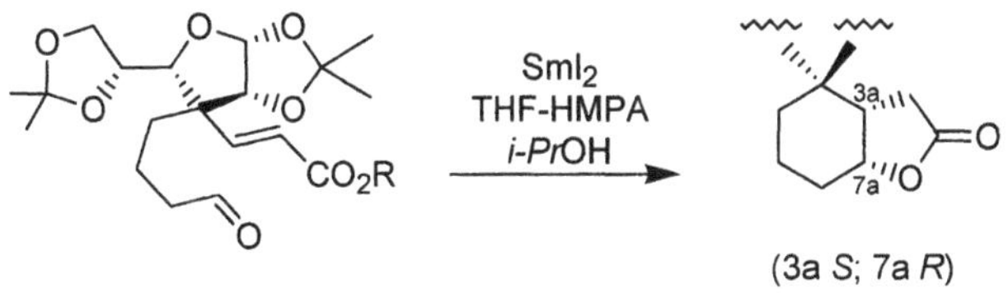

Scheme 21

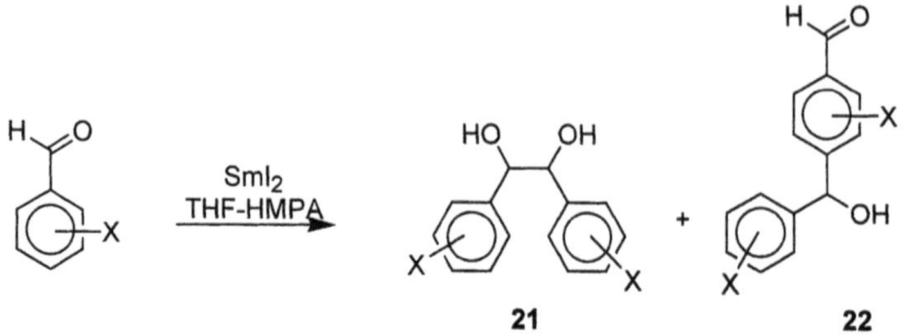

Scheme 22

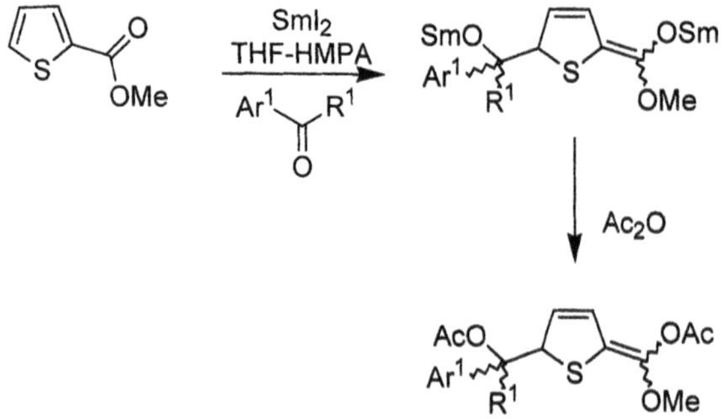

Scheme 23

Scheme 24

Similarly, methyl thiophenecarboxylate reacts at the C5 position with a series of substituted benzaldehydes and acetophenone in the presence of the SmI_2/THF–HMPA system to give a samarium enediolate which can be trapped by a second electrophile [59] (Scheme 23). The samarium-bonded ketyl anion radical does not trap the hydrogen atom or undergo acyloin coupling, presumably due to the hindrance of the ligated HMPA molecules.

Very interesting chemistry has been developed with sulfones and samarium diiodide.

Neither deoxygenation nor desulfonylation occurs upon treatment of various phenylsulfones with SmI_2/THF. However, in the presence of HMPA, reductive desulfonylation was achieved [60]. Thus a series of substituted glycosyl phenylsulfones was converted into the corresponding glycals with SmI_2/THF–HMPA [61] (Scheme 24).

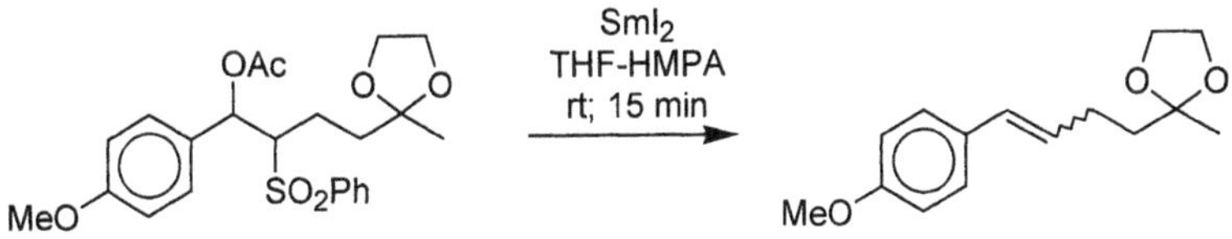

Scheme 25

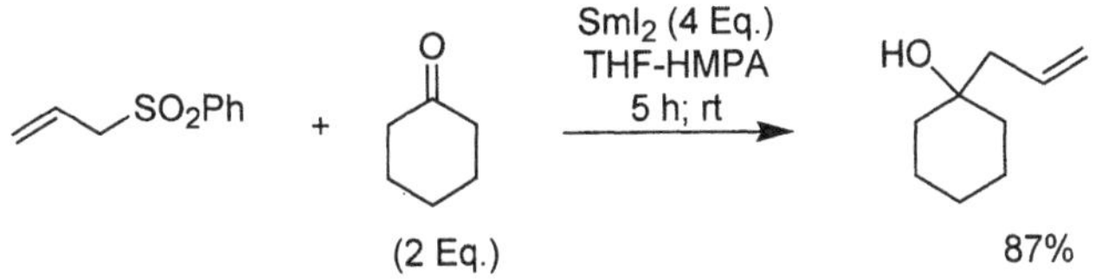

Scheme 26

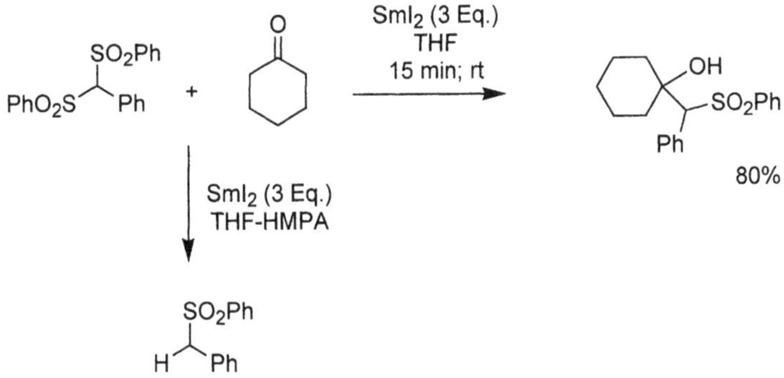

Scheme 27

The Julia alkylidenation was improved using SmI_2/THF-HMPA instead of Na(Hg) for the reductive elimination step. In contrast, a poor result was reported without HMPA [62] (Scheme 25).

In the presence of HMPA, allylic sulfones react with SmI_2 and ketones to give a Barbier-type reaction (Scheme 26). Much lower yields are obtained if the reagents are not thoroughly degassed, and in the absence of HMPA [63].

Completely stereoselective cyclizations using formyl allyl sulfides and sulfones have been described. Optimum conditions involved the slow addition of a solution of the starting material in THF to a 0.1 M solution of SmI_2 containing 8 equiv of HMPA in THF at –10 and –78 °C, respectively [64].

It has also been reported that SmI_2 mediates the in situ reductive addition of geminal bis-phenyl sulfones to ketones in THF. In the presence of HMPA (10 equiv), addition to the carbonyl is completely suppressed, but not desulfonylation [65] (Scheme 27).

Reaction of β-hydroxy or β-acetoxy sulfones with SmI_2 in the presence of HMPA caused effective reductive elimination to provide olefins [66]. In contrast, Kende recorded a poor result in the reductive elimination of the β-hydroxy phenyl sulfones with SmI_2 without HMPA [67]. Finally, it has been reported that contrary to phenyl sulfones, 2-pyridyl sulfones are instantaneously reduced in the presence of SmI_2 without additives [68].

Samarium(II) triflate, a promising reagent for selective organic synthesis, has been prepared by treatment of a THF solution of $Sm(OTf)_3$ with 1 equiv of an organolithium (or organomagnesium) reagent at room tempeature [69,70] (Scheme 28). Generation of the samarium(II) was unsuccessful in the presence of HMPA.

The $Sm(OTf)_3$ reagent mediates the Grignard-type reaction in THF–HMPA. Alkylation, allylation and benzylation of ketones and aldehydes with alkyl, allyl or benzyl halides proceeded via stable organosamarium intermediates [69] (Scheme 29).

Matsuda et al. have studied the hydroxyl group directed intermolecular ketone-olefin coupling reactions, induced by SmI_2, between α-hydroxy ketones and α,β-unsaturated esters or nitriles (Scheme 30). It was noted that reactions

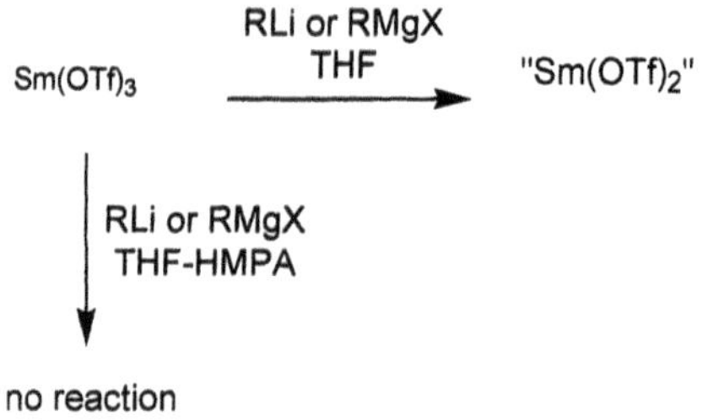

$Sm(OTf)_3$ —(RLi or RMgX, THF)→ "$Sm(OTf)_2$"

$Sm(OTf)_3$ —(RLi or RMgX, THF-HMPA)→ no reaction

Scheme 28

$$RX + 2\,Sm(OTf)_2 \xrightarrow{\text{THF-HMPA}} RSm(OTf)_2 + XSm(OTf)_2$$

Scheme 29

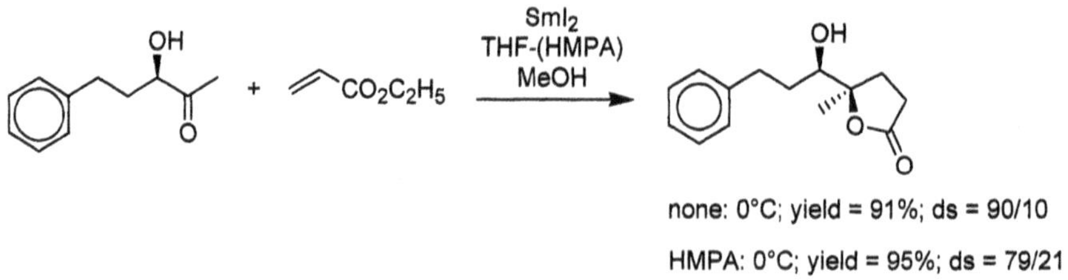

SmI2
THF-(HMPA)
MeOH

none: 0°C; yield = 91%; ds = 90/10

HMPA: 0°C; yield = 95%; ds = 79/21

Scheme 30

run in the presence of HMPA resulted in all cases in some depression in the diastereoselection [71–73]. Various other reactions have been reported which exemplify the crucial role of the solvent HMPA.

Inanaga et al. achieved the hydrodimerization of conjugated esters and amides with the SmI_2/THF–HMPA system in the presence of a proton source [74].(Scheme 31). However, inexplicably, Alper et al. achieved reductions (not reductive dimerization) of the double bond of α,β-unsaturated esters and amides with the same system [75]. In the latter paper, however, the presence of a proton donor is not mentioned.

The reductive coupling of α-bromoacetate to succinic acid diesters mediated by SmI_2 in the presence of HMPA has been reported [76] (Scheme 32).

In contrast, reactions performed without HMPA as additive do not lead to dimerization but to β-keto esters [77] (Scheme 33). Obviously, the intermediate samarium enolate which is formed behaves in a different way according to the presence or the absence of HMPA.

The reductive dimerization of cyclopropane-1,1-dicarboxylic esters using SmI_2/THF has also been reported (Scheme 34). In this case, the addition of

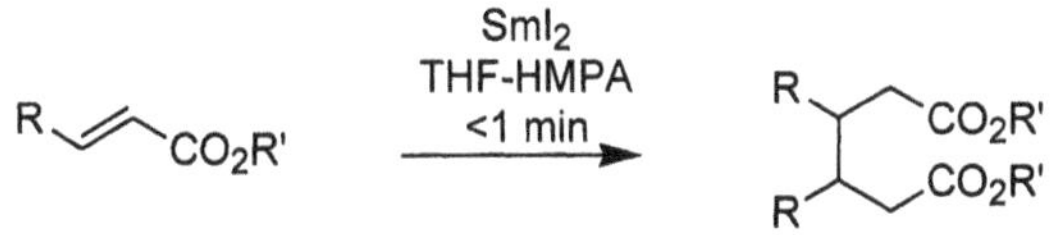

SmI_2, THF-HMPA, <1 min

Scheme 31

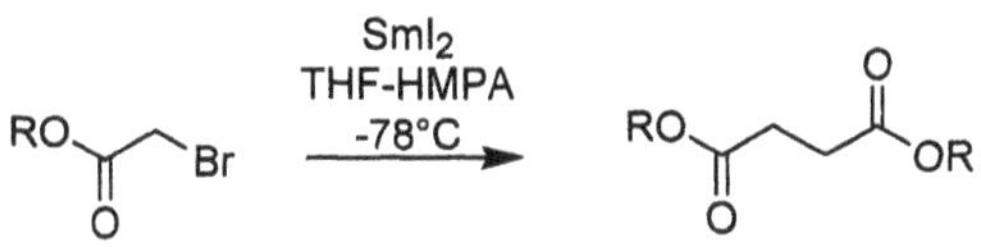

SmI_2, THF-HMPA, -78°C

Scheme 32

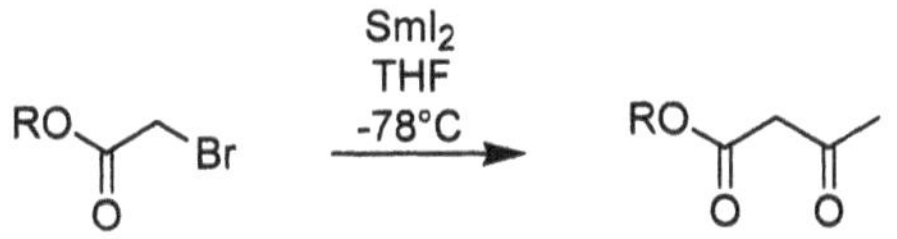

SmI_2, THF, -78°C

Scheme 33

SmI_2, THF

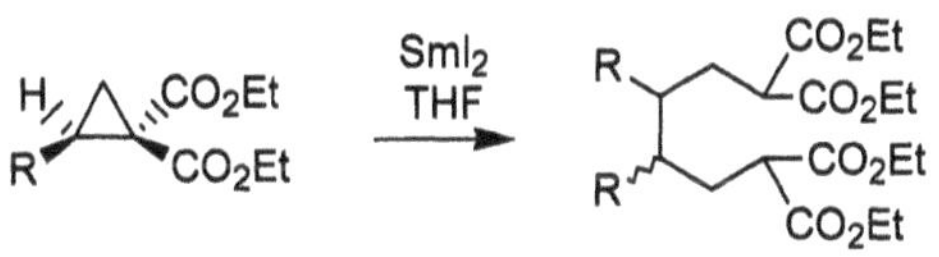

Scheme 34

HMPA did not improve yields of the dimer. On the contrary, yields were higher in the absence of HMPA [78].

A new intramolecular reductive carbon–carbon bond formation has been noted in reactions of 1-(2-formyloxyethyl)-3-formyl-oxycycloalkene with SmI_2 giving a spiro hemiacetal (Scheme 35). The absence of HMPA caused a considerable decrease in the yield of the hemiacetal [79].

Coupling reactions of indoles and carbonyl compounds (either intra- or intermolecularly) have been described. This is a new method for hydroxylation at the C2 position of indole (Scheme 36). The role of HMPA is crucial to prevent reduction or pinacolic coupling of aromatic carbonyl compounds [80].

Stereocontrolled decalin ring annulation reactions through the hydroxyl group directed pinacol coupling using SmI_2 have been reported by Matsuda and et al. Stereocomplementarity was observed depending on the presence or absence of HMPA [81] (Scheme 37). The authors propose that without HMPA only the aldehyde of the substrate is reduced to the ketyl radical during the initial re-

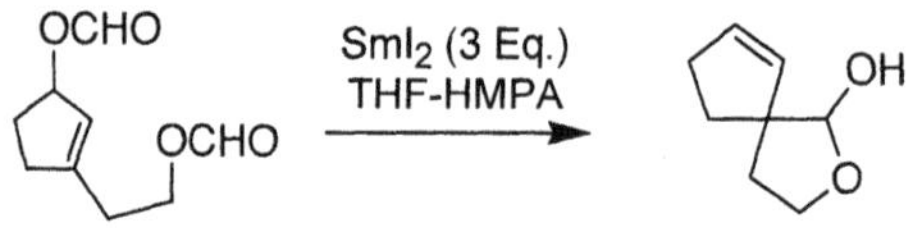

OCHO
OCHO
SmI_2 (3 Eq.)
THF-HMPA
OH
O

Scheme 35

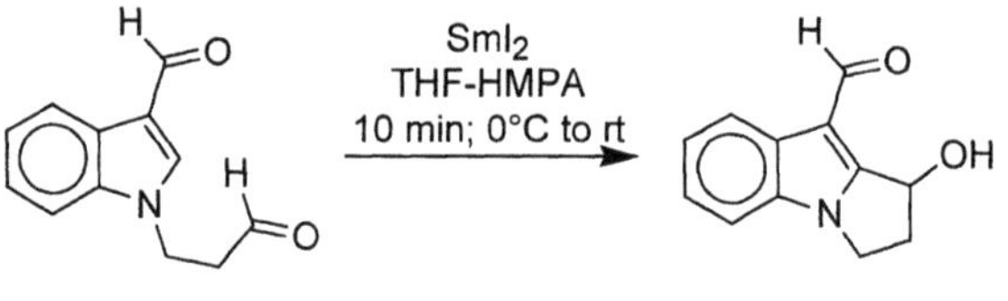

H
O
N
H
O
SmI_2
THF-HMPA
10 min; 0°C to rt
H
O
N
OH

Scheme 36

OH
H
O
O
SmI_2
THF
MeOH; rt
OH
H
OH OH

SmI_2
THF-HMPA
MeOH; rt
OH
H
OH OH

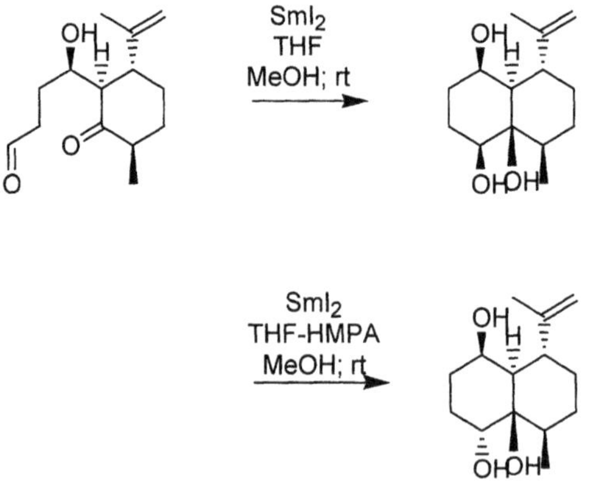

Scheme 37

duction by SmI_2, whereas in the presence of HMPA a ketyl radical pair is generated through a single electron transfer from SmI_2.

Katritzky et al. have prepared α-amino carbanions from tosyl methyl amines. The transformation was achieved by treatment with 2 equiv of SmI_2 [82] in the presence of an electrophile (Scheme 38). In the absence of HMPA the use of isobutyraldehyde as the electrophile gave only the amine dimer (98%). However, the addition of 5% of HMPA resulted in a 67% yield of the desired product (62%) with a minor amount of diamine (17%). The formation of the hydroxy amine probably involves a carbanion intermediate whereas coupling gives the diamine.

Samarium diiodide has also been used for the intramolecular coupling of aldehydes and ketones with *O*-benzyl formaldoxime [83], for the corresponding intermolecular coupling with diphenylhydrazone [84,85] and for the intramolecular coupling of an α,β-unsaturated ester with an oxime ether [86] (Scheme 39). In all these cases the addition of HMPA was found to be essential for a successful reaction.

3.1.2
N,N'-Dimethylpropyleneurea (DMPU)

Because of the toxicity of HMPA, studies on other samarium(II) additives have been carried out by several researchers. DMPU, a well-known and less toxic alternative to HMPA, has sometimes been used as a cosolvent for SmI_2 reactions. Addition of DMPU to a THF solution of SmI_2 results in an immediate color change from blue to purple.

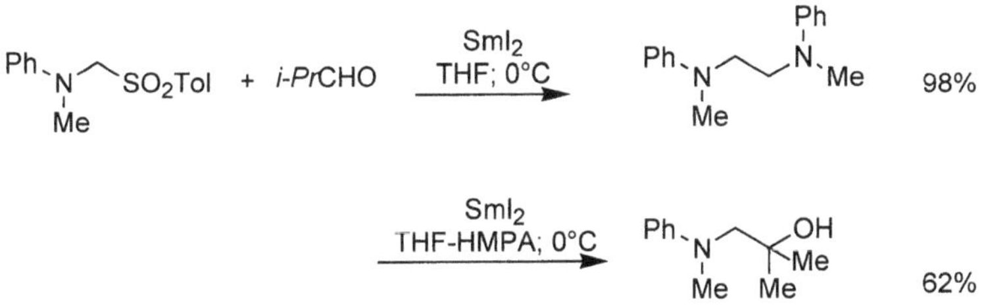

Ph N(Me) SO_2Tol + *i*-PrCHO —SmI_2, THF; 0°C→ Ph N(Me) CH₂CH₂ N(Ph) Me 98%

—SmI_2, THF-HMPA; 0°C→ Ph N(Me) CH₂ C(OH)(Me) Me 62%

Scheme 38

CO_2Me, NOBn, OTBS, Ph —SmI_2, THF-HMPA, rt→ CO_2Me, NHOBn, OTBS, Ph

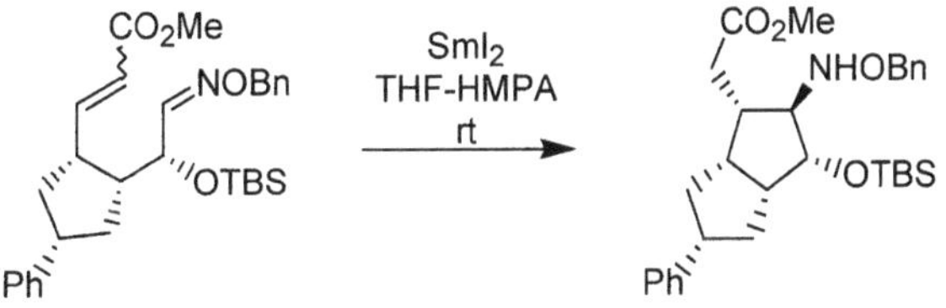

Scheme 39

Curran et al. were the first to use SmI_2/THF–DMPU. This system promoted a tandem radical cyclization [87] (Scheme 40). Reduction of **23** with SmI_2/THF produced **24, 25** and **26** in low yield; addition of HMPA accelerated the reaction and gave **24** in 91% yield. DMPU as additive gave similar results to HMPA with respect to reaction rate and yield but **25** and **26** were also formed albeit in low yield (9 and 4%). In addition, the cyclization in THF–HMPA required less than 2 equiv of SmI_2 (1.3 equiv) but when DMPU was used, a larger excess of SmI_2 was needed, perhaps because the reaction mixture was heterogeneous.

Ring scission of cyclic β-halogeno ethers to olefinic alcohols mediated by SmI_2 has been studied. Scission in the tetrahydrofuran series could be accelerated by addition of HMPA or DMPU with only a small deterioration in diastereoselectivity, but in the tetrahydropyran series there was a drastic change in the stereochemistry of the product when DMPU was used. Obviously, DMPU has a substantial effect on the formation and conformation of complexes in solution [88,89].

A mixture of THF and DMPU has been used as a solvent for SmI_2 in various reactions such as cyclization of alkynyl halides [90,91], tandem iodo-enone cyclization/samarium enolate aldol reaction [92], coupling of β-silylacrylic esters [93], deprotection of arenesulfonamides [94] and pyridine-2-sulfonamides [95], radical ring-opening reactions of cyclopropyl ketones and the trapping of the resulting samarium(III) enolates by a variety of electrophiles [96] (Scheme 41).

From the above studies, the SmI_2/THF–DMPU combination emerges as the most useful, in particular because in this case the problem of second electron delivery is avoided.

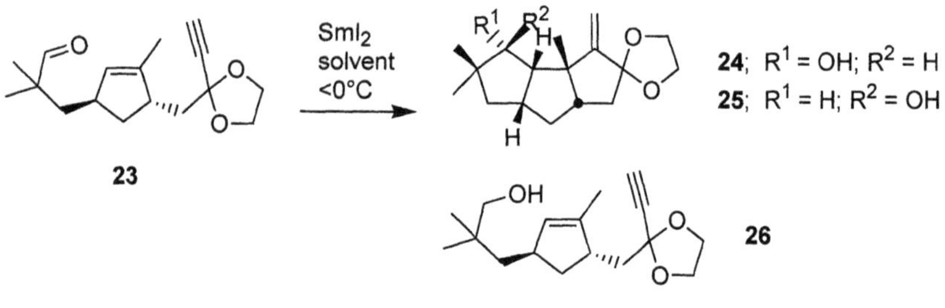

SmI_2, solvent, <0°C

23

24; R^1 = OH; R^2 = H

25; R^1 = H; R^2 = OH

26

Scheme 40

1) SmI_2, THF-DMPU, -78°C to rt

2) AcCl, -78°C to rt

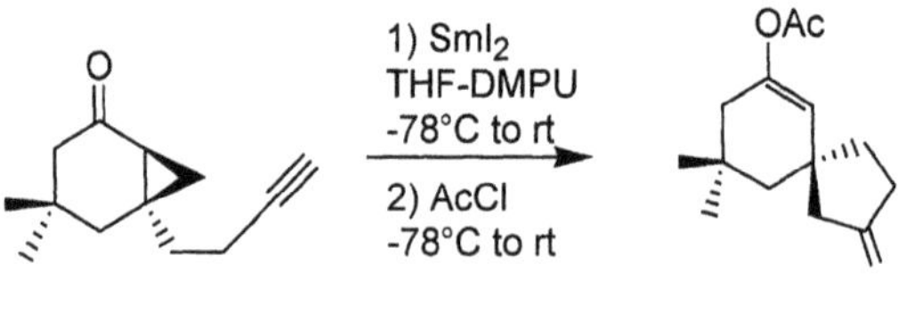

Scheme 41

Scheme 42

The SmI_2/THF–DMPU system has also been used as an alternative to the Na(Hg) amalgam in the Julia–Lythgoe olefination [97] (Scheme 42). It was found that this system gave a positive rate enhancement (from 95% in 5 d with SmI_2/THF to 95% in 35 min). The DMPU reaction also gave considerably higher stereoselectivity (*E*/*Z*=9:1) than was observed using HMPA. Further studies to optimize the reactions have led to the use of 10 to 20 equiv each of DMPU and MeOH, respectively, which gives reasonable reaction times (35 min) and high selectivity for the *E* olefin (99:1).

The role of DMPU has been studied by Curran et al. in the reduction of an *o*-(allyloxy) iodobenzene. It was found that DMPU is much more efficient in MeCN than in THF [98]. The reduced effectiveness of DMPU in THF may be due in part to formation of an insoluble SmI_2/DMPU complex.

3.1.3
Other Nitrogen Ligands

Various other commercially available nitrogen ligands have been screened to increase the reducing power of SmI_2 in THF.

Inanaga et al. have studied the selective conjugate reduction of α,β-unsaturated esters and amides with SmI_2. While the reductive dimerization was promoted by HMPA [74], the use of dimethylformamide (DMF), dimethylacetamide (DMA), tetramethylurea (TMU) and bidentate ligands such as *N,N,N',N'*-tetramethylethylenediamine (TMEDA)and *N,N,N'N'*-tetramethylpropylenediamine (TMPDA) were effective for the conjugate reduction of the substrate. DMA was found to be the most promising. In contrast, in the presence of the tridentate chelating agent *N,N,N'N''N''*-pentamethyldiethylenetriamine, no reaction took place. This result might be due to the complete occupancy of the coordination sphere of samarium [99].

The formation of an aryl radical from the corresponding aryl iodide or aryl bromide using SmI_2 has been carried out in the presence of various other nitrogen ligands (Scheme 43). The results obtained with Et_3N, DBU or TMG were satisfactory or even superior to the ones obtained with HMPA (**27**+**28**: 71%; 4 h; **27**/**28**=81:29). It is worth noting that the ligand/SmI_2 molar ratio has practically no effect on the reaction course if more than 2 equiv of ligand are present [100].

TMEDA has sometimes been used as an additive in the reduction of α,β-epoxy esters. In this case it was associated with HMPA and isopropanol [40], as,

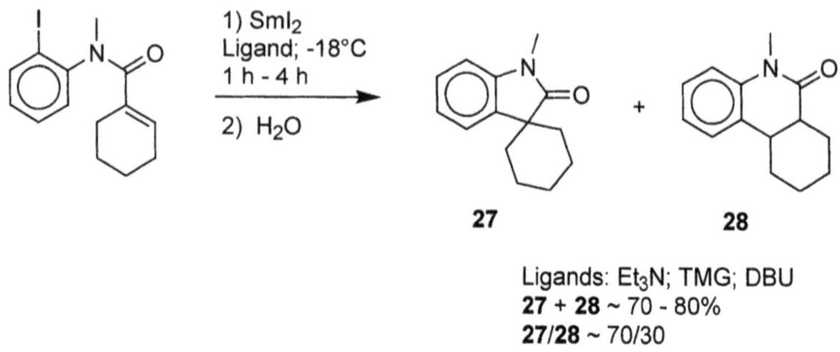

1) SmI_2
Ligand; -18°C
1 h - 4 h
2) H_2O

27 **28**

Ligands: Et_3N; TMG; DBU
27 + **28** ~ 70 - 80%
27/**28** ~ 70/30

Scheme 43

SmI_2 (2.5 Eq.)
THF-HMPA
TMEDA

35%

Scheme 44

i-Bu
i-Bu OH
OH

Scheme 45

for example, in the intermolecular aldol-type reactions of phenacyl bromides with carbonyl compounds [101] and for the reduction of cyclopropyl lactones [102] (Scheme 44).

TMG has been used for the cyclization of a keto vinyl epoxide (Scheme 45). Better yields were achieved than with HMPA; unfortunately, a 1:1 mixture of diastereomers was obtained [55].

The effect of cosolvents (TMP, DBU, PMP, TMU, NMP, DMPU) on the oxidation potential of SmI_2 in THF has been studied. The change in the reducing power of SmI_2 is dependent on the nature and concentration of these cosolvents. For example, the SmI_2/TMU reducing species was determined to have an E_{ox} of −1.99 V (vs. an $Ag/AgNO_3$ electrode) at 30 equiv of cosolvent while the SmI_2/DMPU complex displayed an E_{ox} of −2.21 V at the same cosolvent concentration. The effect of water on the E_{ox} of SmI_2 was also determined; it was found that 55 equiv of water increased the oxidation potential to −1.74 V [103].

3.1.4
Miscellaneous

Kamochi and Kudo found that the reducing ability of SmI_2 is enhanced by addition of a base (KOH, $LiNH_2$ or LiOMe), allowing the fast reduction of carboxylic acids to alcohols [104]. Reduction of nitriles to primary amines has also been realized by the system SmI_2/50% KOH (1:2) in a few minutes at room temperature [105].

BINAPO (the bis-oxide of BINAP) can complex samarium and gives some stereocontrol in the formation of γ-butyrolactones by reaction of ketones with α,β-unsaturated esters and SmI_2 in THF. Enantiomeric excesses (ee's) of up to 90% were obtained [106].

3.2
Proton Sources

All the reactions induced by SmI_2 start with a one-electron transfer giving rise to an anion radical which evolves in various ways: radical chemistry, cleavage into a carbanion and a radical, further reduction to a dianion, etc. Very often marked differences in the product distribution are observed when SmI_2-induced transformations are performed in aprotic or protic conditions, because in the latter case in situ protonation of key intermediates may occur. Kinetic stability of SmI_2/THF solutions in the presence of some water or alcohols have been reported [15,107]. It is then possible to use these solutions for in situ protonation of end products or intermediates. For example, it has been firmly established that the samarium Barbier reaction between an organic halide RI and a ketone involves a reactive organosamarium intermediate $RSmI_2$ since the reaction performed in the presence of SmI_2/THF–*t*-BuOD or EtOD leads to the formation of some RD. Under the same conditions, and in the absence of ketones, the deuteration of RI into RD was also achieved in good yield [108,109].

Apart from its usefulness for some mechanistic investigations the presence of protic additives may disturb the course of a reaction and change the product distribution. It is this aspect which will be considered here.

3.2.1
Water

In 1980, we established that water was the additive of choice for the SmI_2 reduction of ketones and aldehydes [15]. In fact, there is growing evidence that water serves not only as a proton source but can also accelerate certain classes of samarium reductions. Addition of water to THF solutions of SmI_2 induces a color change from the original deep blue to a wine-red similar to the one encountered in a THF–HMPA mixture. The color persists for a few hours if pure and thoroughly deaerated water is used. It appears that water acting as a coordinating ligand (as well as a proton source) greatly increases the reducing ability of SmI_2

(see Sect. 3.1.3 and [103]) and THF–H_2O can thus be considered to be in some reactions a good substitute for THF–HMPA.

Curran et al. have studied reductions of 1,3-diphenylpropanone, ethyl cinnamate, diphenyl sulfoxide, 1-iodododecane, and *o*-allyloxyiodobenzene with SmI_2 in the presence of water [98] (Scheme 46). The accelerating role of water in the reduction of alkyl and aryl iodide has been readily demonstrated [99].

Inanaga et al. reported the use of a 37% aqueous solution of formaldehyde for the reductive cross-coupling of carbonyl compounds with α,β-unsaturated esters with the SmI_2/THF–H_2O system [38] (Scheme 47).

Ohgo et al. have achieved a cross-coupling reaction of aldehydes with α-diketones to give the corresponding adducts [110] (Scheme 48); reactions without HMPA gave better results compared with the reactions using HMPA as additive.

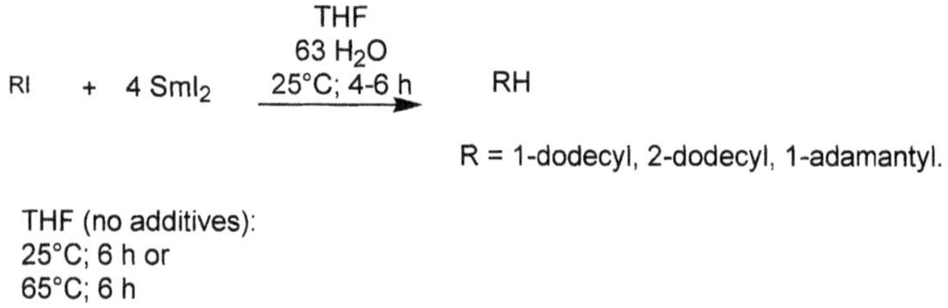

THF
63 H_2O
RI + 4 SmI_2 → (25°C; 4-6 h) RH

R = 1-dodecyl, 2-dodecyl, 1-adamantyl.

THF (no additives):
25°C; 6 h or
65°C; 6 h

Scheme 46

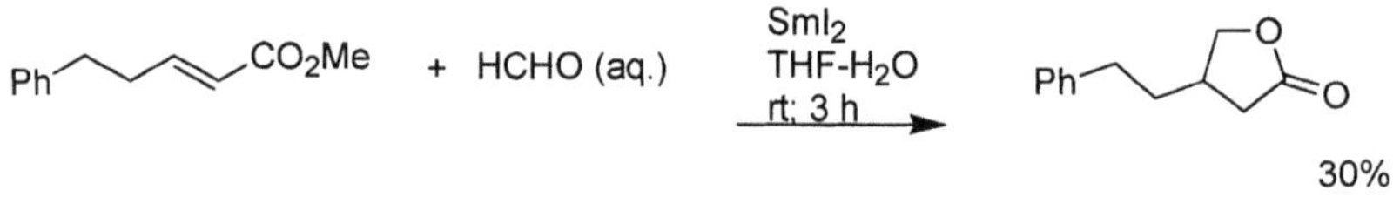

Ph–CH$_2$CH$_2$CH=CH–CO_2Me + HCHO (aq.) → (SmI_2, THF-H_2O, rt; 3 h) Ph 30%

Scheme 47

MeO, O, H, O + H, O → (SmI_2, THF-H_2O)

Ph, OH, O, OMe, OH (*anti*)

Ph, OH, O, OMe, OH (*syn*)

yield: 64%; *anti/syn* = 39/61

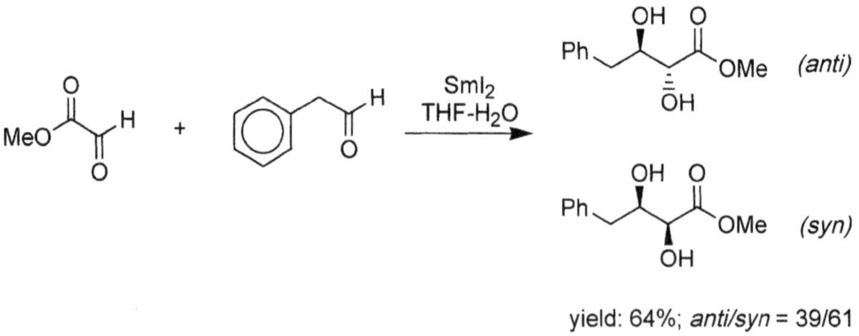

Scheme 48

Unprotected tetrono- and pentonolactone undergo an α-deoxygenation reaction mediated by SmI_2 in THF–H_2O [111] (Scheme 49). The NO reductive cleavage reaction is also accelerated by addition of water [86].

Kamochi and Kudo have reported the use of the SmI_2/THF–H_2O system to reduce aromatic carboxylic acids to alcohols (Scheme 50). Furthermore, aromatic esters amides and nitriles were similarly reduced by this system in good yield [104]. As indicated above, reduction of carboxylic acids to primary alcohols is also effective with SmI_2 in a THF–H_2O–NaOH mixture [105]. In contrast, without water these substrates remain unchanged. With the SmI_2/THF–H_2O system, pyridine was rapidly reduced to piperidine; in similar reactions with pyridine derivatives bearing chloro, amino and cyano substituents, these functionalities were partly eliminated to afford pyridine or piperidine [107].

Curran and Studer transformed in good yields some aromatic dimethylacetals to methylbenzylacetals, by an overall process of reductive demethoxylation [112]. This reaction works by addition of 4 equiv of H_2O at room temperature.

The reduction of allylic halides (in the cepham family) by SmI_2/THF in the presence of some water gave a very fast transformation into exomethylene cephams [113]. The authors made the hypothesis that an intramolecular protonation occurs in a σ-allylsamarium by H_2O coordinated to samarium.

3.2.2
Alcohols

3.2.2.1
Reduction of Carbonyl and C=N Groups

We have shown that THF–MeOH solutions of SmI_2 are excellent reducing agents of ketones or aldehydes to the corresponding alcohols [15]. A mechanistic study

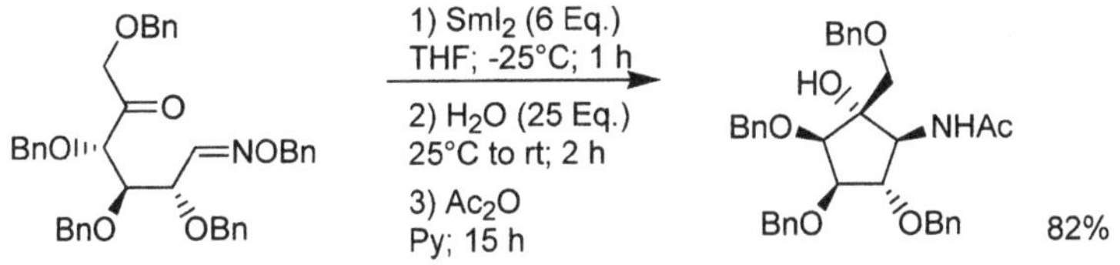

OBn, O, BnO, NOBn, BnO, OBn
1) SmI_2 (6 Eq.) THF; -25°C; 1 h
2) H_2O (25 Eq.) 25°C to rt; 2 h
3) Ac_2O Py; 15 h
BnO, HO, BnO, NHAc, BnO, OBn 82%

Scheme 49

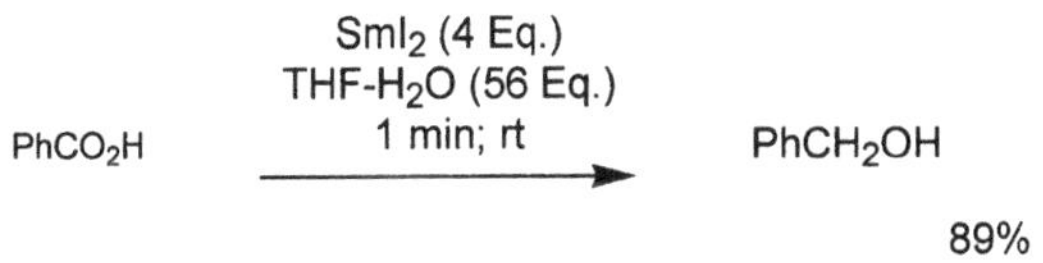

$PhCO_2H$ → SmI_2 (4 Eq.) THF-H_2O (56 Eq.) 1 min; rt → $PhCH_2OH$ 89%

Scheme 50

with MeOD established that 2-octanone was transformed into 2-D-octanol, presumably through the intermediate formation of the carbanion obtained by the reduction of the hydroxy radical issued from the ketyl protonation [114]. Reduction of a mixture of aldehyde and ketone gave a high preference for the aldehyde reduction [15].

The combination of SmI_2 and trimethylsilyl chloride (in replacement of *t*-BuOH) in THF–HMPA was found to accelerate the reduction of sterically hindered or enolizable ketones such as 6-keto- or 20-keto-steroids and 5-cholestenone [115].

Isocyanates R–N=C=S have been transformed by SmI_2 in good yields into thioformamides RNH–(H)C=S in THF–HMPA in the presence of *t*-BuOH at –78 °C [116]. The use of *t*-BuOD gave the C-deuterated thioformamide, establishing that the reaction occurs by two successive one-electron transfers giving this carbanion which is trapped by in situ protonation.

3.2.2.2 Pinacol Formation

In 1983 we recognized that aprotic THF solutions of SmI_2 gave pinacol formation from aldehydes or ketones; the reaction being quite fast with aldehydes or aromatic ketones [117]. However, aromatic carbonyl compounds were coupled in THF by the samarium/I_2/MeOH system, SmI_2 being formed in situ [118].

Hanessian et al. discovered that the intramolecular reductive coupling of various 1,6- and 1,5-dialdehydes gave mainly *cis*-diols [119]. This useful methodology involves the presence of an alcohol (*t*-BuOH or MeOH) acting as an in situ protonating agent for the reaction intermediate. For example, the *myo*-inositol derivative **29** [120] and **30**, an intermediate in the total synthesis of forskolin [121], have been synthesized (Scheme 51). An intramolecular coupling of α-keto aldehyde was also successful, giving a bicyclic system **31** which may be a relay in the total synthesis of taxoids [122].

3.2.2.3 Carbonyl-Ene Couplings

The reductive coupling of ketones or aldehydes with electrophilic alkenes such as conjugated esters (Scheme 52) has been described independently by Inanaga et al. [38] and Fukuzawa et al. [123,124]. The reaction was performed in THF either in the presence of HMPA [38] or an alcohol [123,124]. The mechanistic aspects of this reaction have been discussed [124]. Presumably, the ketyl radical or its protonated form adds to ethyl acrylate with formation of the C–C bond. An efficient enantioselective synthesis of γ-butyrolactones has been derived by using *N*-methylephedrinyl acrylate or crotonate [125]. One example (synthesis of **33**) is indicated in Scheme 52.

The intramolecular reductive ketone-olefin coupling was realized by Molander et al. by the combination SmI_2/THF–*t*-BuOH [126]. For example, the spiro-

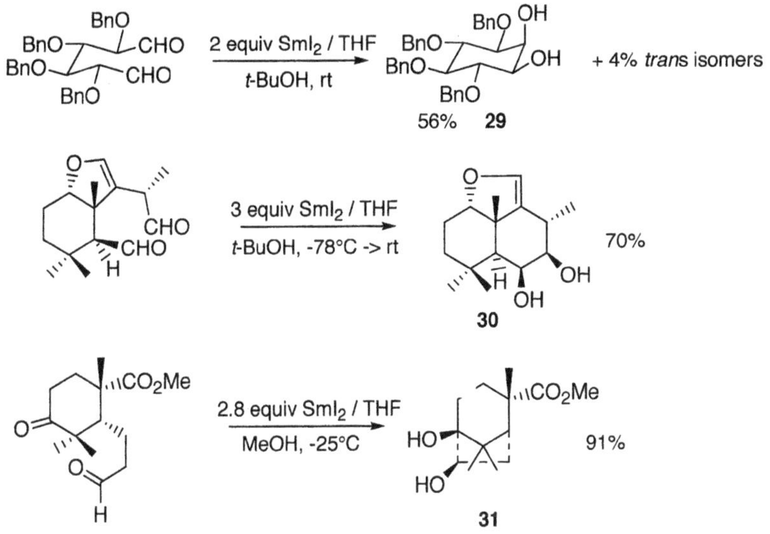

2 equiv SmI_2 / THF, t-BuOH, rt

BnO, BnO, BnO, BnO, CHO, CHO → BnO, OH, BnO, BnO, OH, BnO, OH + 4% *trans* isomers

56% **29**

3 equiv SmI_2 / THF, t-BuOH, -78°C -> rt

CHO, CHO, H → **30** OH, OH, H 70%

2.8 equiv SmI_2 / THF, MeOH, -25°C

CO_2Me → HO, HO, CO_2Me 91% **31**

Scheme 51

+ CO_2Et — 2 equiv SmI_2 / THF, t-BuOH, rt, 2 h → **32** 70%

+ NMe_2 — 2 equiv SmI_2 / THF, t-BuOH, -78°C -> rt → **33** 76%

96% ee (*cis* / *trans* > 99:1)

2 equiv SmI_2 / THF, t-BuOH, -78°C -> rt → HO **34** 87%

ClO_4 - +N Ph **35** — 2 equiv SmI_2 / MeCN, 1 equiv CSA → N Ph **36** 82%

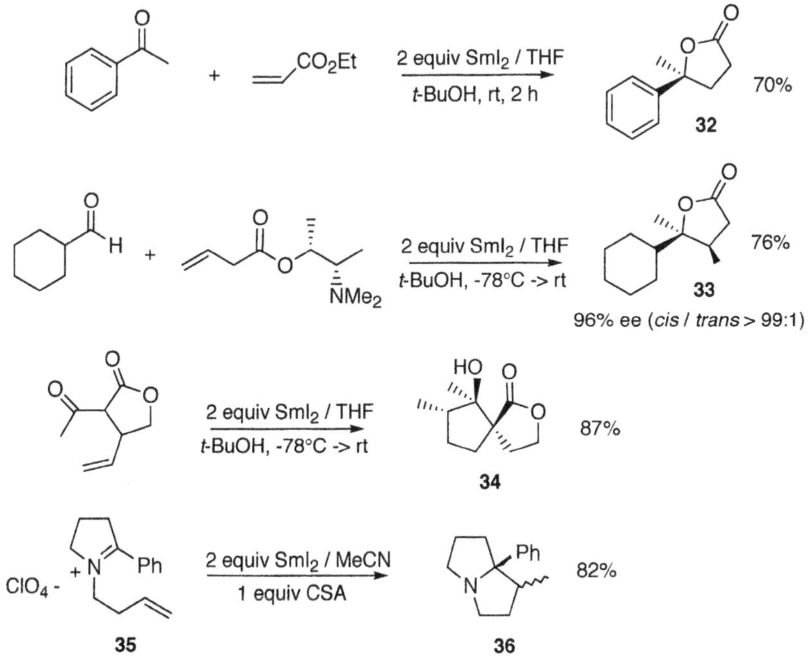

Scheme 52

cyclic lactone **34** was obtained with high diastereoselectivity because of chelation of oxygens on Sm^{3+} (Scheme 52). The protic conditions prevented a retro-aldol process occurring, as observed under aprotic conditions.

Iminium salt **35** was reduced in anhydrous acetonitrile in the presence of at least 1 equiv of camphorsulfonic acid (CSA) (Scheme 52) [127]. In this way, pyrrolidine **36** was obtained in good yield. The cyclization presumably involves a *N*-protonated unsaturated α-amino radical, while the non-protonated form gives less cyclization and more reduction of the C=N bond.

Triple bonds may also undergo coupling reactions to a carbonyl. Inanaga et al. found that the reduction system SmI_2/THF–HMPA–*t*-BuOH efficiently promotes the carbonyl addition on the triple bond, generating allyl alcohol **37** [128] (Scheme 53). A similar procedure was used for the intramolecular coupling reaction [129]. The highest yields were observed for compounds with electron-withdrawing groups on the terminal site of the alkyne, as for the reactions described in Scheme 53.

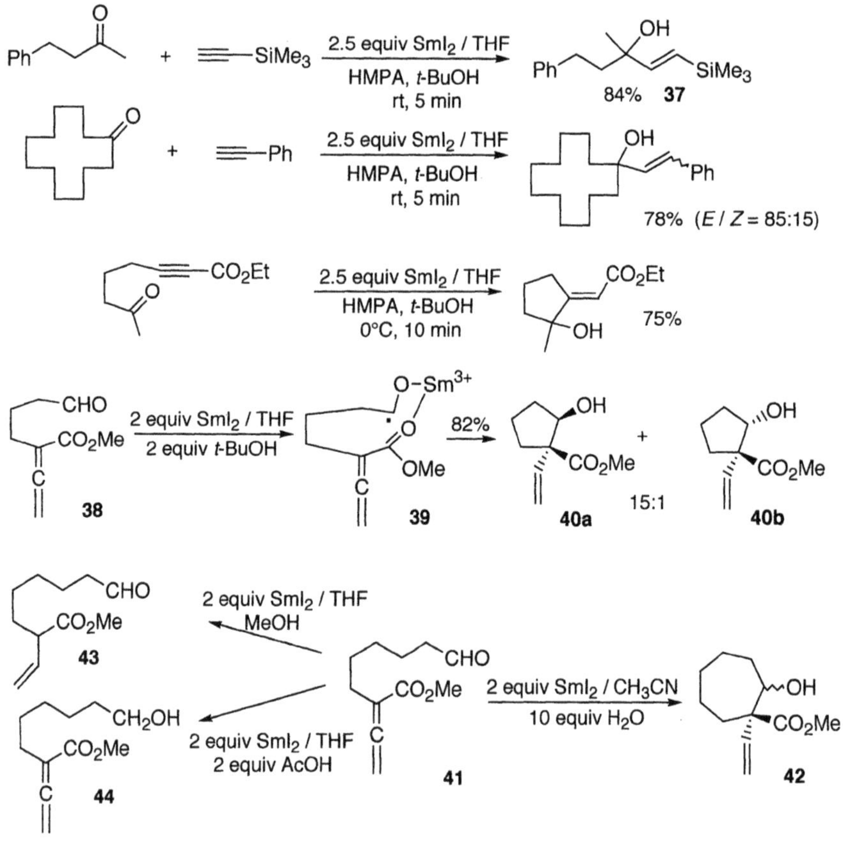

Scheme 53

Intramolecular reductive coupling of aldehydes and allenic esters has also been described [130]. For example, aldehyde **38** was transformed into cyclopentanols **40a** and **40b** (Scheme 53). Presumably ketyl **39** is an intermediate in the reaction. Although being nucleophilic in character it adds to the carbon adjacent to the ester group because of the chelation on samarium, which explains the good diastereoselectivity. The use of homologous aldehydes of **38** should generate rings of larger size than cyclopentane; however, the cyclizations are difficult and need adjustments to the experimental conditions. A six-membered ring analog of **40** was obtained by using methanol instead of *tert*-butanol. The seven-membered homolog **42** was produced in acetonitrile containing 10 equiv of water. Aldehyde **41** was transformed into different products **43** and **44** by minor changes in the experimental conditions.

3.2.2.4
Cleavage of Carbon–Heteroatom Bonds

α-Heterosubstituted ketones are a very widely represented class of compounds amongst organic compounds. Reduction of α-heterosubstituted ketones to ketones is a useful synthetic operation. In 1986 it was shown that SmI_2 is a good reagent for this transformation, which was achieved under mild conditions (–78 °C) in THF–MeOH [131]. For example, α-substituted ketones **45** (Scheme 54) were reduced in 2-dodecanone in good yield for various X groups. Only X=OH gave a poor yield of the desired ketone. The authors envisioned that the initial step of the reaction is formation of an α-ketyl which is rapidly protonated by methanol. A second electron transfer should produce a carbanion which generates an enol by β-elimination. Finally, the enol tautomerizes into the ketone. Under the above experimental conditions α-acetoxy esters were not reduced, a different result to that with α-bromo esters or α-bromo ketones.

The reduction of cyanophosphates **46**, readily prepared from ketones or aldehydes with diethyl phosphorocyanidate and lithium cyanide, led to nitriles in excellent yields [132] (Scheme 54). This reductive cyanation process has also

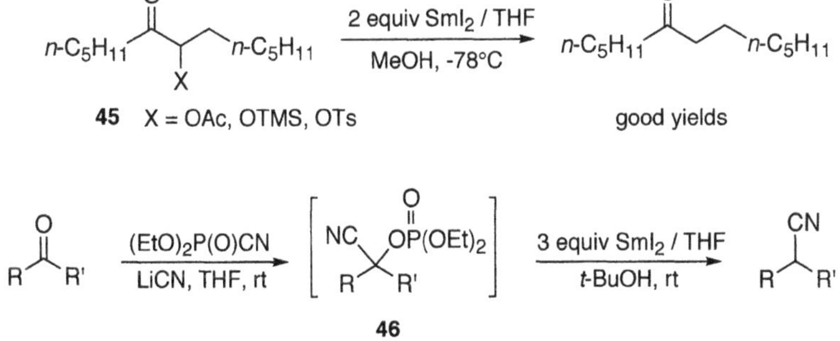

Scheme 54

been applied to conjugated enones or enals, giving β,γ-unsaturated nitriles without products from the migration of the double bond.

Aliphatic α-epoxy ketones were also successfully converted into the corresponding β-hydroxy ketones by SmI_2 in THF–methanol solution [133]. The protonation of the intermediate ketyls is the key step in the process. Radical •C(OH) generated vicinal to an epoxide induces α C–O bond cleavage with ring opening of the epoxide. The alkoxy radical is finally reduced to an alcoholate. Aliphatic α-epoxy ketones have also been treated with diiodosamarium in the presence or absence of methanol [134]. A large excess of methanol (or water) favored the formation of the β-hydroxy ketone, which is absent under aprotic conditions (in this case a small amount of α,β-diketone may be detected).

Deoxygenation reactions have been developed for the preparation of some deoxy sugars. A convenient procedure is to use SmI_2 in THF in the presence of anhydrous ethylene glycol as the proton source (Scheme 55).

Peracetylated aldonolactones gave mixtures of the unsaturated aldonolactone and the deoxy analog [135]. However, good control of the nature of the leaving group and the addition or non-addition of HMPA may fully control the selectivity of the reaction in the direction of the saturated or the unsaturated sugars [136]. Some examples are indicated in Scheme 55. A similar procedure has been used for the anomeric deoxygenation of peracetylated 2-ulosonic acids (as methyl esters) [137]. Deoxygenation in position α to a ketone has been successfully performed in the taxoid family [138].

Reduction of epoxy esters was not selective in the presence of methanol or ethanol but gave good yields of β-hydroxy esters by using *N,N*-dimethylethanolamine (DMEA) [40]. DMEA seems to play the role of a proton source and also acts as a chelator of samarium species, increasing the reduction power of SmI_2 and decreasing the Lewis acidity of Sm(III). See Scheme 56 for some examples.

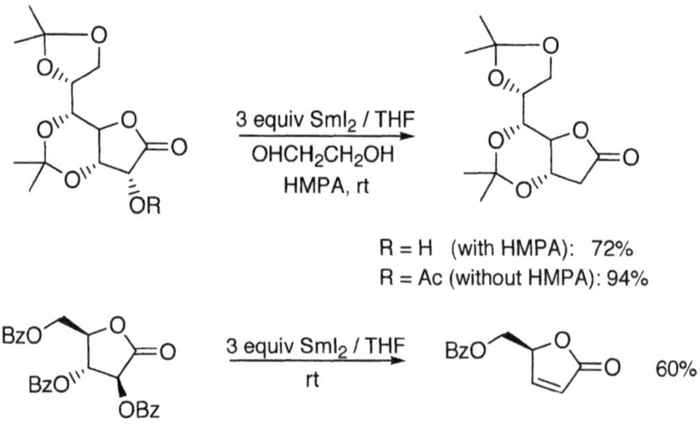

3 equiv SmI_2 / THF, $OHCH_2CH_2OH$, HMPA, rt

R = H (with HMPA): 72%
R = Ac (without HMPA): 94%

3 equiv SmI_2 / THF, rt — 60%

Scheme 55

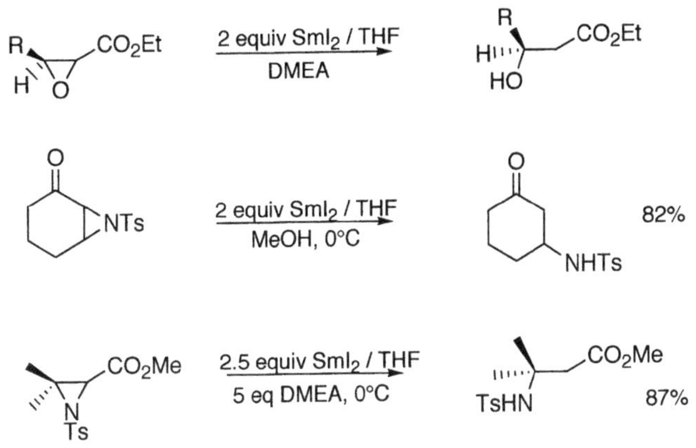

R CO_2Et H O — 2 equiv SmI_2 / THF, DMEA → R H HO CO_2Et

O NTs — 2 equiv SmI_2 / THF, MeOH, 0°C → O NHTs 82%

CO_2Me N Ts — 2.5 equiv SmI_2 / THF, 5 eq DMEA, 0°C → TsHN CO_2Me 87%

Scheme 56

The reduction of 2,3-epoxycycloalkanone hydrazones by SmI_2 in THF in the presence of *t*-BuOH has been investigated by Kang et al. [139]. 2-Acylaziridines were converted into β-amino carboxyl compounds (Scheme 56) by a reaction very similar to the reduction of α,β-epoxy ketones [140]. The main problems were the unselective ring opening of the aziridine ring and the formation of over-reduced amino alcohols. These difficulties were overcome by using 2.5 equiv of SmI_2 in THF at 0 °C in the presence of a suitable proton source. In the case of 2-acylaziridines, methanol was preferred, while for aziridine-2-carboxylates and aziridine-2-carboxamides, DMEA was the best additive. The usefulness of DMEA may be related to several factors, as already pointed out by Inanaga et al. for the reduction of epoxy esters [40] (vide supra).

3.2.2.5
Conclusions

Protic additives other than water strongly modify the course of the reactions mediated by diiodosamarium. As discussed above, the additives protonate some anionic intermediates to new intermediates, driving the reaction in the desired direction. In some specific cases it may coordinate to samarium and modify the redox and Lewis acid properties of some of the samarium complexes. Chiral protic additives may also act as asymmetric protonating agents of prochiral intermediates. For example, Takeuchi et al. reduced benzil to the corresponding samarium enediolate and treated it with various chiral amino alcohols. The resulting benzoin could be obtained with ee's up to 91% by using quinidine as the proton source [141]. The catalytic use of a C_2-symmetric diol for asymmetric protonation of some samarium enolates was achieved in the presence of trityl alcohol (an achiral proton source for regeneration of the catalyst). In this way ee's up to 93% were observed [142]. Enantioselective protonation of samarium enolates generated from 2-methoxy-2-substituted cyclohexanones by SmI_2 in the presence of some chelating chiral diols gave up to 94% ee of 2-substituted cyclohexanones [143,144]. The in situ asymmetric protonation of allenylsamariums obtained by reduction of propargylic esters has recently been described (see Sect. 3.3.6).

3.2.3 *Acids*

The combination of SmI_2 with a protic or aprotic acid ($POCl_3$, HCl, H_2SO_4 or H_3PO_4) has been investigated by Kamochi and Kudo for the reduction of aromatic acids and derivatives [145]. Excellent results were obtained by the addition of an amount of 85% H_3PO_4. Thus, at room temperature, acids or esters gave primary alcohols, while amides $ArCONH_2$ were transformed into aldehydes and aromatic nitriles into primary amines. All these reactions were performed in THF in a few minutes. In some cases there were some differences to the corresponding reduction in the presence of a base which was described in Sect. 3.1.4. A mechanism was proposed by the authors to explain the influence of acids, involving protonation of various intermediates.

Curran and Studer recently used 1 equiv of trifluoroacetic acid or BF_3–OEt_2 as a useful additive to promote the reductive coupling of aromatic dimethylacetals to 1,2-diaryl-1,2-dimethoxyethanes [112]. If the acids were replaced by water, a reductive demethoxylation occurred (see Sect. 3.2.1). The added acid may activate the acetal, giving formation of an oxocarbenium ion which is then reduced by SmI_2 to a radical which couples or is further reduced. It is interesting to note that a Lewis acid such as BF_3–OEt_2 is compatible with the use of SmI_2.

3.3 Metal Salts

3.3.1 *Iron Salts*

In 1980 we noticed that a samarium Barbier reaction between *n*-butyl iodide and 2-octanone which needed 8 h at 65 °C could be run in 3 h at room temperature in the presence of a catalytic amount (2 mol%) of ferric chloride [15]. This experiment was inspired by the known catalytic effect of the Fe^{3+} ion on the Ln^{2+}/Ln^{3+} conversion [146]. The beneficial influence of the Fe(III) derivative in reactions mediated by SmI_2 has been used by several authors in various processes. Thus, Molander et al. described a general procedure for five- and six-membered ring annulation starting from 2-(ω-iodoalkyl) cycloalkanones [147].

Diiodosamarium induced an intramolecular Barbier reaction, under mild conditions (room temperature), in the presence of a catalytic amount of iron tris(dibenzoylomethane [$Fe(DBM)_3$]. In this way many bicyclic tertiary alcohols were prepared. These conditions were also convenient for the cyclization

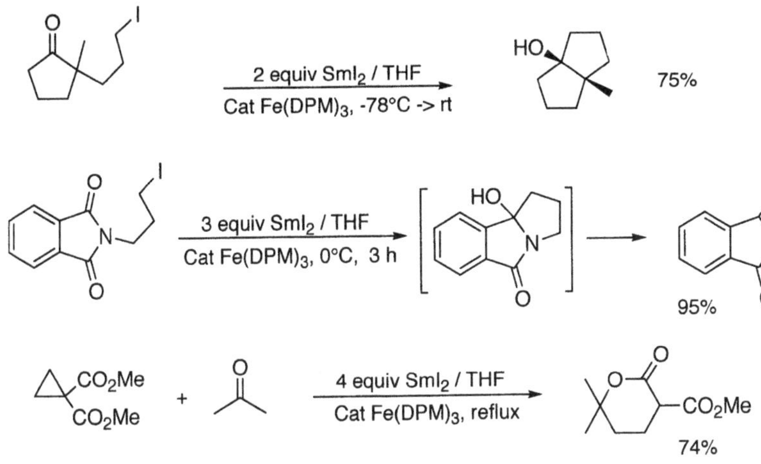

2 equiv SmI_2 / THF
Cat $Fe(DPM)_3$, -78°C -> rt
75%

3 equiv SmI_2 / THF
Cat $Fe(DPM)_3$, 0°C, 3 h
95%

4 equiv SmI_2 / THF
Cat $Fe(DPM)_3$, reflux
74%

Scheme 57

of ω-iodoalkyl acyl derivatives (esters, lactones or amides) into ketones [148]. A comparison was made between addition of HMPA or of a catalytic amount of $Fe(DBM)_3$ or $Fe(acac)_3$. Both methods are often of the same efficiency. This type of reaction was named intramolecular nucleophilic acyl substitution of halo-substituted esters or lactones. One example is shown in Scheme 57. A similar cyclization involving *N*-iodoalkyl cyclic imides was later described by Ha et al. (see Scheme 57) [149] using SmI_2/*t*-BuOH. Iron(III) catalysis was also useful in the reductive cleavage of cyclopropanes activated by ester functions [150]. This cleavage may also be realized in the presence of an aliphatic ketone which traps the intermediate samarium homoenolate, one example is depicted in Scheme 57.

Iron(III) catalysis may be considered as a good alternative to HMPA activation in many reactions induced by SmI_2. Recently, we reinvestigated this area to discover whether other metal salts could be equivalent or superior. Diiodonickel was the result of this investigation (vide infra).

3.3.2
Nickel Salts

We screened the catalytic behavior of various transition-metal salts or complexes in the classical Barbier reaction between 2-octanone and 1-iodobutane in the presence of 2 equiv of SmI_2 in THF at room temperature [151]. Many metal halides significantly improved the yield (compared to the blank reaction): FeX_3 (X= Br, Cl, I), $CuCl_2$, CuI, NiX_2 (X=Br, I), AgX (X=Br, I), $CoBr_2$. Amongst them NiI_2 proved to be a superior catalyst allowing an almost quantitative reaction in less than 10 min. Surprisingly, acetylacetonates of Ni, Fe, Co and Cu are quite inefficient in this reaction.

Diiodonickel (1 mol%) accelerated significantly many of the reductions performed by SmI_2 in THF. For example, some epoxides were deoxygenated in good yields to the corresponding alkenes at room temperature in 20 min, 2-octanone gave the pinacol in 10 min (instead of 24 h without catalyst) at room temperature, and cyclohexane carbonyl chloride coupled quantitatively into 2-hydroxy-1,2-dicyclohexylethanone in 1 min (reaction performed in THP).

An unexpected feature of the SmI_2/NiI_2 system is the high reactivity of esters under intermolecular Barbier conditions. As for Grignard reagents, the reaction gives tertiary alcohols; one example is mentioned in Scheme 58. The coupling between a ketone and a conjugated ester, which is promoted by HMPA [38], may be easily obtained by NiI_2 catalysis (Scheme 58). Nucleophilic acylation of esters by acid chlorides has been realized by the Barbier procedure (Scheme 58) [152]. The reaction intermediate **A** can be either hydrolyzed into mixed α-ketols or quenched by other electrophiles (aldehydes or anhydrides). Recently, NiI_2 catalysis has been successfully applied to the cyclization of iodoalkyl α,β-unsaturated esters or amides (for one example, see Scheme 58); iron(III) catalysts were equally efficient [153]. NiI_2 has also been used to selectively activate 1-iodo-3-chloropropane for the Barbier condensation on various keto esters, the reaction occurring on the carbon bearing the iodine [154].

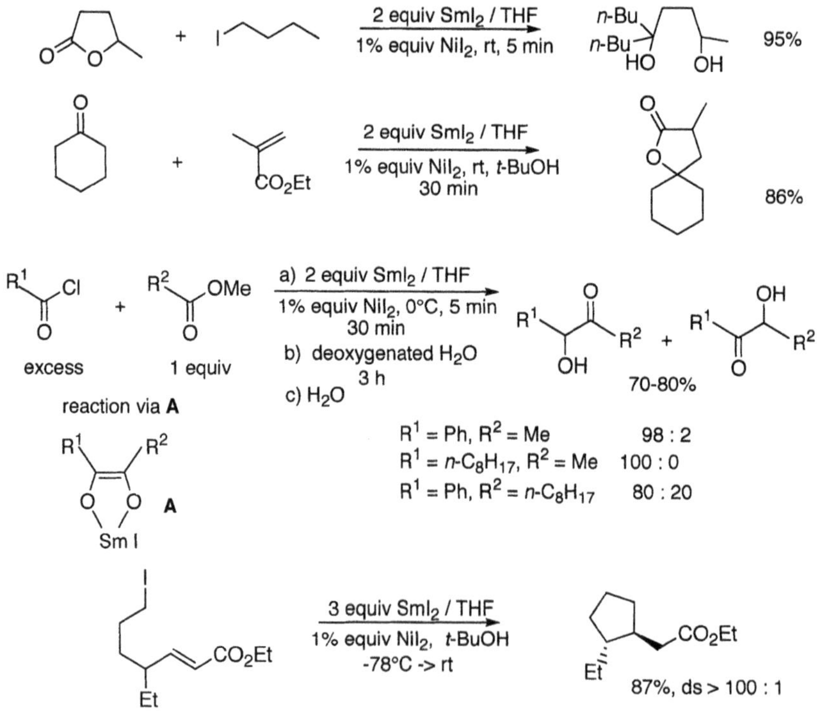

2 equiv SmI_2 / THF, 1% equiv NiI_2, rt, 5 min — 95%

2 equiv SmI_2 / THF, 1% equiv NiI_2, rt, t-BuOH, 30 min — 86%

a) 2 equiv SmI_2 / THF, 1% equiv NiI_2, 0°C, 5 min, 30 min; b) deoxygenated H_2O, 3 h; c) H_2O

excess + 1 equiv; reaction via **A**; 70-80%

R^1 = Ph, R^2 = Me	98 : 2
R^1 = n-C_8H_{17}, R^2 = Me	100 : 0
R^1 = Ph, R^2 = n-C_8H_{17}	80 : 20

3 equiv SmI_2 / THF, 1% equiv NiI_2, t-BuOH, -78°C -> rt — 87%, ds > 100 : 1

Scheme 58

3.3.3
Cobalt Salts

In 1991, Inanaga et al. described the reduction at room temperature of some disubstituted alkynes by the combination SmI_2/proton source in the presence of some transition-metal catalysts (3% equiv) in THF [155]. By a good choice of catalyst ($CoCl_2$, 4 PPh_3) and proton source (MeOH, *i*-PrOH or AcOH) it was possible to orientate the reaction towards the exclusive formation of the *Z*-alkene. When the same reaction was performed in the presence of HMPA then the *E*-alkene was produced. Iron(III) and Ni(II) catalysts were found to be less efficient. It was assumed that the reactive species were the corresponding transition-metal hydrides obtained by reduction of the initial complexes.

3.3.4
Copper Salts

When copper salts were added to organosamarium species produced in situ from an organic halide and SmI_2, a transmetallation process gave organocoppers able to react in catalyzed conjugate additions [156,157].

The cross-coupling of alkylsamarium reagents with alkyl halides may be catalyzed by copper salts such as CuX (X=Br, Cl or I) or Li_2CuCl_4 [158]. The most effective catalyst for the desired cross-coupling was CuBr or Li_2CuCl_4. The reaction was performed at room temperature in THF with 8 equiv of HMPA. Some examples are listed in Scheme 59. In order to optimize cross-coupling products R^1–R^2, the reaction was performed in two steps, with initial formation of the organosamarium from R^1X and then addition of the catalyst and of R^2X.

3.3.5
Chromium Salts

Alkylation of ketones by *gem*-dibromoalkanes in the presence of a stoichiometric amount of SmI_2/Sm (1:1) and a catalytic amount of $CrCl_3$ (10% equiv) was studied by Utimoto et al. [159]. It gave in moderate yields (40–70%) various

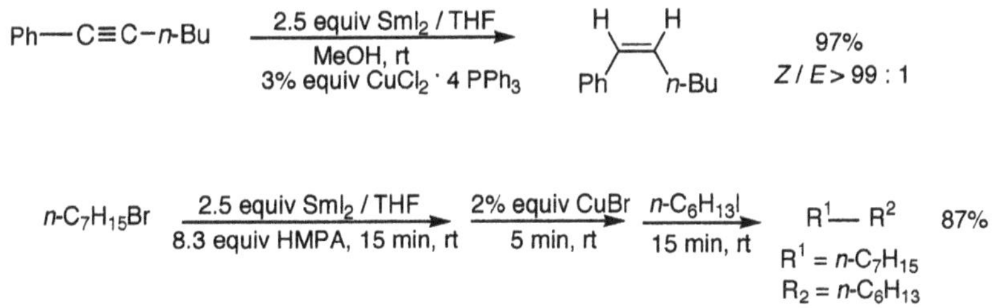

Ph—C≡C–*n*-Bu → (2.5 equiv SmI_2 / THF; MeOH, rt; 3% equiv $CuCl_2 \cdot 4\,PPh_3$) → (*Z*)-PhCH=CH–*n*-Bu 97%, *Z*/*E* > 99 : 1

n-$C_7H_{15}Br$ → (2.5 equiv SmI_2 / THF; 8.3 equiv HMPA, 15 min, rt) → (2% equiv CuBr; 5 min, rt) → (*n*-$C_6H_{13}I$; 15 min, rt) → R^1—R^2 87%
R^1 = *n*-C_7H_{15}
R_2 = *n*-C_6H_{13}

Scheme 59

alkenes as *E/Z* mixtures. The reaction presumably involves a *gem*-dimetallic reagent [$C(SmX_2)_2$ or $C(SmX_2)(CrX_2)$]. The reaction can be applied to easily enolizable ketones such as β-tetralone.

3.3.6
Palladium Complexes

In 1986 Inanaga et al. found that allylic or propargylic acetates could be reduced in THF by diiodosamarium in the presence of a catalytic amount of a Pd(0) complex and 1 equiv of 2-propanol [160,161]. Some examples are indicated in Scheme 60. The mechanism of the reaction likely involves π-allyl (or σ-allenyl) palladium intermediates which are reduced first to radicals and then to carbanions (see Scheme 60). A final protonation by the alcohol generates the products.

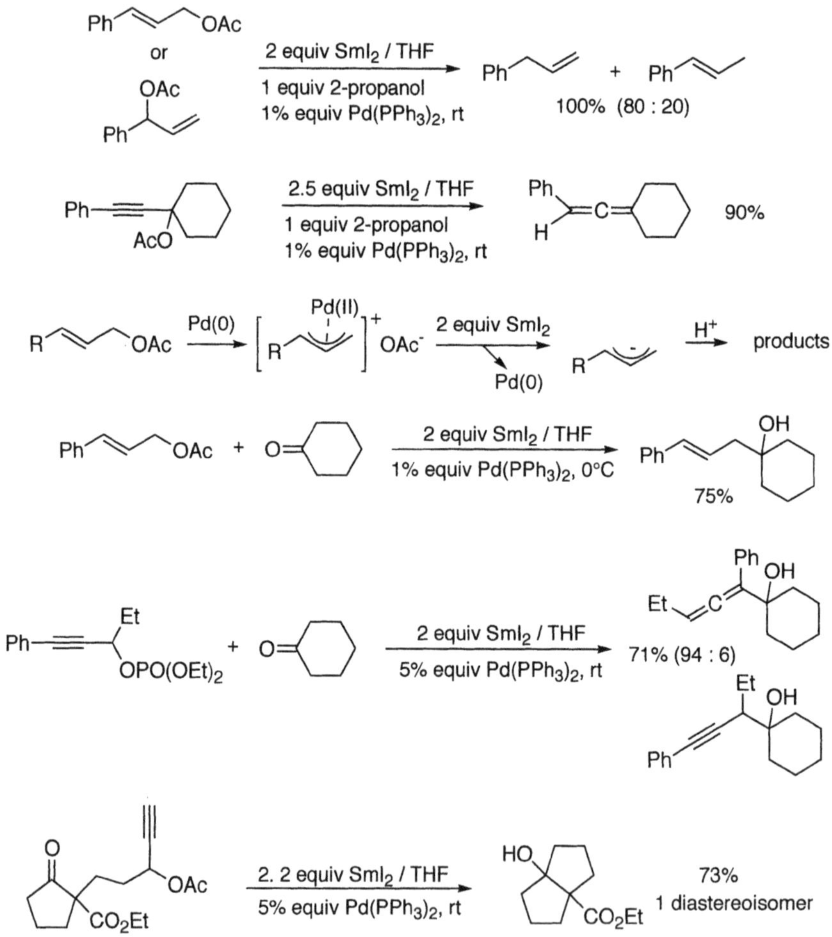

Scheme 60

Some limitations to this reaction have been noticed [162]. The reductive coupling of allylic acetates with carbonyl compounds (which act as electrophiles in replacement of a proton donor) provided homoallylic alcohols [163]. Similarly, coupling of propargylic esters to carbonyl compounds gave allenyl alcohols or acetylenic alcohols according to the structure of the reactants [164,165]. Some examples can be found in Scheme 60. The intramolecular coupling reaction using ω-keto-alkynyl esters allowed the synthesis of five- or six-membered ring systems [166]. A highly regio- and stereoselective synthesis of allylsilanes was achieved by reductive silylation of allylic phosphates with a SmI_2/HMPA/Pd(0) system and TMSCl [167].

A regiodivergent reduction of allylic esters with SmI_2 in the presence or absence of a Pd(0) catalyst gave either α- or γ-protonated products by tuning the proton source (alcohols, H_2O) and the ester functionality (Scheme 61) [168].

A changeover of regioselectivity was observed by Mikami et al. in the reduction of secondary propargylic phosphates by the SmI_2/Pd(0)/proton source system [169]. *tert*-Butanol and dimethyl (*R*,*R*)-tartrate gave allene and acetylene, respectively (Scheme 61). This process was modified in an asymmetric synthesis of chiral allenes [170].

2,3-Naphthoquinodimethanes were easily generated from *ortho*-bis(α-acetoxypropargyl) benzene derivatives by the SmI_2/Pd(0)/proton donor system [171]. The reduction occurs by a 6π-electron cyclization of an *O*-diallenylbenzene intermediate.

3.3.7
Lithium Salts

The Barbier reaction discussed in Sect. 3.3.2 was not accelerated by addition of LiCl. In the literature there are only a few examples where lithium salts are ben-

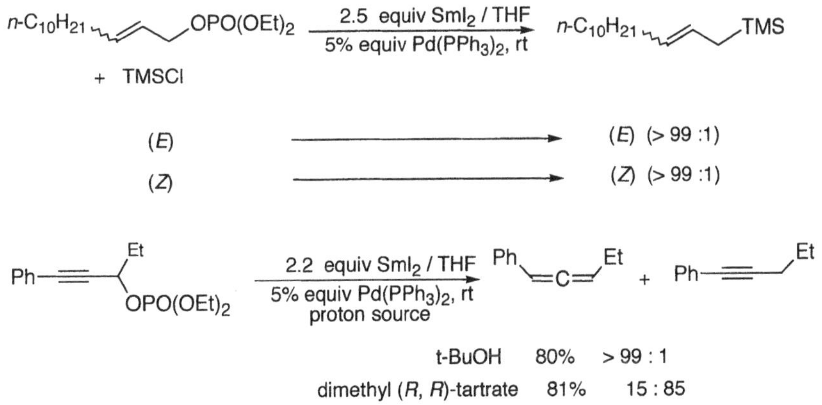

n-$C_{10}H_{21}$–CH=CH–CH$_2$OPO(OEt)$_2$ + TMSCl → (2.5 equiv SmI_2 / THF, 5% equiv $Pd(PPh_3)_2$, rt) → n-$C_{10}H_{21}$–CH=CH–CH$_2$TMS

(*E*) → (*E*) (> 99 :1)

(*Z*) → (*Z*) (> 99 :1)

Ph–C≡C–CH(Et)OPO(OEt)$_2$ → (2.2 equiv SmI_2 / THF, 5% equiv $Pd(PPh_3)_2$, rt, proton source) → PhCH=C=CHEt + Ph–C≡C–CH$_2$Et

proton source	yield	ratio
t-BuOH	80%	> 99 : 1
dimethyl (*R*, *R*)-tartrate	81%	15 : 85

Scheme 61

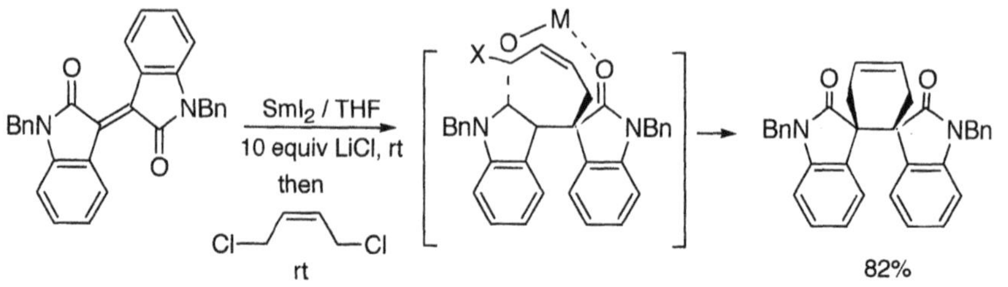

Scheme 62

eficial for reactions mediated by SmI_2. However, it is worth mentioning that a reductive dialkylation of isoindigo by *cis*-1,4-dichloro-2-butene only works in the presence of an excess of LiCl (KCl is inefficient) as indicated in Scheme 62 [172]. The mechanism was not elucidated; a chelate may be involved.

Recently, Flowers et al. initiated a study on the influence of additives on the reactivity of SmI_2, in parallel with electrochemical investigations (see Sect. 3.1.1). These authors found that the oxidation potentials of SmI_2 in THF containing 12 or more equivalents of LiBr or LiCl were, respectively, –1.98 and –2.11 V (instead of –1.33 V). They established that the corresponding molecular species were not $SmBr_2$ and $SmCl_2$. These new reagents are highly reactive and facilitate the pinacol coupling of cyclohexanone at room temperature in a few minutes at room temperature [173].

4 Conclusions

In this chapter we have tried to present the main changes which occur for reactions mediated by diiodosamarium when THF is replaced by other solvents or when some additives are introduced into the THF solution. Many useful organic transformations have been achieved by this approach but, due to lack of space, this area has not been covered extensively. However, we hope to have demonstrated that introduction into THF of various additives in stoichiometric or catalytic amounts or replacement of THF by another solvent may drastically modify the behavior of SmI_2. The most promising approach involves the use of molecular catalysts since one may expect a wide range of chemo- or stereoselectivities by suitable changes in the structure of the catalysts.

5 References

1. Matignon CA, Cazes E (1906) Ann Chim Phys 8:417
2. Jantsch G, Skalla N, Jawurek L (1931) Z Anorg Chem 201:207
3. Namy J-L, Girard P, Kagan HB (1977) New J Chem 1:5
4. Girard P, Namy J-L, Kagan HB (1980) J Am Chem Soc 102:2693

5. Kagan HB (1985) In: Marks TJ, Fragala IL (eds) Fundamental aspects of organo-f-elements chemistry. Reidel, Dordrecht, p 49
6. Kagan HB, Namy J-L (1986) Tetrahedron 42:6573
7. Kagan HB (1990) New J Chem 14.453
8. Soderquist JA (1991) Aldrichimica Acta 24:15
9. Molander GA (1992) Chem Rev 92.15
10. Molander GA, Harris CR (1996) Chem Rev 96:307
11. Skrydstrup T (1998) Angew Chem Int Ed Engl 36:345
12. Molander GA (1994) Organic Reactions 46:211
13. Inanaga J (1990) Rev Heteroatom Chem 3:75
14. Namy J-L, Girard P, Kagan HB, Caro P (1981) New J Chem 5:479
15. Matsukawa M, Inanaga J, Yamaguchi M (1987) Tetrahedron Lett 28:5877
16. Otsubo K, Kawamura M, Inanaga J, Yamaguchi M (1987) Chem Lett 1487
17. The stoichiometry of bonded THF molecules to SmI_2 is 5, the removal of coordinated THF has been discussed [18]
18. Evans W, Gummersheimer TS, Ziller JW (1995) J Am Chem Soc 117:8999
19. Murakami M, Hayashi M, Ito Y (1992) J Org Chem 57:793
20. Murakami M, Hayashi M, Ito Y (1995) Appl Organomet Chem 9:385
21. Booth SE, Benneche T, Undheim K (1995) Tetrahedron 51:3665
22. Namy J-L, Colomb M, Kagan HB (1994) Tetrahedron Lett 35:1723
23. Hamann-Gaudinet B, Namy J-L, Kagan HB (1997) Tetrahedron Lett 38:6585
24. Matsukawa M, Inanaga J, Yamaguchi M (1987) Tetrahedron Lett 28:5877
25. Imamoto J, Takeyama T, Yokoyama M (1984) Tetrahedron Lett 25:3225
26. Zoretic PA, Yu BC, Caspar L (1989) Synthetic Commun 1859
27. Collin J, Namy J-L, Dallemer F, Kagan HB (1991) J Org Chem 56:3118
28. Ruder SM (1992) Tetrahedron Lett 33:2621
29. Hamann B, Namy J-L, Kagan HB (1996) Tetrahedron 52:14225
30. Chebolu A, Whittle RR, Sen A (1985) Inorg Chem 24:3082
31. Inanaga J, Ujikawa O, Yanaguchi M (1991) Tetrahedron Lett 32:1737
32. Ying T, Bao W, Zhang Y, Xu W (1996) Tetrahedron Lett 37:3885
33. Akane N, Kanegawa Y, Nishiyama Y, Ishii Y (1992) Chem Lett 2431
34. Akane N, Hutano T, Kusui H, Nishiyama Y, Ishii Y (1994) J Org Chem 59:7902
35. Kunishima M, Hioki K, Ohara T, Tani S (1992) J Chem Soc Chem Commun 219
36. Kunishima M, Hioki K, Tani S, Kato A (1994) Tetrahedron Lett 35:7253
37. Kunishima M, Hioki K, Kono K, Kato A, Tani S (1997) J Org Chem 62:7542
38. Otsubo K, Inanaga J, Yamaguchi M (1986) Tetrahedron Lett 27:5763
39. Fukuzawa S-I, Ilida M, Nakawishi A, Fiyinami T, Sakai SJ (1987) J Chem Soc Chem Commun 920
40. Otsubo K, Inanaga J, Yamaguchi M (1987) Tetrahedron Lett 28:4437
41. Inanaga J, Ishikawa M, Yamaguchi M (1987) Chem Lett 1485
42. Ujikawa O, Inanaga J, Yamaguchi M (1989) Tetrahedron Lett 30:2837
43. Handa Y, Inanaga J, Yamaguchi M (1989) J Chem Soc Chem Commun 298
44. Hou Z, Wakatsuki Y (1994) J Chem Soc Chem Commun 1205
45. Shabangi M, Flowers II RA (1997) Tetrahedron Lett 38:137
46. Imamoto T, Yamanoi Y, Tsuruta H, Yamaguchi K, Yamazaki M, Inanaga J (1995) Chem Lett 949
47. Molander GA, Mc Kie JA (1992) J Org Chem 57:3132
48. Skene WG, Scaiano JC, Cozens FL (1996) J Org Chem 61:7918
49. Hasegawa E, Curran DP (1993) Tetrahedron Lett 34:1717
50. Curran DP, Totleben MJ (1992) J Am Chem Soc 114:6050
51. Totleben MJ, Curran DP, Wipf P (1992) J Org Chem 57:1740
52. Wipf P, Venkatraman S (1993) J Org Chem 58:3455
53. Molander GA, Mckie JA (1994) J Org Chem 59:3186

54. Aurrecoechea JM, Iztueta E (1995) Tetrahedron Lett 36:7129
55. Molander GA, Shakya SR (1996) J Org Chem 61:5885
56. Tadano K-I, Isshiki Y, Minami M, Ogawa S (1993) 58:6266
57. Shiue J-S, Liu C-C, Fang J-M (1993) Tetrahedron Lett 34:335
58. Shiue J-S, Lin M-H, Fang J-M (1997) J Org Chem 62:4643
59. Yang S-M, Fang J-M (1997) Tetrahedron Lett 38:1589
60. Künzer H, Stahnke M, Sauer G, Wieckert R (1991) Tetrahedron Lett 32:1949
61. De Pouilly P, Chénéde A, Mallet J-M, Sinaÿ P (1992) Tetrahedron Lett 33:8065
62. Ihara M, Zuzuki S, Taniguchi T, Tokunaga Y, Fukumoto K (1994) Synlett 859
63. Clayden J, Julia M (1994) J Chem Soc Chem Commun 2261
64. Kan J, Nara S, Ito S, Matsuda F, Shirahama H (1994) J Org Chem 59:5111
65. Chandrasekhar S, Yu J, Falck JR (1994) Tetrahedron Lett 35:5441
66. Ihara M, Suzuki S, Taniguchi T, Tokunaza Y, Fukumoto K (1995) Tetrahedron 51:9873
67. Kende A.S, Mendoza JS (1990) Tetrahedron Lett 31:7105
68. Mazéas D, Skrydstrup T, Doumeix O, Beau J-M (1994) Angew Chem Int Ed Engl 33:1383
69. Fukuzawa S-I, Mutoh K, Tsuchimoto T, Hiyama T (1996) J Org Chem 61:5400
70. Hanamoto T, Sugimoto Y, Sugino A, Inanaga J (1994) Synlett 377
71. Kawatsura M, Matsuda F, Shirahama H (1994) J Org Chem 59:6900
72. Matsuda F (1995) J Synth Org Chem Jpn 53:987
73. Kawatsura M, Hosaka K, Matsuda F, Shirahamatt (1995) Synlett 729
74. Inanaga J, Handa Y, Tabuchi T, Otsubo K (1991) Tetrahedron Lett 32:6557
75. Cabiera A, Alper H (1992) Tetrahedron Lett 33:5007
76. Balaux E, Ruel R (1996) Tetrahedron Lett 37:801
77. Park HS, Lee IS, Kim YH (1995) Tetrahedron Lett 36:1673
78. Yamashita M, Okuyama K, Ohhara T, Kawasaki I, Ohta S (1996) Synlett 547
79. Shibuya K, Nagaoka H, Yamada Y (1991) J Chem Soc Chem Commun 1545
80. Shiue J-S, Fang J-M (1993) J Chem Soc Chem Commun 1277
81. Kawatsura M, Kishi E, Kito M, Sakai T, Shirahama H, Matsuda F (1997) Synlett 479
82. Katritzky AR, Feng D, Qi M (1997) J Org Chem 62:6222
83. Hanamoto T, Inanaga J (1991) Tetrahedron Lett 32:3555
84. Sturino CF, Fallis AG (1994) J Am Chem Soc 116:7447
85. Sturino CF, Fallis AG (1994) J Org Chem 59:6514
86. Marco-Coutelles J, Gallego P, Rodriguez-Fernandez M, Khiar N, Destabel C, Bernabé M, Martinez-Grau A, Chiara JL (1997) J Org Chem 62:7397
87. Fevig TL, Elliott RL, Curran DP (1988) J Am Chem Soc 110:5064
88. Crombie L, Rainbow LJ (1988) Tetrahedron Lett 29:6517
89. Crombie L, Rainbow LJ (1994) J Chem Soc Perkin Trans. I 673
90. Bennett SM, Larouche D (1991) Synlett 805
91. Zhou Z, Larouche D, Bennett SM (1995) Tetrahedron 51:11623
92. Curran DP, Wolin RL (1991) Synlett 317
93. Fleming I, Ghosh SK (1992) J Chem Soc Chem Commun 1775
94. Vedejs E, Lin S (1994) J Org Chem 59:1602
95. Goulaouic-Dubois C, Gruggisberg A, Hesse M (1995) J Org Chem 60:5969
96. Bateg RA, Harling JD, Motherwell WB (1996) Tetrahedron 52:11421
97. Keck GE, Savin KA, Weglarz MA (1995) J Org Chem 60:3194
98. Hasegawa E, Curran DP (1993) J Org Chem 58:5008
99. Inanaga J, Sakai S, Handa Y, Yamaguchi M, Yokyama Y (1991) Chem Lett 2117
100. Cabri W, Candiani I, Colombo M, Franzoi L, Bedeschi A (1995) Tetrahedron Lett 36:949
101. Aoyagi Y, Yoshimura M, Tsuda M, Tsuchibuchi T, Kawamata S, Teteno H, Asano K, Nakamura H, Obokata M, Ohta A, Kodama Y (1995) J Chem Soc Perkin Trans I 689
102. Molander GA, Alonso-Alijal C (1997) Tetrahedron 53:8067
103. Shabangi M, Sealy JM, Fuchs JR, Flowers RA II (1998) Tetrahedron Lett 39:4429

104. Kamochi Y, Kudo T (1993) Chem Lett 1495
105. Kamochi Y, Kudo T (1991) Chem Lett 893
106. Mikami K, Yamaoka M (1998) Tetrahedron Lett 39.4501
107. Kamochi Y, Kudo T (1993) Heterocycles 36:2383
108. Namy J-L, Collin J, Bied C, Kagan HB (1992) Synlett 733
109. Curran DP, Fevig TL, Jasperse CP, Totleben MJ (1992) Synlett 943
110. Miyoshi N, Takeuchi S, Ohgo Y (1993) Chem Lett 2129
111. Hanessian S, Girard C (1994) Synlett 861
112. Studer A, Curran DP (1996) Synlett 255
113. Cabri W, Candiani I, Bedsechi A (1993) Tetrahedron Lett 34:6931
114. Girard P, Namy J-L, Kagan HB (1981) Tetrahedron, Suppl 31:175
115. Honda T, Katoh M (1997) J Chem Soc Chem Commun 369
116. Park HS, Lee I.S, Kim YH (1996) J Chem Soc Chem Commun 1805
117. Souppe J, Namy J-L, Kagan HB (1983) Tetrahedron Lett 24:765
118. Yamada R, Negoro N, Yanada K, Fujita T (1997) Tetrahedron Lett 38:3271
119. Chiara JL, Cabri W, Hanessian S (1991) Tetrahedron Lett 32:1125
120. Guidot JP, Le Gall T, Mioskowski C (1994) Tetrahedron Lett *36*:6671
121. Anies C, Pancrazi A, Lallemand JY (1994) Tetrahedron Lett 35:7771
122. Arseniyadis S, Yashunsky DV, Pereira de Fruitas R, Munoz Dorado M, Toromanoff E, Potier P (1993) Tetrahedron Lett 34:1137
123. Fukuzawa S, Nakanishi A, Fujinami T, Sakai S (1986) J Chem Soc Chem Commun 624
124. Fukuzawa S, Nakanishi A, Fujinami T, Sakai S (1988) J Chem Soc Perkin Trans I 1669
125. Fukuzawa S, Seki K, Tatsukawa M, Mutoh K (1997) J Am Chem Soc 119:1482
126. Molander GA, Kenny C (1987) Tetrahedron Lett 28:4367
127. Martin SF, Yang CP, Laswell WL, Rüeger H (1988) Tetrahedron Lett 29:6685
128. Inanaga J, Katsuki J, Ujikawa O, Yamaguchi M (1991) Tetrahedron Lett 32:4921
129. Shim SC, Hwang JT, Kang HY, Chang MH (1990) Tetrahedron Lett 31:4765
130. Gillmann T (1993) Tetrahedron Lett 34:607
131. Molander GA, Hahn G (1986) J Org Chem 51:1135
132. Yoneda R, Harusawa S, Kurihara T (1989) Tetrahedron Lett 30:3681
133. Molander GA, Hahn G (1986) J Org Chem 51:2596
134. Hagesawa E, Ishiyama K, Fujita T, Kato T, Abe T (1997) J Org Chem 62:2396
135. Inanaga J, Katsuki J, Yamaguchi M (1991) Chem Lett 1025
136. Hanessian S, Girard C, Chiara JL (1992) Tetrahedron Lett 33:573
137. Hanessian S, Girard C (1994) Synlett 863
138. Py S, Pan JW, Khuong-Huu F (1994) Tetrahedron 50:6881
139. Kang HY, Hong WS, Lee SH, Choi KI, Koh HY (1997) Synlett 33
140. Molander GA, Stengel PJ (1997) Tetrahedron 53:8887
141. Takeuchi S, Miyoshi N, Hirata K, Hayashida H, Ohgo Y (1992) Bull Chem Soc Jpn 65:2001
142. Nakamura Y, Takeuchi S, Ohira A, Ohgo Y (1996) Tetrahedron Lett 37:2805
143. Nakamura Y, Takeuchi S, Ohgo Y, Yamaoka M, Yoshida A, Mikami K (1997) Tetrahedron Lett 38:2709
144. Mikami K, Yamaoka M, Yoshida A, Nakamura Y, Takeuchi S, Ohgo Y (1998) Synlett 607
145. Kamochi Y, Kudo T (1992) Tetrahedron 48:4301
146. Clifford AF, Beachell HC (1948) J Am Chem Soc 70:2730
147. Molander GA, Etter JB.(1986) J Org Chem 51:1778
148. Molander GA, Mc Kie JA (1993) J Org Chem 58:7216
149. Ha DC, Yun CS, Yu E (1996) Tetrahedron Lett 37:7577
150. Imamoto T, Hatajima T, Yoshizawa T (1994) Tetrahedron Lett 35:7805
151. Machrouhi F, Hamann B, Namy J-L, Kagan HB (1996) Synlett 633
152. Machrouhi F, Namy J-L, Kagan HB (1997) Tetrahedron Lett 38:7183
153. Molander GA, Harris CR (1997) J Org Chem 62:7418

154. Molander G, Alonso-Alija C (1998) J Org Chem 63:4366
155. Inanaga J, Yokohama Y, Baba Y, Yamaguchi M (1991) Tetrahedron Lett 32:5559
156. Totleben MJ, Curran DP, Wipf P (1992) J Org Chem 57:1740
157. Wipf P, Venkatraman S (1993) J Org Chem 58:5455
158. Bukowitz WF, Wu Y (1997) Tetrahedron Lett 38:3171
159. Matsubara S, Horiuchi M, Takai K, Utimoto K (1995) Chem Lett 259
160. Tabuchi T, Inanaga J, Yamaguchi M (1986) Tetrahedron Lett 27:601
161. Tabuchi T, Inanaga J, Yamaguchi M (1986) Tetrahedron Lett 27:5237
162. Marco-Contelles J, Destabel C, Chiara JL (1996) Tetrahedron Asymmetry 7:105
163. Tabuchi T, Inanaga J, Yamaguchi M (1986) Tetrahedron Lett 27:1195
164. Tabuchi T, Inanaga J, Yamaguchi M (1987) Chem Lett 2275
165. Mikami K, Yoshida A, Matsumoto S, Fong F, Matsumoto Y, Sugino A, Hanamoto T, Inanaga J (1995) Tetrahedron Lett 36:907
166. Aurrecoechea JM, Anton RF (1994) J Org Chem 59:702
167. Hanamoto T, Sugino A, Kikuwa T, Inanaga J (1997) Bull Soc Chim Fr 134:391
168. Yoshida A, Hanamoto T, Inanaga J, Mikami K (1998) Tetrahedron Lett 39:1777
169. Yoshida A, Mikami K (1997) Synlett 1375
170. Mikami K, Yoshida A (1997) Angew Chem Int Ed Engl 36:858
171. Inanaga J, Sugimoto Y, Hanamoto T (1992) Tetrahedron Lett 33:7035
172. Link JT, Overman LE (1996) J Am Chem Soc 118:8166
173. Fuchs JR, Mitchell ML, Shabangi M, Flowers RA II (1997) Tetrahedron Lett 38:8157

Chiral Heterobimetallic Lanthanoid Complexes: Highly Efficient Multifunctional Catalysts for the Asymmetric Formation of C-C, C-O, and C-P Bonds

Masakatsu Shibasaki* and Harald Gröger

Graduate School of Pharmaceutical Sciences, The University of Tokyo, Hongo, Bunkyo-ku, Tokyo 113, Japan
* *e-mail:* mshibasa@mol.f.u-tokyo.ac.jp

The use of alkali metal-containing, heterobimetallic lanthanoid complexes as catalysts in asymmetric synthesis is reviewed. This new and innovative type of chiral catalyst, which was recently developed by Shibasaki et al., contains a Lewis acid as well as a Brønsted base moiety, thereupon showing a similar mechanistic effect as observed in enzyme chemistry. The heterobimetallic complexes have been successfully applied as highly stereoinducing catalysts in many different types of asymmetric reactions, including the stereoselective formation of C-C, C-O, and C-P bonds.

Keywords: Heterobimetallic catalysts, Lanthanoid complexes, Asymmetric synthesis, Homogenous catalysis

Abbreviations 200

1 Introduction 200

2 Structural Requirements for an Efficient Bimetallic Catalyst: What does the Catalysts Look Like? 201

3 Asymmetric Catalytic C-C Bond Formation Using Heterobimetallic Lanthanoid Complexes 203

3.1 Nitroaldol Reaction 203
3.1.1 Development of an Efficient Catalysis in Model Reactions 203
3.1.2 Enantioselective Catalytic Nitroaldol Reaction: Adducts with One Stereogenic Center 206
3.1.3 Diastereoselective Catalytic Nitroaldol Reaction Starting from Chiral Aldehydes 207
3.1.4 Diastereoselective and Enantioselective Nitroaldol Reaction 209
3.1.5 Tandem Inter-Intramolecular Catalytic Asymmetric Nitroaldol Reaction 210
3.1.6 Recent Improvements (Second Generation of LnLB Catalyst) and Summary 211
3.2 Direct Asymmetric Aldol Reaction 213
3.3 Michael Addition Reaction 213
3.4 Diels-Alder Reaction 217

4 Catalytic Asymmetric C-O Bond Formation Using Heterobimetallic Lanthanoid Complexes: Epoxidation 218

5 Catalytic Asymmetric C-P Bond Formation Using Heterobimetallic Lanthanoid Complexes 220

5.1 Hydrophosphonylation of Aldehydes 220
5.2 Hydrophosphonylation of Imines 223
5.2.1 Hydrophosphonylation of Acyclic Imines 223
5.2.2 Hydrophosphonylation of Cyclic Imines 225

6 Summary 230

7 References and Notes 230

Abbreviations

Ln	lanthanoid metal ion (III)
M	alkali metal ion
L	lithium
S	sodium
P	potassium
B	1,1'-binaphthol (BINOL)
LnMB	lanthanoid-alkali metal-binaphthoxide complex of type LnM_3tris(binaphthoxide)
LLB	lanthanum-lithium-binaphthoxide complex $LaLi_3$tris(binaphthoxide)

1 Introduction

One of the most fascinating aspects in the history of asymmetric catalysis with its countless successful applications in the stereoselective synthesis of a broad variety of functional groups is the structural variety of the complexes which are able to be used as catalysts [1, 2]. Many catalysts have been developed based on different ideas and concepts of mechanistic effect. However, in spite of the abundance of such catalysts which have been successfully applied in asymmetric catalysis, not a handful of them possess multifunctional abilities catalyzing different type of enantioselective reactions. The development of such a type of chiral catalyst, the catalytic effect of which is not limited to one reaction but to different types of asymmetric synthetic organic transformations, remained an attractive challenge for a long time.

In the following, such a desired new and innovative multifunctional catalytic system is reviewed: The chiral heterobimetallic lanthanoid complexes, developed by Shibasaki et al., have recently been shown to catalyze a broad spectra of organic reactions including many "classical" carbon-carbon bond formations like the nitroaldol reaction, Michael addition, aldol reaction and so on, but also an oxidation reaction as well as the asymmetric formation of carbon-phosphorus bonds with excellent stereoselectivity in all cases (Fig. 1) [3–5]. This new concept for catalytic asymmetric reactions, based on the idea of using chiral heterobimetallic lanthanoid complexes which function as both Brønsted base and Lewis acid, just like an enzyme, makes possible a variety of efficient catalytic asymmetric reactions.

2
Structural Requirements for an Efficient Bimetallic Catalyst: What does the Catalysts Look Like?

For a better understanding of many successful applications of the rare earth-alkali metal containing LnM_3tris(binaphthoxide) complexes (LnMB, Ln=rare earth, M=alkali metal) to catalytic asymmetric synthesis, intense investigations also focused on the determination of the structure. It has been shown that these complexes, which can be readily prepared from the corresponding rare earth trichlorides and/or rare earth isopropoxides [5], possess a structure as presented in Fig. 2. This structure was supported by various NMR spectroscopic, MS spectrometric, X-ray crystallographic and other analytic investigations of a variety of LnMB complexes.

Beginning with LDI-TOF mass spectral analysis, these investigations revealed that the structure was in fact a heterobimetallic complex consisting of one lanthanum, three lithium and three BINOL moieties [6]. Although LDI-TOF mass

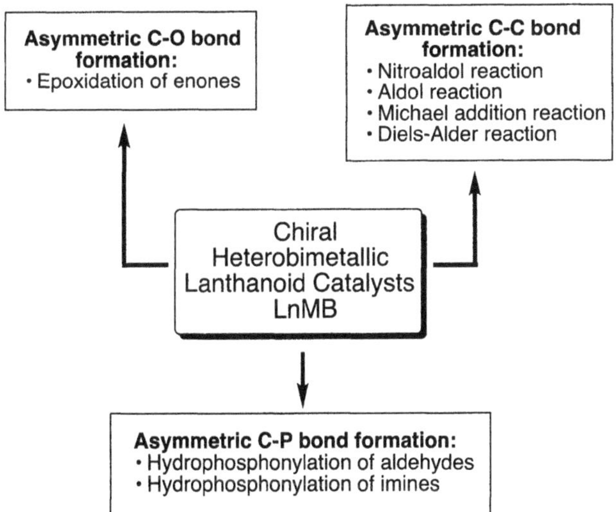

Figure 1. Applications of heterobimetallic lanthanoid complexes in asymetric catalysis

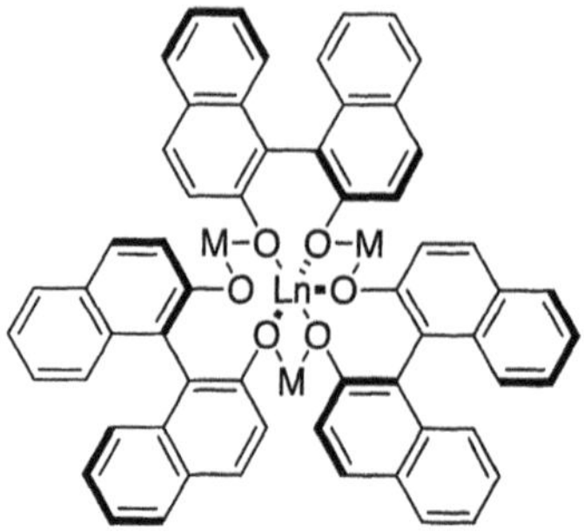

Figure 2. The structure of (*R*)-LnMB. (M=alkali metal, Ln=lanthanoid metal)

spectrometry has a mass accuracy of about ±0.1%, the proposed structural framework was strongly supported by the similarity of the mass spectra of various other rare earth complexes, since rare earth elements have their own atomic weights and isotope abundance distributions. Successful attempts have also been made to obtain X-ray grade crystals of rare earth complexes of type LnLB and LnSB [6–8]. Several rare earth-sodium-BINOL complexes (LnSB, S=sodium) could be crystallized from THF starting from rare earth trichlorides including $LaCl_3$, disodium *(R)*-binaphthoxide, NaOtBu, and H_2O. The elemental analyses suggested a stoichiometry of $LnNa_3$tris(binaphthoxide)-6THF-$2H_2O$, which was confirmed by X-ray crystallographic analyses of La, Pr, Nd, and Eu complexes as being correct. The X-ray crystallographic analyses showed that these crystals had almost the same structure (Fig. 2), with the differences being in the distance of the atoms around the center from the central rare earth metal [6, 7]. Each rare earth *(S)*-BINOL complex has a chiral center at the rare earth metal so that the *(S)*-BINOL complex can exist as a diastereomer. Nevertheless, it is interesting to note that La, Pr, Nd, and Eu complexes starting from *(S)*-BINOL exist only in the Λ-configuration, perhaps due to the greater thermodynamic stability of this configuration compared to the Δ-form.

All the above-mentioned LnSB crystals showed similar mass patterns compared to those of Ln-Li-BINOL complexes, though they contained sodium instead of lithium. The oligomeric structure of the catalysts in the reaction mixture was supported by a slightly positive asymmetric amplification in the asymmetric nitroaldol reaction [9]. In addition, quite recently Shibasaki et al. reported the first X-ray crystallographic structure of a lithium-containing heterobimetallic catalyst of type LnLB, namely the $SmLi_3$tris(binaphthoxide) complex SmLB [8].

Concerning the heterobimetallic potassium lanthanoid catalysts *(R)*-LnPB, a very recent NMR spectroscopic as well as FAB and ESI mass spectrometric study of isolated *(R)*-YbPB complexes provided a clear insight into the assembly of this heterobimetallic complex in solution. At first, the assumption that the proposed structure of *(R)*-YbPB (according to Fig. 2) corresponds with those of isolated crystals *(R)*-YbPB was confirmed [10]. Moreover, the application of these isolated crystals *(R)*-YbPB in an asymmetric hydrophosphonylation reaction (see also Sect. 5.2.2.) gave (nearly) the same good result as the corresponding catalyst solution, indicating that the determined structure of the crystals corresponds with the actual catalytically active species. Such a knowledge of the composition of the heterobimetallic catalysts, namely the potassium derivative *(R)*-YbPB, in solution represents one of the rare cases in catalytic asymmetric synthesis in which the "real catalytically active species" could be isolated, applied, and analyzed by NMR spectroscopical and mass spectrometrical methods.

In summary, it is noteworthy that all of these heterobimetallic asymmetric complexes of type LnMB are stable in organic solvents such as THF, CH_2Cl_2 and toluene which contain small amounts of water, and also are insensitive to oxygen. Using a catalytic amount of LnMB complexes various kinds of asymmetric C-C bond formation reactions, an asymmetric epoxidation, and asymmetric hydrophosphonylations of either imines or aldehydes (catalyzed by LnPB, P=potassium) proceed efficiently to give the corresponding desired products in high stereoselectivity. In detail, these asymmetric reactions will be discussed in the following sections.

3 Asymmetric Catalytic C-C Bond Formation Using Heterobimetallic Lanthanoid Complexes

3.1 Nitroaldol Reaction

3.1.1 *Development of an Efficient Catalysis in Model Reactions*

In 1992, Shibasaki et al. reported for the time an application of chiral heterobimetallic lanthanoid complexes (LnLB) as chiral catalysts in asymmetric catalysis, namely the catalytic asymmetric nitroaldol reaction (Henry reaction), which is one of the most classical C-C bond forming processes [11]. Additionally, this work represents the first enantioselective synthesis of β-nitroalcohol compounds by the way of enantioselective addition of nitroalkanes to aldehydes in the presence of a chiral catalyst. The chiral BINOL based catalyst was prepared starting from anhydrous $LaCl_3$ and an equimolar amount of the dialkali metal salt of BINOL in the presence of a small amount of water [9].

Starting from prochiral aldehydes **1–3**, the desired products b were obtained in good chemical yields and with enantioselectivities up to 90% ee (Scheme 1) [11]. The amount of the catalyst is not shown in Scheme 1 due to the unknown structure of the catalyst (at this time).

Investigations concerning the influence of the rare earth metal component showed pronounced differences both in the reactivity and in the enantioselec-

RCHO + CH_3NO_2 (10 equiv) → La-Li-(S)-BINOL complex, THF, -42 °C, 18 h → R–CH(OH)–CH_2NO_2

1: R = $PhCH_2CH_2$
2: R = iPr
3: R = cyclohexyl

4: 79% (73% ee), R = $PhCH_2CH_2$
5: 80% (85% ee), R = iPr
6: 91% (90% ee), R = cyclohexyl

Scheme 1. The first catalytic asymmetric nitroaldol reaction catalyzed by chiral lanthanoid complexes.

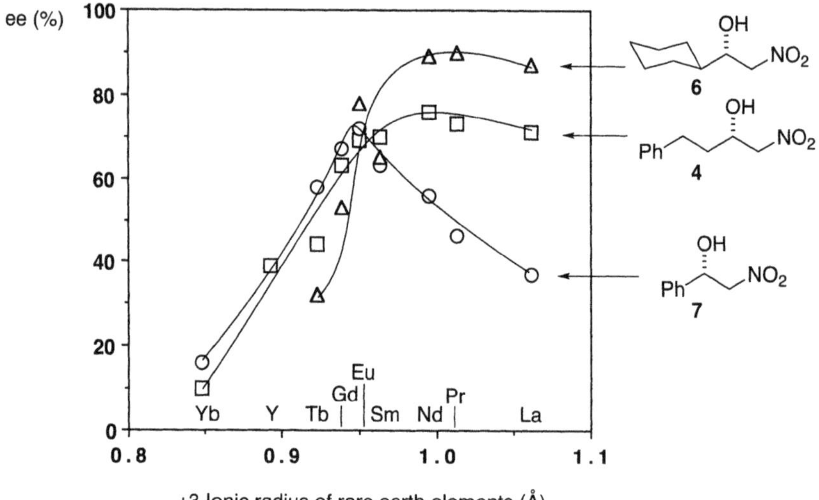

Figure 3. Effects of the ionic radii of rare earth elements on the enantioselectivity.

tivity among the various rare earth metals used [12]. When benzaldehyde and nitromethane were used as starting materials, the corresponding Eu complex gave 7 in 72% ee (91%) in contrast to 37% ee (81%) in the case of the La complex (–40 °C, 40 h). The unique relationship between the ionic radii of rare earth metals and the enantioselectivities of several nitroaldols **4, 6, 7** are depicted in Fig. 3.

Consequently, small changes in the structure of the catalyst (ca. 0.1 Å in ionic radius of the rare earth cation) cause drastic changes in the optical purity of the produced nitroaldols. Although in general nitroaldol reactions are regarded as equilibrium processes, no detectable retro-nitroaldol reaction was observed in the Ln-BINOL complex catalyzed asymmetric nitroaldol reactions.

The proposed mechanism for the asymmetric nitroaldol reaction catalyzed by heterobimetallic lanthanoid complexes is shown in Scheme 2 [5]. In the initial step, the nitroalkane component is deprotonated and the resulting lithium nitronate coordinates to the lanthanoid complex under formation of the inter-

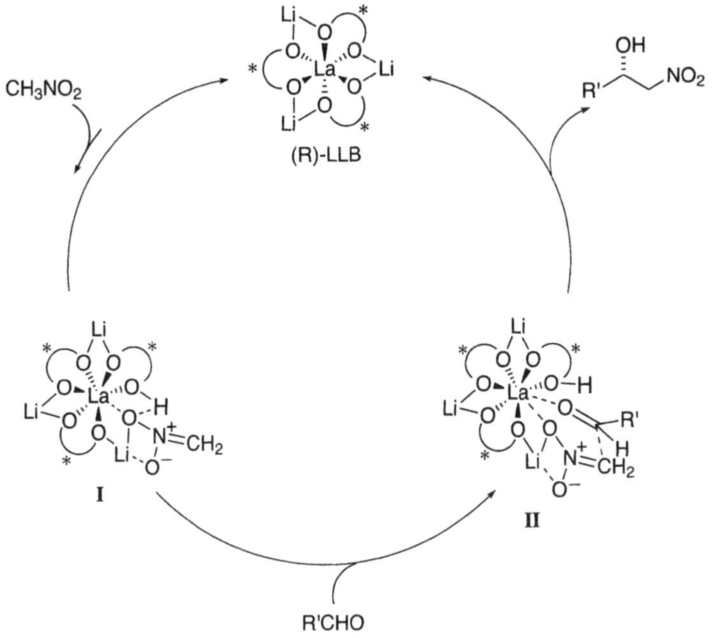

Scheme 2. A possible mechanism for catalytic asymmetric nitroaldol reactions with nitromethane.

mediate I. [13] Subsequent addition of the aldehyde by coordination of the C=O double bond to the lanthanoid (III) center ion leads to intermediate II, in which the carbonyl function should be attacked by the nitronate via a six-membered transition state (in an asymmetric environment). A proton exchange reaction step will then generate the desired optically active nitroalkanol adduct with regeneration of the "free" rare earth complex LnLB.

In addition to the early results of the general and effective catalytic asymmetric nitroaldol reaction (Scheme 1), which proceeds efficiently in the presence of 3.3 mol% of LLB, the knowledge of the structure of the LnLB complexes (see Sect. 2) led to an extension of this catalytic method to a wide range of further applications which are described below.

The structural modification of the BINOL ligand system also plays an important role concerning stereoselection in the asymmetric Henry reaction. Improved enantioselectivites were obtained using a number of *(R)*-BINOL derivatives in which the 6,6'-positions were substituted [14]. Their utility as asymmetric catalysts was assessed with enantioselectivities up to 88% ee accompanied by chemical yields up to 85% in the nitroaldol reaction of nitromethane with hydrocinnamaldehyde **1**. Whereas the substitution at the 6,6'-position of BINOL proved to be effective in obtaining superior asymmetric catalysts, the use of complexes derived from 3,3'-disubstituted BINOL derivatives [15, 16] gave racemic **4** and BIPOL derived catalyst [17] gave **4** in only 39% ee.

Recently, Okamoto et al. showed that the reactivity and selectivity also depends on the alkali metal component in the heterobimetallic catalysts [18]. Using the bulkier 2-nitropropane as starting material in a model reaction with benzaldehyde, almost no reaction occurred at –30 °C in the presence of the lithium containing catalyst LLB, whereas higher temperatures as well as the use of HMPA as a co-solvent only led to racemic product. However, in the presence of the corresponding potassium-containing catalyst LPB the desired reaction proceeded with 46% ee. In contrast, the use of LLB was connected with superior enanioselectivity and chemical yield (compared to LPB) when replacing 2-nitropropane by the less bulkier nitromethane (LLB: 91% yield; 48% ee; LPB: 71% yield; 6% ee) [18].

3.1.2
Enantioselective Catalytic Nitroaldol Reaction: Adducts with One Stereogenic Center

A first example of an efficient application of the LnLB catalyzed nitroaldol reaction as key step in a multi-step syntheses was presented by Shibasaki et al. in the asymmetric approach to three kinds of optically active β-blockers **10, 13,** and **16** (Scheme 3) [12, 19–21].

Using **14** and 10 mol equiv of nitromethane at –50 °C in the presence of 3.3 mol% of *(R)*-LLB catalyst, a 76% yield of nitroaldol **15** in 92% ee was obtained. Reductive alkylation of the nitroaldol **15** to **16** was accomplished by a PtO_2 catalyzed hydrogenation. Thus, *(S)*-(–)-pindolol **16** was synthesized in only four steps from 4-hydroxyindole [20]. Interestingly, the nitroaldols **9, 12,** and **15** were found to have *(S)*-configuration when *(R)*-LLB was used. The nitronates thus appear to react preferentially with the Si face of the aldehydes, in the oppo-

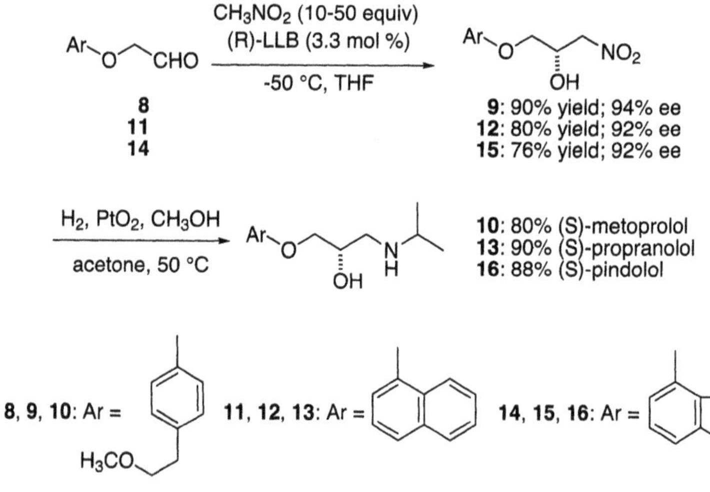

CH_3NO_2 (10-50 equiv)
(R)-LLB (3.3 mol %)
-50 °C, THF
8, **11**, **14**
9: 90% yield; 94% ee
12: 80% yield; 92% ee
15: 76% yield; 92% ee
H_2, PtO_2, CH_3OH
acetone, 50 °C
10: 80% (S)-metoprolol
13: 90% (S)-propranolol
16: 88% (S)-pindolol
8, 9, 10: Ar =
11, 12, 13: Ar =
14, 15, 16: Ar =

Scheme 3. Catalytic asymmetric synthesis of β-blockers using (*R*)-LLB as a catalyst.

site sense to the enantiofacial selectivity which might have been expected on the basis of the previous results (cf. Scheme 1). These results suggested that the presence of an oxygen atom at the β-position had a pronounced influence on the enantiofacial selectivity.

The LLB type catalysts were also successfully applied in the asymmetric nitroaldol reaction of quite unreactive α,α-difluoro aldehydes. However, catalytic asymmetric nitroaldol reaction of a broad variety of α,α-difluoro aldehydes proceeded satisfactorily when using the heterobimetallic asymmetric catalysts with modified, 6,6'-disubstituted BINOL ligands [22]. The best results were obtained with the samarium (III) complex (5 mol%) generated from 6,6'-bis{(triethylsilyl)ethynyl}BINOL with enantioselectivities up to 95% ee. The *(S)*-configuration of one representative nitroaldol adduct showed that the nitronate reacted preferentially on the Si face of aldehyde in the presence of *(R)*-LLB (20 mol%; 74% yield; 55% ee). It is noteworthy that the enantiotopic face selection for α,α-difluoro aldehydes is reverse to that for nonfluorinated aldehydes. The stereoselectivity for α,α -difluoro aldehydes is identical with that of β-oxa-aldehydes, suggesting that the fluorine atoms at the α-position have a great influence on enantioface selection.

3.1.3
Diastereoselective Catalytic Nitroaldol Reaction Starting from Chiral Aldehydes

The diastereoselective catalytic nitroaldol reaction has been investigated starting from optically active α-amino aldehyde, e.g. **17**. The adducts of type **18** are attractive intermediates for the synthesis of unnatural erythro-amino-2-hydroxy acids, which are important components of several biologically active compounds. For example, the promising HIV-protease inhibitor KNI-272 [23, 24] contains *(2S,3S)*-3-amino-2-hydroxy-4-phenylbutanoic acid (erythro-AHPA, **19**) as a subunit. A conventional diastereoselective synthesis in the presence of achiral bases led to limited internal induction with *erythro/threo* ratios of **18** in the range between 62:38 and 74:26. The use of the achiral complex $La(O^iPr)_3$ gave the product **18** in an 89:11 *erythro/threo* ratio [25]. However, this limitation of diastereoselection has been overcome using catalytic amounts of lithium-containing heterobimetallic complexes LnLB (Scheme 4).

In the presence of *(R)*-LLB (3.3 mol%), the treatment of *N*-phthaloyl-L-phenylalanal **17** with nitromethane at –40 °C gave practically a single stereoisomer of *(2R,3S)*-2-hydroxy-4-phenyl-3-phthaloylamino-1-nitrobutane **18** in 92% yield (>99:1 *erythro*-selectivity) [25]. Interestingly, reaction of the *(S)*-aldehyde **17** with nitromethane, using the *(S)*-LLB complex as a catalyst, led to a reduced diastereo- and enantioselectivity (96% yield; *erythro/threo* 74:26; 90% ee(*erythro*)). The conversion of the nitroaldol adduct **18** into **19** was achieved in one pot (80% yield).

A further example of a diastereoselective nitroaldol reaction using heterobimetallic lanthanoid complexes as catalysts was recently reported by Okamoto et al. [18] in connection with a novel approach to 1α,24*(R)*-dihydroxyvitamin D_3

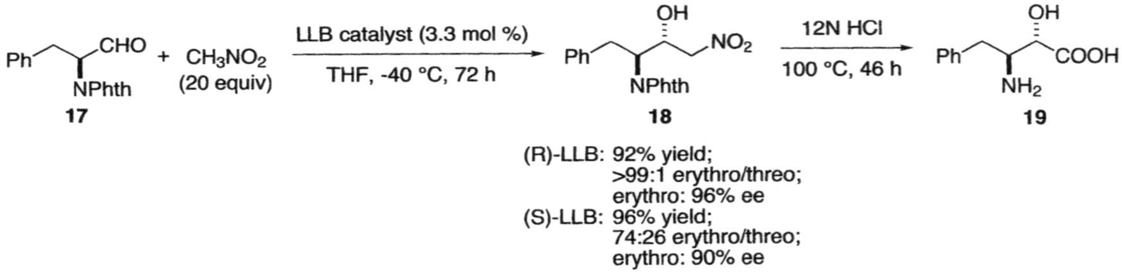

Ph–CH(NPhth)–CHO **17** + CH_3NO_2 (20 equiv) → [LLB catalyst (3.3 mol %), THF, -40 °C, 72 h] → **18** → [12N HCl, 100 °C, 46 h] → **19**

(R)-LLB: 92% yield; >99:1 erythro/threo; erythro: 96% ee

(S)-LLB: 96% yield; 74:26 erythro/threo; erythro: 90% ee

Scheme 4. Diastereoselective nitroaldol reaction as key step in the synthesis of *erythro*-AHPA **19**.

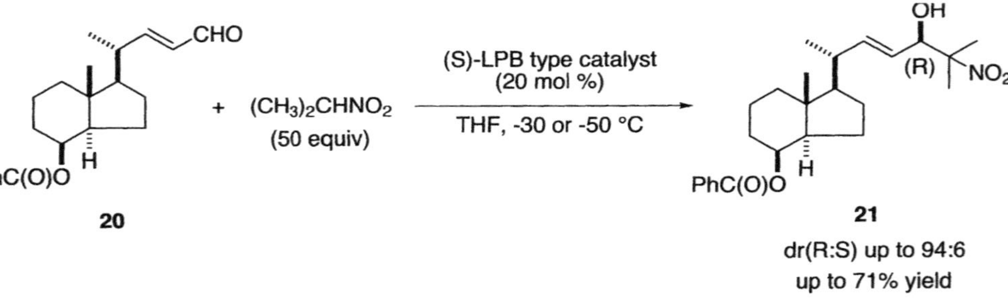

Scheme 5. Catalytic diastereoselective nitroaldol reaction promoted by the LPB type catalyst.

[26], which is an active analogue of vitamin D_3 and induces keratinocyte differentiation [27, 28]. Herein, several rare earth metal complexes were used to catalyze the nitroaldol reaction of the CD-ring 24-aldehyde precursor **20** with 2-nitropropane (Scheme 5). In accordance with the results of the corresponding model reaction with benzaldehyde when using 2-nitropropane as starting material (see Sect. 3.1.1.), the best results were achieved in the presence of the potassium containing lanthanoid complex of type LPB with *(S)*-6,6'-{(triethylsilyl)ethynyl}BINOL as ligand. The desired nitroaldol adduct **21** was formed in yields up to 71% and with diastereomeric ratios dr up to 94:6 (Scheme 5). It is noteworthy, that a conjugate double bond in the aldehyde component was needed for good asymmetric induction.

3.1.4
Diastereoselective and Enantioselective Nitroaldol Reaction

LnLB type catalysts are also able to promote diastereoselective and enantioselective nitroaldol reactions starting from prochiral materials. In preliminary work, LLB gave unsatisfactory results in terms of both diastereoselectivity (*syn*/*anti* ratio 63 : 37 to 77 : 23) and enantioselectivity (<78% ee) [19]. However, an effective asymmetric induction was obtained in the presence of LL(B-a) type catalysts containing 6,6'-substituted BINOL (Scheme 6; for the structure of B-a, see Scheme 7).

The application of the catalysts of type LL(B-a) (3.3 mol%) to diastereoselective nitroaldol reactions led to high *syn*-selectivity and enantioselectivity [29]. In all cases, much higher *syn*-selectivity (*syn*/*anti* ratio up to 94:6) and enanti-

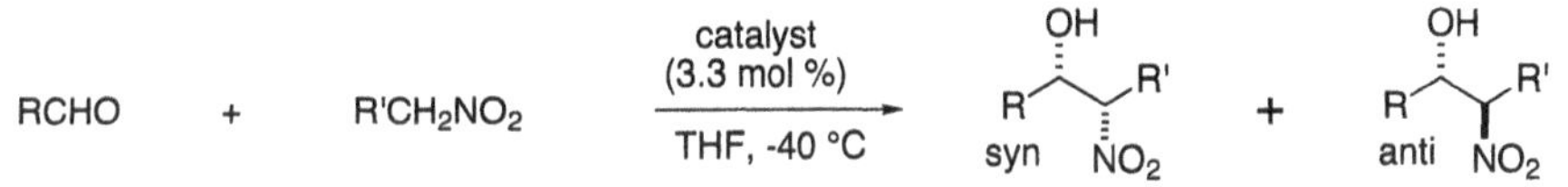

RCHO + $R'CH_2NO_2$ —catalyst (3.3 mol %), THF, -40 °C→ syn + anti (OH, R, R', NO_2)

1 R = $PhCH_2CH_2$
2: R = $CH_3(CH_2)_4$

23: R' = Et
24: R' = CH_2OH

25(syn), **26** (anti): R = $PhCH_2CH_2$, R' = Et
27(syn), **28** (anti): R = $PhCH_2CH_2$, R' = CH_2OH
29(syn), **30** (anti): R = $CH_3(CH_2)_4$, R' = CH_2OH

Reagent	Catalyst	Time [h]	Product	Yield [%]	*syn*/*anti*	ee (*syn*) [%]
1 + 23	LLB	138	**25 + 26**	89	85 : 15	87
1 + 23	LL(B-a)	138	**25 + 26**	85	93 : 7	95
1 + 24	LLB	111	**27 + 28**	62	84 : 16	66
1 + 24	LL(B-a)	111	**27 + 28**	97	92 : 8	97
22 + 24	LLB	93	**29 + 30**	79	87 : 13	78
22 + 24	LL(B-a)	93	**29 + 30**	96	92 : 8	95

Scheme 6. Diastereoselective and enantioselective nitroaldol reaction.

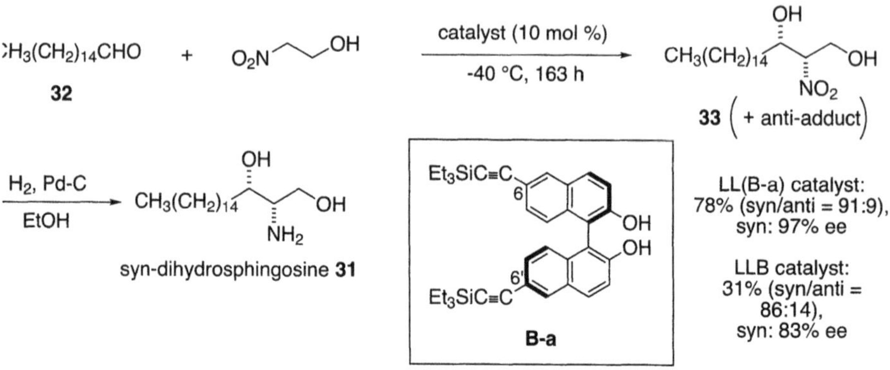

Scheme 7. Catalytic asymmetric synthesis of *syn*-dihydrosphingosine.

oselectivity (up to 97% ee) was obtained using the catalysts with 6,6'-substituted BINOL instead of LLB (for comparison representative results are given in Scheme 6). The optical purities of the minor anti-adducts **26**, **28**, and **30** were lower than those of the *syn*-adducts **25**, **27**, and **29**, indicating that the former were not generated by epimerization of the nitro group. In fact, treatment of the *syn*-adducts with catalysts such as LLB and its derivatives resulted in near-quantitative recovery of the starting materials with unchanged optical purities.

The *syn*-selective asymmetric nitroaldol reaction was successfully applied to the catalytic asymmetric synthesis of *threo*-dihydrosphingosine **31**, which elicits a variety of cellular responses by inhibiting protein kinase C (Scheme 7) [30]. Nitroaldol reaction of hexadecanal **32** with 3 equiv of nitroethanol catalyzed by LL(B-a) gave the corresponding nitroaldol adduct **33** in high syn-selectivity (91:9) and 78% yield, with the *syn*-adduct **33** being obtained with up to 97% ee [29]. In this case, under similar conditions the LLB-catalyzed reaction proceeded only slowly to give an 86:14 ratio of the syn and *anti*-adducts in 31% yield (with lower optical purity: 83% ee). The hydrogenation of **33** in the presence of 10% Pd on charcoal afforded *threo*-dihydrosphingosine **31** in 71% yield.

3.1.5
Tandem Inter-Intramolecular Catalytic Asymmetric Nitroaldol Reaction

The asymmetric catalytic nitroaldol reaction was furthermore successfully extended to the field of asymmetric tandem reactions [31]. Tandem reactions are especially useful to construct compounds with several chiral centers in a one-pot synthesis starting from simple achiral components in the presence of a chiral catalyst. The first tandem inter-intramolecular catalytic asymmetric nitroaldol reaction has been realized in the reaction of the cyclopentanedione derivative **34** with nitromethane using a catalytic amount of LnLB according to Scheme 8 [31].

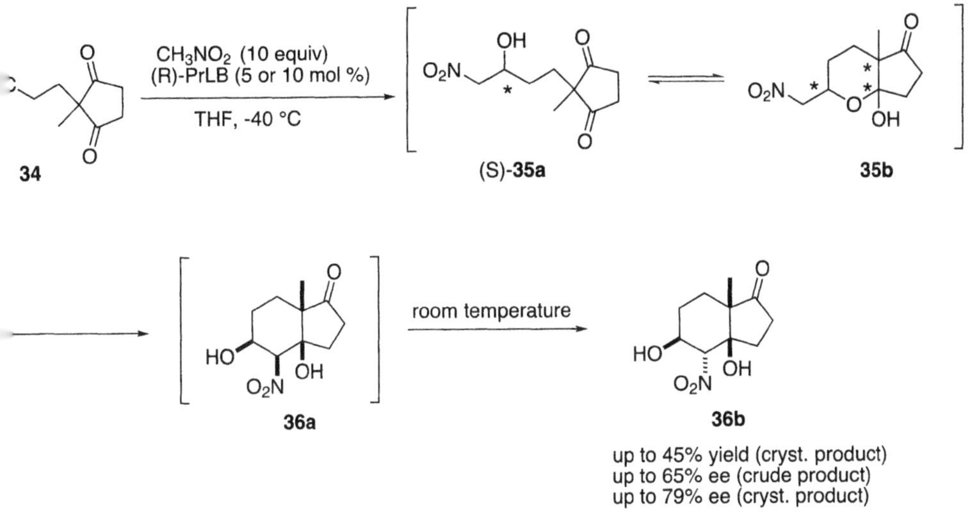

CH_3NO_2 (10 equiv)
(R)-PrLB (5 or 10 mol %)
THF, -40 °C

34 (S)-35a 35b

room temperature

36a 36b

up to 45% yield (cryst. product)
up to 65% ee (crude product)
up to 79% ee (cryst. product)

Scheme 8. Tandem inter-intramolecular catalytic asymmetric nitroaldol reaction.

In addition to temperature effects, the optical purity of the product **36b** strongly depends on the lanthanoid center ion. In the presence of *(R)*-PrLB complex (5 mol%) as the most efficient catalyst, the hexahydro-1-indanone derivative **36b** was formed with enantioselectivities up to 65% ee (for comparison: LLB (10 mol%; -20 °C): 39% ee; YbLB (10 mol%; -20 °C): 7% ee) [31]. After crystallization, **36b** was isolated with up to 79% ee and 41% yield.

3.1.6
Recent Improvements (Second Generation of LnLB Catalyst) and Summary

The catalytic asymmetric nitroaldol reactions promoted by LLB or its derivatives require at least 3.3 mol% of asymmetric catalysts for efficient conversion. However, even in the case of 3.3 mol% of catalyst, reactions are rather slow. Attempts were made to reduce the required catalytic amount and accelerate the reactions, which led to a second-generation heterobimetallic lanthanoid catalyst (LLB-II), prepared from LLB, 1 mol equiv of H_2O, and 0.9 mol equiv of butyllithium. The use of only 1 mol% of LLB-II efficiently promoted catalytic asymmetric nitroaldol reactions and additionally LLB-II (3.3 mol%) accelerated these reactions [32]. A comparison of the efficiency of LLB (or LL(B-a)) and the second-generation catalysts LLB-II (or LL(B-a)-II) is given in Scheme 9. The structure of LLB-II has not yet been unequivocally determined. However, it appears that it is a complex of LLB and LiOH.

Using a second-generation LnLB catalyst consisting of 6,6'-bis{(trimethylsilyl)ethynyl}BINOL and Sm, an efficient catalytic asymmetric synthesis of arb-

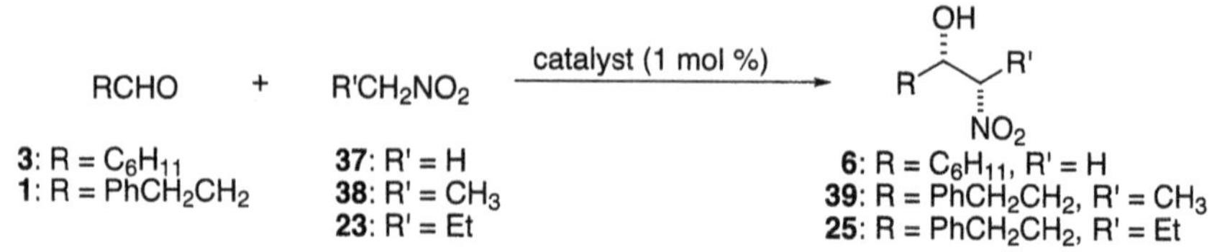

RCHO + R'CH$_2$NO$_2$ → (catalyst (1 mol %)) → product (OH, NO$_2$)

3: R = C$_6$H$_{11}$
1: R = PhCH$_2$CH$_2$

37: R' = H
38: R' = CH$_3$
23: R' = Et

6: R = C$_6$H$_{11}$, R' = H
39: R = PhCH$_2$CH$_2$, R' = CH$_3$
25: R = PhCH$_2$CH$_2$, R' = Et

Reagent	Catalyst	Time [h]	Temp. [°C]	Product	Yield [%]	*syn/anti*	ee (*syn*) [%]
3 + **37**	LLB	24	-50	6	5.6	–	88
3 + **37**	LLB-II	24	-50	6	73	–	89
1 + **38**	LL(B-a)	113	-30	39	25	70 : 30	62
1 + **38**	LL(B-a)-II	113	-30	39	83	89 : 11	94
1 + **23**	LL(B-a)	166	-40	25	trace	–	–
1 + **23**	LL(B-a)-II	166	-40	25	84	95 : 5	95

Scheme 9. Comparison of the catalytic activity of LLB and second-generation LLB (LLB-II) or LL(B-a) and LL (B-a)-II. [LLB-II: LLB + H_2O (1 mol equiv) + BuLi (0.9 mol equiv.); LL(B-a)-II: LL(B-a) + H_2O (1 mol equiv) + BuLi (0.9 mol equiv)]; For the structure of the BINOL derivative ligand B-a, see Scheme 7

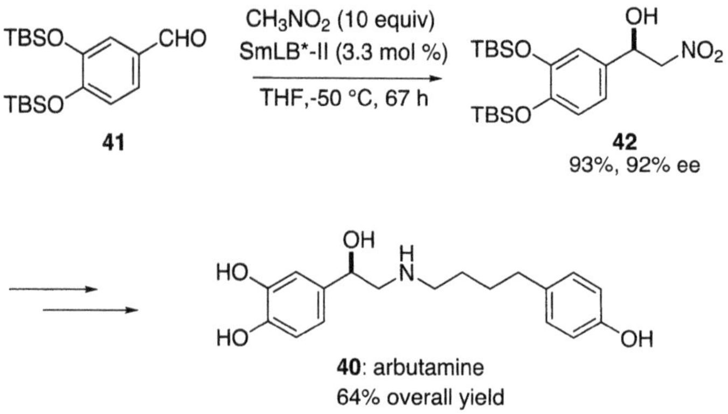

TBSO, TBSO, CHO **41** → (CH$_3$NO$_2$ (10 equiv), SmLB*-II (3.3 mol %), THF,-50 °C, 67 h) → TBSO, TBSO, OH, NO$_2$ **42** 93%, 92% ee → → HO, HO, OH, H N, OH

40: arbutamine
64% overall yield

Scheme 10. A catalytic asymmetric synthesis of arbutamine. [SmLB*-II = SmLi$_3$tris(*R*)-6,6'-bis(trimethylsilylethynyl)binaphthoxide) + H_2O (1.0 mol equiv to Sm) + BuLi (0.6 mol equiv to Sm)].

utamine **40**, which is a useful β-agonist [33, 34], was achieved (Scheme 10) [8]. In the key step, the nitroaldol adduct **42** was formed in 93% yield and with 92% ee [8].

In conclusion, since the first example of a catalytic asymmetric nitroaldol reaction (Henry reaction) was reported in 1992 by Shibasaki et al., this reaction has been developed into a highly efficient synthetic method for the stereoselective synthesis of nitroalkanols. Using alkali metal-containing heterobimetallic lanthanoid complexes as catalysts, a broad variety of nitroalkanol derivatives containing one, two, or more stereogenic centers have been constructed in a highly stereoselective manner. Until now, this new and innovative catalytic concept has been applied to the synthesis of several biologically active and pharmaceutically interesting compounds (or their precursors).

3.2 Direct Asymmetric Aldol Reaction

Although the development of a range of catalytic asymmetric aldol-type reactions has proven to be a valuable contribution to asymmetric synthesis [35–37], in all of these reactions pre-conversion of the ketone moiety to a more reactive species such as an enol silyl ether, enol methyl ether, or ketene silyl acetal has been an unavoidable necessity. However, quite recently Shibasaki et al. reported that a direct catalytic asymmetric aldol reaction, which is known in enzyme chemistry, is also possible in the presence of heterobimetallic lanthanoid catalysts [38]. Using *(R)*-LLB (20 mol%), which shows both Lewis acidity and Brønsted basicity similar to the corresponding aldolases, the desired optically active aldol adducts were obtained with up to 94% ee. A variety of aldehydes and unmodified ketones can be used as starting materials (Scheme 11).

Moreover, although it is known that aldol reactions that utilize acetone as a starting material are generally difficult to control, this reaction proceeds well in the presence of LLB (20 mol%) and 10 equiv of acetone to give the aldol adduct with up to 74% ee and in 82% yield. The postulated mechanism of this direct catalytic aldol reaction is presented in Scheme 12.

Therein, the lanthanum center ion (III) should function as a Lewis acid activating the aldehyde, whereas the lithium binaphthoxide moiety act as Brønsted base moiety. The synergetic effect of both groups, as can be seen in intermediate III, appeared to be responsible that the reaction proceeds without any activation of the starting materials, especially the ketone component.

3.3 Michael Addition Reaction

Catalytic asymmetric Michael reactions are one of the most important synthetic methods for obtaining asymmetric centers by enantioselective construction of carbon-carbon bonds [39, 40]. The first lanthanoid complex catalyzed Michael

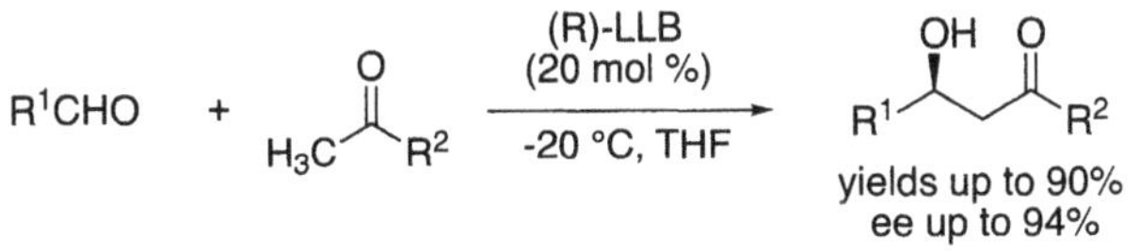

R^1CHO + H_3C–C(=O)–R^2 —(R)-LLB (20 mol %), -20 °C, THF→ R^1–CH(OH)–CH_2–C(=O)–R^2

yields up to 90%
ee up to 94%

Scheme 11. LLB-catalyzed direct aldol reaction.

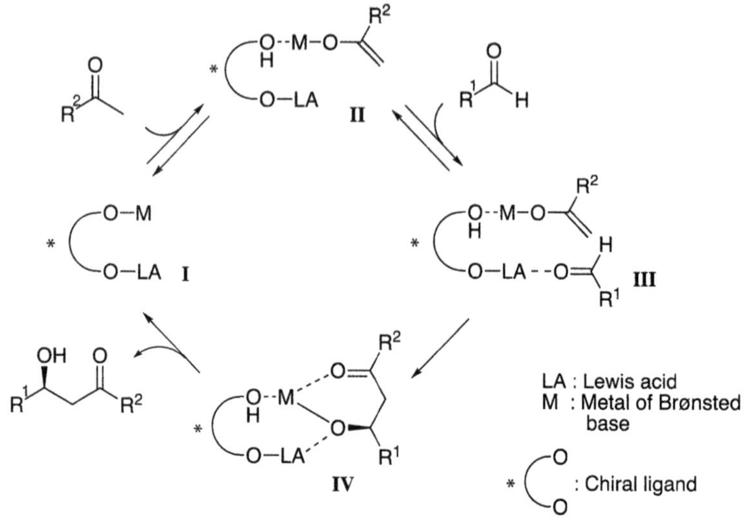

Scheme 12. Catalytic Cycle of Direct Catalytic Asymmetric Aldol Reactions.

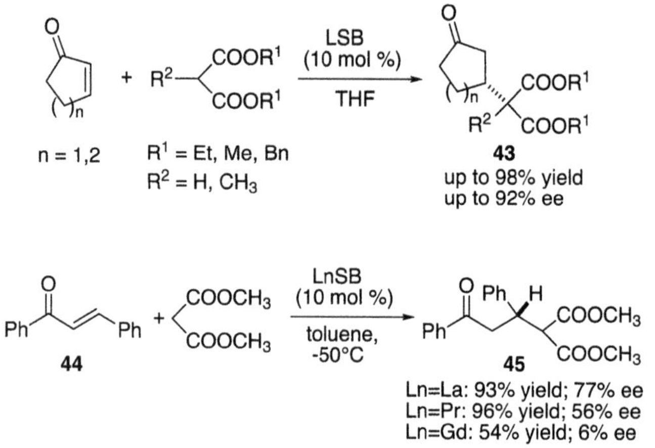

Scheme 13. LnSB-catalyzed asymmetric Michael addition.

addition reaction was realized using malonates as Michael donor in the catalytic asymmetric Michael reaction of various enones [41, 42]. Although ineffective as an asymmetric catalyst for nitroaldol reactions, LSB (L=lanthanum, S=sodium) was found to be effective in this case, giving Michael adducts **43** in up to 92% ee and almost quantitative yield (Scheme 13) [41]. In general, the use of THF as a solvent led to the best results, whereas in the case of the corresponding LSB-catalyzed reaction of *trans*-chalcone **44** the use of toluene was essential to give the product in good enantiomeric excess. Center metal effects were also investigated for this Michael reaction, indicating that LSB was the best catalyst for catalytic asymmetric Michael reactions.

In order to clarify the nature of the interaction between the enone and the asymmetric catalyst, the complexation was studied by ^{1}H-NMR spectroscopy after mixing cyclohexenone and the asymmetric bimetallic complexes and observing the chemical shift of the α-proton of cyclohexenone [41]. Interestingly, it was found that complexation with LSB induced a small downfield shift on the α-proton of cyclohexenone whereas PrSB, a moderately effective asymmetric catalyst for Michael reactions, induced a large upfield shift. In strong contrast, in the case of either EuSB or LLB, which gave only near-racemic Michael adducts, the ^{1}H-NMR spectra showed no changes in chemical shift of the α-proton of cyclohexenone. These NMR studies indicated that the carbonyl group of the enone coordinated to lanthanum and/or praseodymium metal in the LnSB molecule, while the enone did not coordinate to either LLB and/or EuSB. These changes of chemical shift were observed even in the presence of dimethyl malonate. The chemical phenomena described above might be understood by considering the differing dihedral angles of binding of the BINOL moiety to the center metal in each case.

Additionally, computational simulations of the enantioselection process using Rappé's universal force field (UFF) [43–45] were carried out which clearly indicated that the *(R)*-LSB complex complexes better as a pro-*(R)* adduct than as pro-*(S)* adduct (ΔE=4.9 kcal/mol) [41]. The proposed catalytic cycle is shown in Scheme 14. Thus, the basic LSB complex also acts as a Lewis acid, controlling the orientation of the carbonyl function and so activating the enone to attack. It ap-

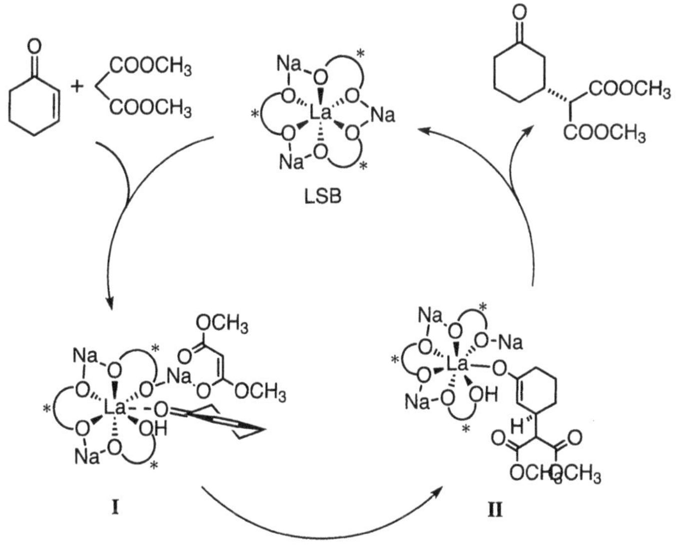

Scheme 14. Proposed catalytic cycle for the asymmetric Michael reaction promoted by LSB.

pears that the multifunctional nature of the LSB catalyst makes possible the formation of Michael adducts with high ees even at room temperature. As final step the resultant sodium enolates **II** of the optically active Michael adducts appear to abstract a proton from an acidic OH so as to regenerate the LSB catalyst. In both catalytic asymmetric Michael reactions and nitroaldol reactions, enones and/or aldehydes appear to coordinate to the rare earth metal. The reason why LSB is more effective for catalytic asymmetric Michael reactions whereas LLB is more effective for catalytic asymmetric nitroaldol reactions is still unclear at present. However, it seems that slight differences in bond lengths in chelate structures such as **I** and **II**, as well as slight differences in "bite" angle for the BINOL moiety caused by varying the alkali metal component may be responsible for this effect.

Another type of an LSB catalyzed asymmetric Michael reaction, in which the asymmetric center is induced on the side of the adduct originating from the Michael donor, was also reported by Shibasaki et al. (Scheme 15) [46]. In a preliminary study, it was found that the reaction of **46** with 3-buten-2-one in THF using 10 mol% of LSB gave **47** with 23% ee, while carrying out the reaction in toluene afforded **47** with 75% ee. However, when the amount of LSB was reduced to 5 mol%, the enantiomeric excess of **47** declined to a more modest 25% ee. To offset this decline while still maintaining the lower level of catalyst slow addition of **46** was essential. Accordingly, the use of syringe pump methods gave **47** with high enantiomeric excess (89% ee). The reaction was further improved when using CH_2Cl_2 as a solvent. Thus, the asymmetric Michael reaction catalyzed by 5 mol% of LSB in CH_2Cl_2 gave **47** in 89% yield and with 91% ee, without the need for slow addition previously encountered (Scheme 15). In addition, in this case the catalytic asymmetric Michael reaction for **47** was not so affected by the choice of rare earth metal.

In conclusion, as shown in Scheme 15, slow addition of β-keto ester and the use of CH_2Cl_2 as solvent are generally quite effective methods for preventing reduction of enantiomeric excess for the various Michael adducts. On the other

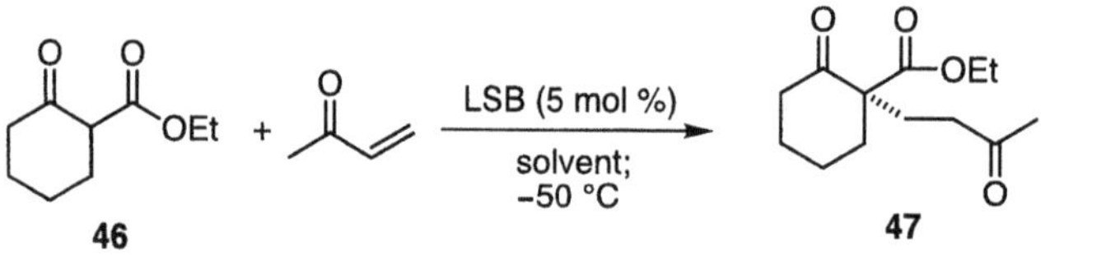

OEt + → LSB (5 mol %) solvent; -50 °C → OEt

46 **47**

solvent	yield	ee
toluene	83%	25%
toluene[a)]	76%	89%
CH_2Cl_2	89%	91%

a) Slow addition of nucleophile with syringe-pump methods over 8h

Scheme 15. LSB-catalyzed Michael addition of **46** with methyl vinyl ketone.

hand, malonates give the adducts with high ees regardless of the solvent used [41]. These results can be rationalized by comparison of the pKa of a β-keto ester with that of a malonate; the former is significantly more acidic than the latter. Thus, the concentration of the resulting Na-enolate can be expected to be greater in the case of the β-keto ester, and moreover this Na-enolate will react with an enone more slowly than the Na-enolate derived from a malonate. We suggest that this combination of more rapid formation and longer lifetime increases the likelihood of dissociation of the Na-enolate from the chelated ensemble, thus giving a product of lower ee. On the other hand, in less polar CH_2Cl_2 the Na-enolate would, even in this case, remain part of the ensemble, thereby affording the product with high ee (Fig. 4). Furthermore, it appears that slow addition of the β-keto ester also acts to limit undesired ligand exchange between BINOL moieties and the Michael donor.

In both types of catalytic asymmetric Michael reactions, the use of either second-generation LSB or 6,6'-substituted BINOL derived LSB type catalysts did not result in significantly improved results.

3.4 Diels-Alder Reaction

A further application of the heterobimetallic lanthanoid catalysts of the LLB type to the field of catalytic asymmetric Diels-Alder reactions [47, 48] was also achieved by Shibasaki et al. [49]. In general, LLB type complexes are multifunctional asymmetric catalysts, showing both Brønsted basicity and Lewis acidity. Nevertheless, in this study the use of LLB type catalysts acting as asymmetric Lewis acids alone was examined and led to the development of an LLB (type) catalyzed asymmetric Diels-Alder reaction [49]. Representative results for the catalytic asymmetric Diels-Alder reactions using **48** and cyclopentadiene in toluene as a solvent are shown in Scheme 16.

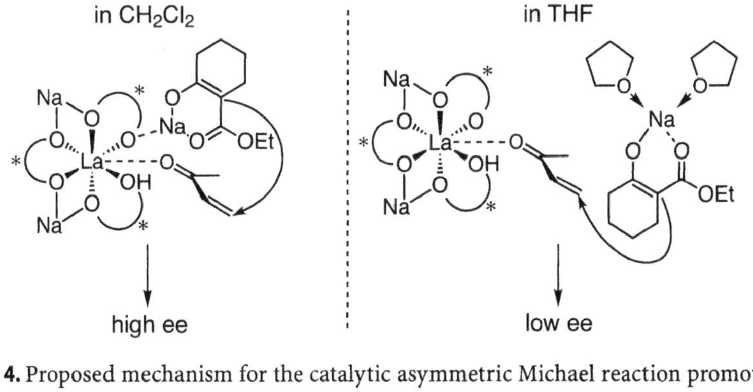

Figure 4. Proposed mechanism for the catalytic asymmetric Michael reaction promoted by LSB.

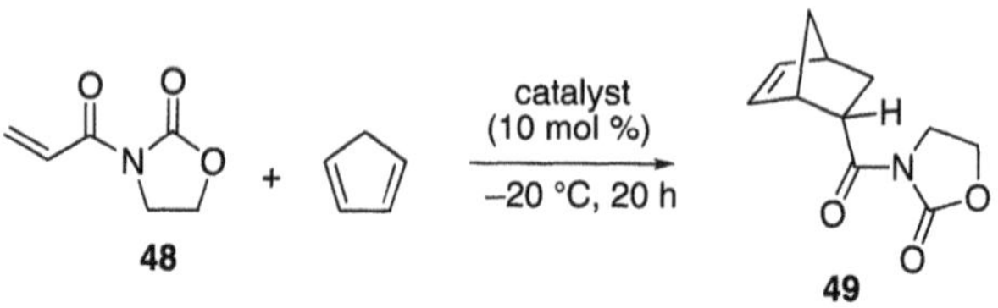

catalyst (10 mol %)
−20 °C, 20 h

48 49

	(R)-LLB	(R)-LLB*[a]
yield:	82%	100%
endo/exo:	15:1	36:1
ee:	63%	86%

a) LLB*=LaL_3tris(6,6'-dibromobinaphth-oxide)

Scheme 16. Catalytic asymmetric Diels-Alder reaction.

Compared to the result with LLB, the use of a 6,6'-dibromosubstituted BINOL derived LLB type catalyst LLB* led to significantly improved yield, *endo:exo* ratio and enantioselectivity (86% ee). Interestingly, the addition of 12-crown-4 to the reaction medium resulted in the formation of the adduct **49** with much lower enantiomeric excess. This result appears to suggest that the lithium cation(s) play a key role in activation of the dienophile.

4 Catalytic Asymmetric C-O Bond Formation Using Heterobimetallic Lanthanoid Complexes: Epoxidation

Catalytic asymmetric epoxidations are one of the most important asymmetric processes [1]. In addition to previous successful achievements of other groups with allylic alcohols [50, 51] and unfunctionalized olefins [52–55], the first efficient catalytic asymmetric epoxidation of a variety of enones with broad generality has been developed when using chiral lanthanoid complexes [56].

In the presence of the sodium-containing heterobimetallic catalyst *(R)*-LSB (10 mol%), the reaction of enone **52** with TBHP (2 equiv) was found to give the desired epoxide with 83% ee and in 92% yield [56]. Unfortunately LSB as well as other bimetallic catalysts were not useful for many other enones. Interestingly, in marked contrast to LSB an alkali metal free lanthanoid BINOL complex, which was prepared from $Ln(O\text{-}i\text{-}Pr)_3$ and *(R)*-BINOL or a derivative thereof (1 or 1.25 molar equiv) in the presence of MS 4A (Scheme 17), was found to be applicable to a range of enone substrates. Regarding enones with an aryl-substituent in the α-keto position, the most effective catalytic system was revealed when using a lanthanum-*(R)*-3-hydroxymethyl-BINOL complex La-**51** (1–5 mol%) and cumene hydroperoxide (CMHP) as oxidant. The asymmetric epoxidation proceeded with excellent enantioselectivities (ees between 85 and 94%) and yields up to 95%.

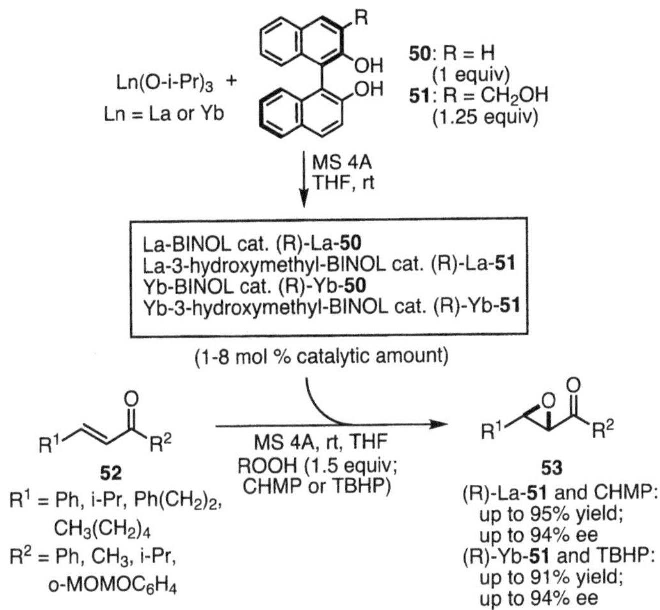

Scheme 17. Catalytic asymmetric epoxidation of enones.

In contrast to this result, the enones with an alkyl moiety in the α-keto position were best converted to the corresponding epoxides when replacing lanthanum by ytterbium in the corresponding catalytically active complex of the above mentioned type. Using TBHP (1.5 equiv) as oxidant in the presence of 5 mol% of Yb-**51** gave enantioselectivities up to 94%. However, the use of either Yb-**50** catalyst or a La-CMHP system afforded the product of type **53** with less satisfactory results. It seems likely that the difference in ionic radius between lanthanum and ytterbium, as well as the difference in Lewis acidities, accounts for the observed center metal effects.

Although the structure of the catalytically active species could not have been unequivocally determined, it was found that an almost 1:1 ratio of $Ln(O\text{-}i\text{-}Pr)_3$ (Ln=La or Yb) and BINOL gave the maximum enantiomeric excesses. In addition, an asymmetric amplification of the catalytic asymmetric epoxidation has been obtained (Fig. 5) [56], which led to the conclusion that an oligomeric structure of these lanthanoid-BINOL catalysts may play a key role in these catalytic asymmetric epoxidations of enones. Regarding the mechanism, a Ln-alkoxide moiety in the catalysts appears to act as a Brønsted base, activating a hydroperoxide moiety so as to make possible a Michael reaction. At the same time another Ln metal ion seems to act as a Lewis acid, both activating and controlling the orientation of the enone. Furthermore, it is noteworthy that this catalytic asymmetric epoxidation can be carried out at room temperature using 1–8 mol% of an asymmetric catalyst to give epoxides with good enantiomeric excesses.

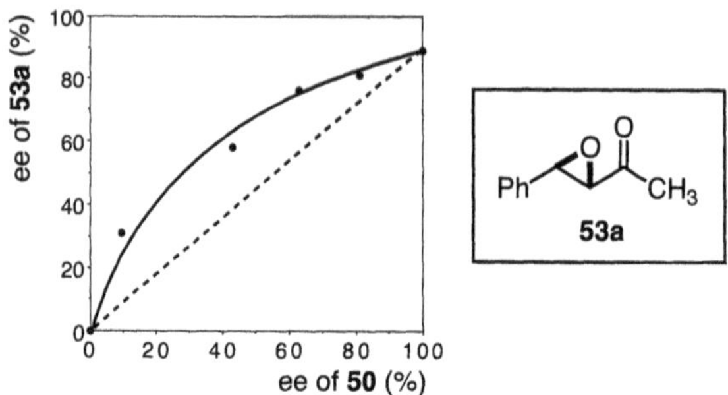

Figure 5. Asymmetric amplification in the Yb-**50** catalyzed epoxidation reaction to **50a**.

5 Catalytic Asymmetric C-P Bond Formation Using Heterobimetallic Lanthanoid Complexes

The application of the heterobimetallic lanthanoid complexes of the LnMB type led to a breakthrough in establishing a highly efficient asymmetric catalytic route to α-hydroxy as well as α-amino phosphonic acid esters, which have attracted much attention due to their wide ranging biological activity [57–64]. The heterobimetallic catalysis described below represents the first and until now the only highly efficient asymmetric catalytic approach to both α-hydroxy and α-amino phosphonates by the attractive way of asymmetric catalytic hydrophosphonylation.

5.1 Hydrophosphonylation of Aldehydes

For the first time an LLB catalyzed enantioselective hydrophosphonylation of aldehydes has been independently reported by Shibuya et al. and Rath and Spilling in the early 1990s [65, 66]. Due to the unknown catalyst structure of LLB at this time, and to the fact that the purity of LLB catalyst utilized by this groups appeared to be rather low, only low to modest ees have been achieved using LLB. According to Shibuya et al. and Rath and Spilling, the hydrophosphonylation of benzaldehyde gave the corresponding α-hydroxy phosphonates in less than 30% ee [65, 66]. However, under similar conditions Shibasaki et al.'s reinvestigation of the catalytic asymmetric hydrophosphonylation of benzaldehyde and cinnamaldehyde with 1.3 equiv of dimethyl phosphite in the presence of 10 mol% of LLB gave the corresponding α-hydroxy phosphonates in strongly improved 76% ee (79% yield) and in 72% ee (78% yield), respectively [67]. It is noteworthy that with slow addition of the aldehydes, the enantiomeric excesses of both products rose up to 83% ee (73% yield) and 79% ee (88% yield) respectively.

The reaction strongly depends on the solvent, increasing the ee from 4 to 16 to 79% ee when changing the solvent from diethyl ether to dichloromethane to THF, which has been shown to be the solvent of choice. The LLB catalyzed hydrophosphonylation was most efficient in the case of aromatic aldehydes of low reactivity or arylated α,β-unsaturated aldehydes as starting materials, with enantioselectivities in the range of 84–95% ee (Scheme 18). A comparison of the catalytic ability of LLB with its relative aluminum complex ALB showed that LLB is superior for most of the aldehyde systems. Interestingly, however, for electron-deficient aromatic aldehydes the aluminum-lithium-complex ALB gave better results (e.g. in the case of *p*-nitrobenzaldehyde: 36% ee vs 71% ee). Thus, LLB and ALB can be used in a complementary manner for the catalytic asymmetric hydrophosphonylation of aldehydes. In contrast to other reactions, herein the use of neither 6,6'-bis((triethylsilyl)ethynyl)BINOL nor the second-generation LLB catalyst gave a significant improvement in the results.

The above-mentioned effects of slow addition of the aldehydes on the enantioselection can be best explained as follows. Heterobimetallic catalysts such as LLB are believed to activate both nucleophiles and electrophiles. For the hydrophosphonylation of comparatively unreactive aldehydes the activated phosphite can react only with aldehydes which are pre-coordinated to lanthanum. However, in the case of reactive aldehydes such as benzaldehyde and cinnamaldehyde, the Li-activated phosphite may be able to undergo a competing reaction with the unactivated aldehyde. If such aldehydes are added in one portion, the ee of the product will be reduced. Slow addition of aldehyde, in contrast, has the effect of maximizing the ratio of activated to unactivated aldehyde present in solution, by allowing time for the catalytic cycle to complete and regenerate the catalyst, thereby facilitating aldehyde activation. Reactive aldehydes should, therefore, be added slowly in order to avoid the side reaction which proceeds without activation of the aldehyde by LLB (Scheme 19).

A detailed investigation into the hydrophosphonylation of substituted benzaldehydes as well as heteroaromatic aldehydes has been recently reported by Shibuya et al. [68] Therein this group found an interesting effect of the *p*-substituent on enantioselectivity. In agreement with the results reported by Shibasaki

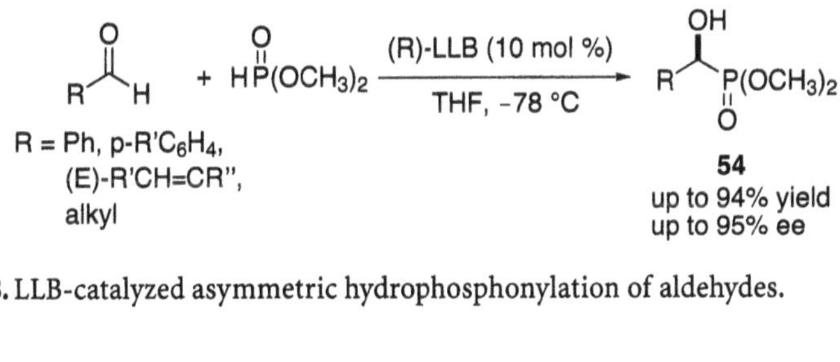

RCHO + HP(O)(OCH3)2 —(R)-LLB (10 mol %), THF, -78 °C→ RCH(OH)P(O)(OCH3)2

R = Ph, p-R'C6H4, (E)-R'CH=CR", alkyl

54
up to 94% yield
up to 95% ee

Scheme 18. LLB-catalyzed asymmetric hydrophosphonylation of aldehydes.

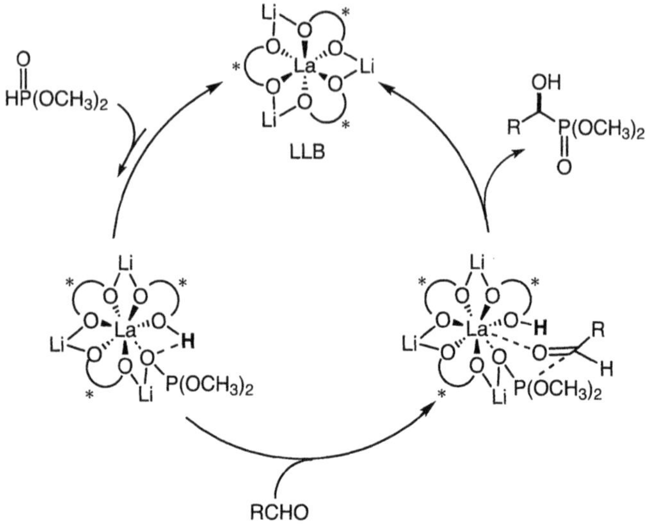

Scheme 19. A proposed catalytic cycle for asymmetric hydrophosphonylation.

et al. [67], good results were obtained when using *p*-methoxybenzaldehyde in the presence of LLB with up to 82% ee and excellent yields [68]. A significant influence of the lanthanoid metal center ion and the alkali metal component was also found, underlining that the choice of the rare earth metal and alkali metal component in the Shibasaki catalyst is a crucial factor in inducing good enantioselectivity. Interestingly, compared to LLB the use of either EuLB or SmLB as asymmetric catalyst gave the reverse enantioselection (in case of *p*-methoxy benzaldehyde), although with low ee (8 and 9%, respectively).

The role of the aromatic ring within a substrate was investigated in the LLB-catalyzed hydrophosphonylation of a range of heteroaromatic aldehydes **55** (X= O,S) with diethyl phosphite. Good yields were obtained for all products **56**, although enantioselectivities strongly varied from 18 to 73% ee (Scheme 20). The enantioselectivities obtained from the reactions with aldehydes bearing a thiophene nucleus were found to be higher than those of the corresponding furancarboxaldehyde derivatives. Regarding the series of reactions with aldehydes which possess the same heterocycle, the numerical value of the ee of the products increased with an increase in the numerical value of the super-delocalizability $S_r^{(E)}$ [69, 70] at the carbonyl oxygen within the substrates (see Scheme 20). These results reveal that the electrophilic coordination of the lanthanum atom in a presumed LLB-phosphite complex to the carbonyl atom of the aldehyde might be one critical factor governing the enantioselectivity.

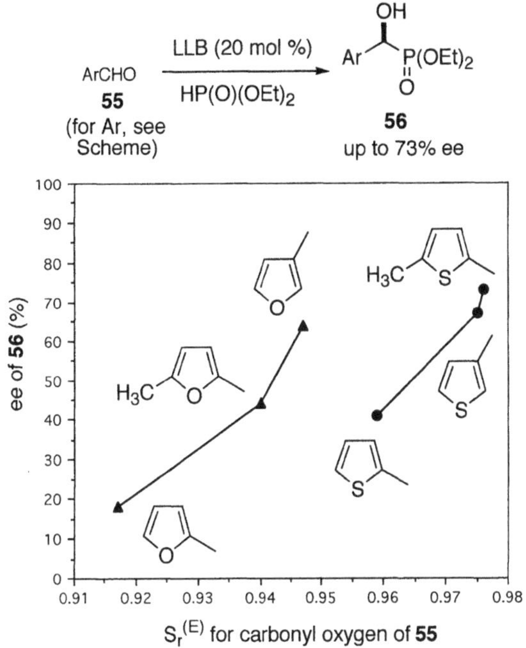

Scheme 20. LLB-catalyzed asymmetric hydrophosphonylation of heteroaromatic aldehydes.

5.2
Hydrophosphonylation of Imines

5.2.1
Hydrophosphonylation of Acyclic Imines

α-Amino phosphonic acids are interesting compounds for the use in the design of enzyme inhibitors [59, 60]. As in the case of α-hydroxy phosphonic acids, the absolute configuration of the α-carbon strongly influences the biological properties. Although several (especially diastereoselective) methods for the synthesis of optically active α-amino phosphonic acids have been known for a long time [71, 72], the first catalytic asymmetric hydrophosphonylation of imines has been reported recently by Shibasaki et al. [73] using potassium-containing LnK_3tris(binaphthoxide) complexes (LnPB) as most efficient catalysts.

Interestingly, almost the first results in asymmetric hydrophosphonylation with acyclic imines revealed that the lanthanum-potassium-BINOL catalyst LPB was more effective than the analog sodium and lithium complexes LSB and LLB, both of which have been shown to be highly efficient in asymmetric C-C bond formations (see Sect. 3). As a representative example, in the presence of LPB (20 mol%) and 5 equiv of dimethyl phosphite the hydrophosphonylation of imine **57a** proceeds at room temperature under formation of the desired product in 91% ee, whereas the use of LLB and LSB gave decreased enantioselectivities (LSB: 49% ee; LLB: 38% ee) and yields. Moreover, attempts to improve the efficiency of this asymmetric catalysis were successful. In the presence of only 5 or 10 mol% catalytic amount of LPB and 1.5 equiv of phosphite, another α-amino phosphonate (bearing a benzhydryl group as *N*-substituent) was obtained in excellent 92 or 96% ee and good yields up to 82% when using **57b** as imine component (Scheme 21).

The broad generality of this asymmetric hydrophosphonylation method using catalytic amounts of LPB was shown by the effective conversion of several types of imines to the corresponding optically active α-amino phosphonates in satisfactory to excellent enantioselectivities up to 96% ee and yields up to 87%. Concerning the substitution pattern of the imine component, the authors focused on the use of alkyl and alkenyl groups as C-substituents, whereas the benzhydryl group was used as *N*-substituent in most of the cases due to the possibility to cleave this group from the products in order to obtain the pure α-amino phosphonic acids with a primary amino function. In case of two imines, the use of GdPB or PrPB gave the best results. Once again, these results confirm impressively the ability of the LnMB type catalysts as a flexible asymmetric catalytic system which can be easily modified and "designed" according to the needs for the desired reaction by changing lanthanoid center ion, alkali metal component and/or BINOL derivative. However, in this reaction neither the use of second-generation LPB catalyst nor of LPB derivatives derived from modified BINOLs gave improved results.

The proposed mechanism of this catalytic asymmetric hydrophosphonylation is shown in Scheme 22. The first step of this reaction is the deprotonation of dimethyl phosphite by LPB to generate the potassium dimethyl phosphite. This potassium phosphite immediately coordinates to a rare earth to give **I** due to the strong oxophilicity of rare earth metals [74]. **I** then reacts with an imine to give an optically active potassium salt of the α-amino phosphonate, which leads via proton-exchange reaction to an α-amino phosphonate and LPB, thereby completing the catalytic cycle and giving the desired asymmetric hydrophosphonylation.

$HP(O)(OCH_3)_2$, (R)-LPB (5 - 20 mol %)

57 → **58**

a: R = $CH(p\text{-}CH_3OC_6H_4)_2$
b: R = $CHPh_2$

up to 82% yield
up to 96% ee

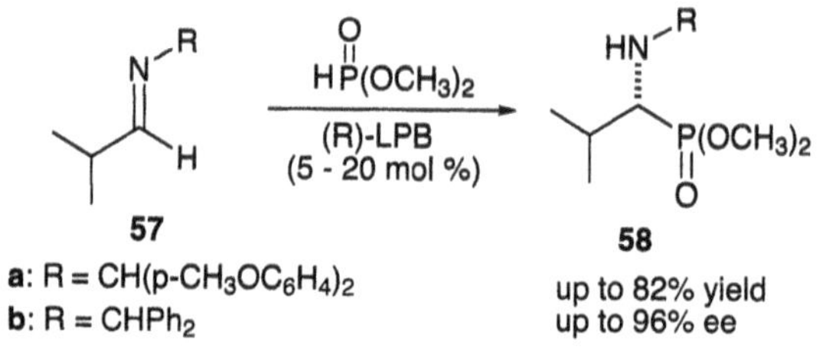

Scheme 21. Asymmetric hydrophosphonylation of acyclic imines catalyzed by (*R*)-LnPB.

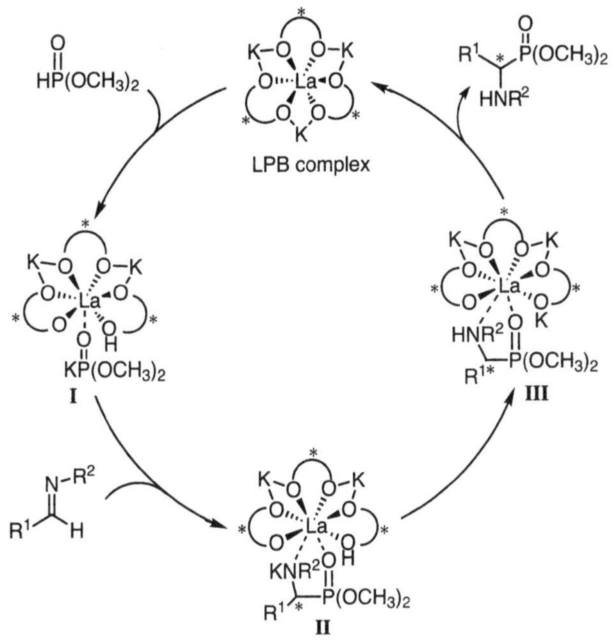

Scheme 22. Proposed catalytic cycle.

In conclusion, the lanthanoid complex catalyzed hydrophosphonylation of acyclic imines represents an efficient method to produce optically active α-amino phosphonates in modest to high enantiomeric excess.

5.2.2
Hydrophosphonylation of Cyclic Imines

In spite of the high level of interest in the asymmetric synthesis of α-amino phosphonic acids, less is known in the case of cyclic α-amino phosphonates. Although in recent years descriptions of promising pharmaceutical applications for cyclic compounds (and acylated derivatives thereof) have been published [61–64], until now no efficient general asymmetric route has been available to prepare this class of α-amino phosphonates. Several attempts at a diastereoselective synthetic route which were made by the addition of a stoichiometric amount of chiral phosphites to cyclic imines, namely thiazolines, gave only limited diastereoselection ratios of *dr*=2:1 or below [75, 76].

The first efficient enantioselective synthetic route to cyclic α-amino phosphonates, namely the pharmaceutically interesting heterocyclic phosphonates of type **1** [63, 64], by an asymmetric addition reaction of dialkyl phosphites to heterocyclic imines was recently developed by Shibasaki et al. (Scheme 23) [10, 77].

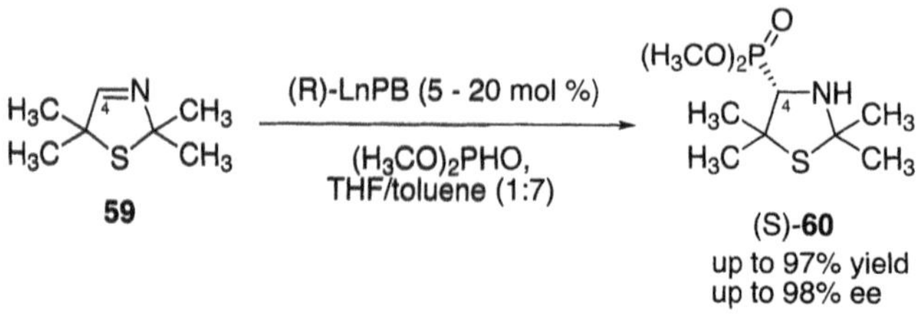

59 → (S)-60: (R)-LnPB (5 - 20 mol %), $(H_3CO)_2PHO$, THF/toluene (1:7); up to 97% yield, up to 98% ee

Scheme 23. Enantioselective hydrophosphonylation catalyzed by (*R*)-LnPB complexes.

Therein, potassium-containing heterobimetallic lanthanoid complexes have been chosen as catalysts and thiazolines of type **59** were used as the imine substrates. To produce the optically active α-amino phosphonate *(S)*-**60**, the model compound 2,2,5,5-tetramethyl-3-thiazoline **59** was treated with 5 equiv of dimethyl phosphite in the presence of several kinds of chiral lanthanoid-potassium-binaphthoxide complexes [*(R)*-LnPB]. At first, 20 mol% of LaK_3tris(binaphthoxide) (*(R)*-LPB) in THF/toluene (1:7) at room temperature was used, which has been shown to be the most efficient catalytic system in the asymmetric hydrophosphonylation of acyclic imines (see Sect. 5.2.1). However, only a modest enantioselectivity of 61% ee accompanied by a 53% chemical yield was observed in the formation of *(S)*-**60** using this method. The efficiency of the reaction was improved by increasing the reaction temperature to 50 °C. Investigations of the influence of further lanthanoid metal components in the catalyst structure revealed that, in the presence of Sm, Gd, Dy, and Yb, the ee values rose to 97% ee accompanied by good to high chemical yields [10, 77]. The characteristic relationship between the ionic radii of the rare earth metal ions (III) and the enantioselectivity is summarized in Fig. 6. The functional course of this graph supports the idea that *(R)*-LnMB catalysts represent a multifunctional, highly flexible catalyst system by varying their incorporated components.

In contrast to the asymmetric hydrophosphonylation of acyclic imines, the rare earth metals with lower ionic radii in the range of [Yb(III)] to [Gd(III)] were connected with the highest enantioselectivities of approximately 95% ee (whereas high optical purities with acyclic imines were achieved with La(III), a lanthanoid metal with a relatively large ionic radii). The asymmetric hydrophosphonylation of **59** with *(R)*-LPB was found to be limited, with only 64% ee. In agreement with the hydrophosphonylation of acyclic imines, potassium was needed as the alkali metal component in the complex *(R)*-LnMB to obtain acceptable results. In order to identify a highly efficient catalytic system which could maintain a high ee and yield even in the presence of small catalytic amounts, the influence of reduced catalytic amounts has been investigated. In the case of the *(R)*-SmPB catalyzed asymmetric hydrophosphonylation, a steady decrease in chemical yields was observed with smaller catalytic amounts, whereas nearly unchanged high chemical yields were obtained when reducing the concentration of the *(R)*-YbPB catalyst from 20 to 15 to 10 mol%. A further decrease in the

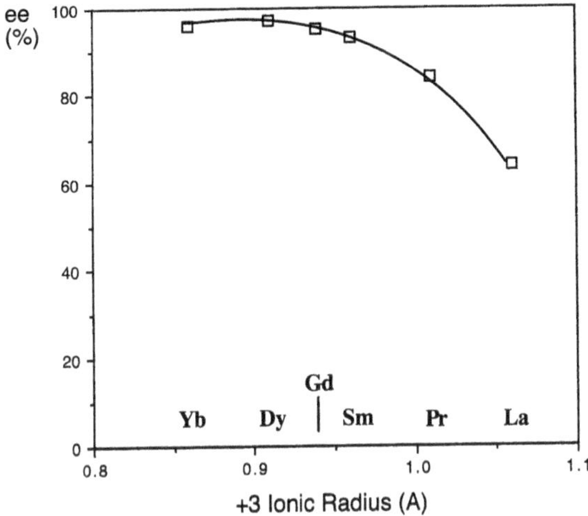

Figure 6. Effect of Lanthanoid Ionic Radii on Enantioselectivity (corresponding to the reaction shown in Scheme 23).

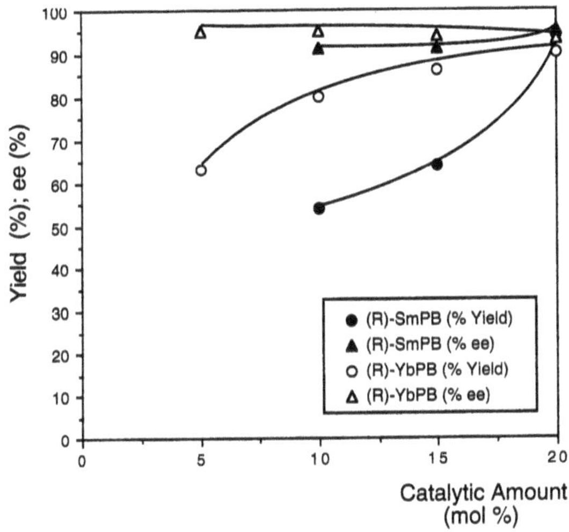

Figure 7. Influence of the catalytic amount on asymmetric hydrophosphonylation of imine **59.**

concentration of the catalyst to 5 mol% *(R)*-YbPB gave *(S)*-**60** in a still satisfactory 63% yield. In all cases the enantioselectivity of the *(R)*-YbPB-catalyzed reaction was approximately 95% ee. The functional dependence of the chemical yield and enantiomeric excess from the catalytic amount is shown schematically in Fig. 7.

Considering the superior asymmetric catalysis properties of La-Li-6,6'-disubstituted BINOL complexes in the enantioselective nitroaldol reaction, investigations were carried out with corresponding ytterbium catalysts. The use of the 6,6'-bis(methoxy)binaphthoxide derivative as ligand gave *(S)*-**60** in improved 81% chemical yield while maintaining high enantioselectivity (96% ee; catalyic amount: 5 mol%) [10]. Moreover, the flexibility of the optimized catalytic conditions with regard to other substituted thiazolines was investigated using 5 mol% catalytic amount of *(R)*-YbPB in connection with a reaction temperature of 50 °C and THF/toluene (1:7) as a solvent. After a 48 h reaction time all corresponding α-amino phosphonates were synthesized with high enantioselectivities (up to 96% ee) nearly independently of the substituents at the thiazoline ring system. In addition, at these reaction conditions (48 h reaction time) hydrophosphonylation of the model component **59** gave the corresponding product in further improved 88% chemical yield (compared to 63% after 40 h).

The proposed mechanistic course is shown in Scheme 24 [10]. As the first step, an interaction between the P=O bond's oxygen and the Yb(III) center ion might occur, which results in the formation of a lanthanoid/phosphite complex **I**. The preferred coordination of phosphite (instead of the theoretically also possible imine bond at the nucleophile N-atom) to Yb(III) is due to the high oxophilicity of lanthanide(III) ions. However, in structure **I** the phosphorus atom doesn't show any nucleophilic abilities, which are necessary for a nucleophilic attack on the C=N double bond. Consequently, a tautomeric rearrangement of structure **I** should take place, leading to a Ln(III)-coordinated phosphite form **IIb**. Therein, the λ^3 phosphorus atom shows enhanced nucleophilic character caused by the lone pair, which now allows a nucleophilic attack on the C=N double bond of the thiazolines. The nucleophilicity of the phosphorus atom should be further increased by (partial) deprotonation and an additional coordination of the resulting anion to the potassium in the intermediate **IIa**. Consequently, the P-nucleophile reacts (in a high enantioselective manner) with the C=N double bond of the thiazoline **59** to form the chiral potassium salt of the 4-thiazolidinylphosphonate **III**. A subsequent proton-exchange reaction step produces the α-amino phosphonate *(S)*-**60** and *(R)*-LnPB, which are connected to each other in structure **IV**. The final step of the catalytic cycle is achieved by a dissociation of the α-amino phosphonate *(S)*-**60** from the rare earth complex **IV**, which regenerates the "free" catalyst *(R)*-LnPB.

As shown in Fig. 8, two possibilities are conceivable for the addition step of the imine to the lanthanoid-phosphite complex **IIa**. To determine whether structure **V** or **VI** seems to be the more reasonable in the addition reaction of a dimethyl phosphite with the C=N double bond of cyclic imines, several hydrophosphonylation experiments using different types of phosphites were carried out in the presence or absence of the Lewis acid boron trifluoride. As it was shown that a high level of Lewis acid activation of the imine is required independently from the nucleophilicity of the phosphorus nucleophile which was used, a transition state of type **VI** appeared to exist as the dominant transition state structure.

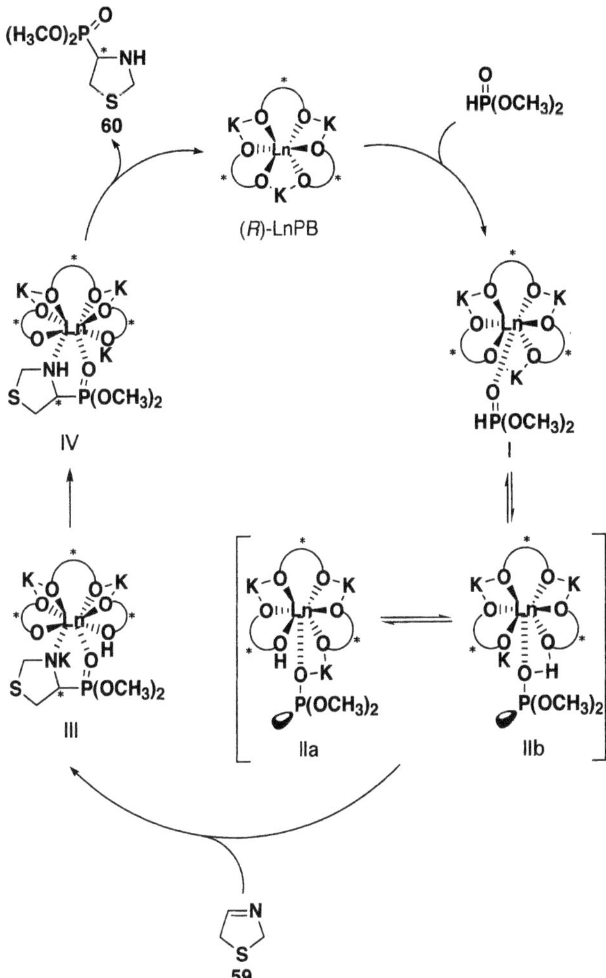

Scheme 24. Proposed reaction mechanism (for graphical reasons, the substituents at the thiazoline **59** are not shown in this scheme).

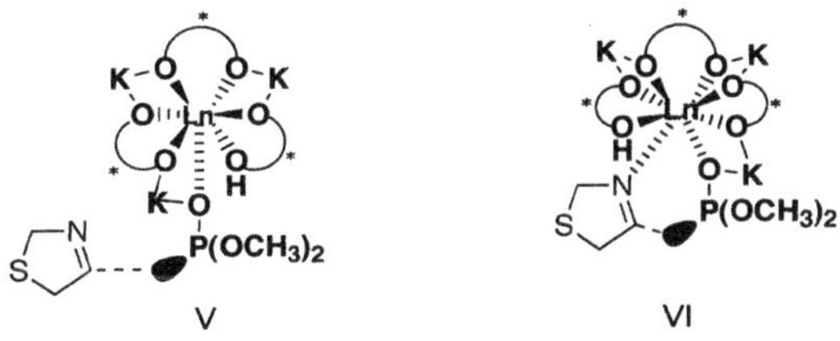

Figure 8. Two possibilities for coordination of imine in the addition step.

6 Summary

In conclusion, chiral heterobimetallic lanthanoid compexes LnMB, which were recently developed by Shibasaki et al., are highly efficient catalysts in stereoselective synthesis. This new and innovative type of chiral catalyst contains a Lewis acid as well as a Brønsted base moiety and shows a similar mechanistic effect as observed in enzyme chemistry. A broad variety of asymmetric transformations were carried out using this catalysts, including asymmetric C-C bond formations like the nitroaldol reaction, direct aldol reaction, Michael addition and Diels-Alder reaçtion, as well as C-O bond formations (epoxidation of enones). Thereupon, asymmetric C-P bond formation can also be realized as has been successfully shown in case of the asymmetric hydrophosphonylation of aldehydes and imines. It is noteworthy that all above-mentioned reactions proceed with high stereoselectivity, resulting in the formation of the desired optically active products in high to excellent optical purity.

7 References and Notes

1. Noyori R (1994) In: Asymmetric catalysis in organic synthesis. Wiley, New York
2. Ojima I (1993) In: Catalytic asymmetric synthesis. VCH, New York
3. Steinhagen H, Helmchen G (1996) Angew Chem 208:2489; Angew Chem Int Ed Engl 35:2337
4. Shibasaki M, Sasai H (1996) Pure & Appl Chem 68:523
5. Shibasaki M, Sasai H, Arai T (1997) Angew Chem 109:1290; Angew Chem Int Ed Engl 36:1236
6. Sasai H, Suzuki T, Itoh N, Tanaka K, Date T, Okamura K, Shibasaki M (1993) J Am Chem Soc 115:10,372
7. Sasai H, Arai T, Satow Y, Houk KN, Shibasaki M (1995) J Am Chem Soc 117:6194
8. Takaoka E, Yoshikawa N, Yamada YMA, Sasai H, Shibasaki M (1997) Heterocycles 46:157
9. Sasai H, Suzuki T, Itoh N, Shibasaki M (1993) Tetrahedron Lett 34:851
10. Gröger H, Saida Y, Sasai H, Yamaguchi K, Martens J, Shibasaki M (1998) J Am Chem Soc (in press)
11. Sasai H, Suzuki T, Arai S, Arai T, Shibasaki M (1992) J Am Chem Soc 114:4418
12. Sasai H, Suzuki T, Itoh N, Arai S, Shibasaki M (1993) Tetrahedron Lett 34:2657
13. Although the lithium nitronate is first generated, there also appears to be a significant possibility that the aldehyde coordinates to La first
14. A 6,6'-disubstituted BINOL derivative was also used in the asymmetric Diels-Alder reaction by:Terada M, Motoyama Y, Mikami K (1994) Tetrahedron Lett 35:6693
15. Lingfelter DS, Hegelson RC, Cram DJ (1981) J Org Chem 46:393
16. Maruoka K, Itoh T, Araki Y, Shirasaka T, Yamamoto H (1988) Bull Chem Soc Jpn 61:2975
17. Yamamoto K, Noda K, Okamoto Y (1985) J Chem Soc Chem Commun 1065
18. Oshida JI, Okamoto M, Azuma S, Tanaka T (1997) Tetrahedron: Asymmetry 8:2579
19. Sasai H, Itoh N, Suzuki T, Shibasaki M (1993) Tetrahedron Lett 34:855
20. Sasai H, Yamada YMA, Suzuki T, Shibasaki M (1994) Tetrahedron 50:12,313
21. Sasai H, Suzuki T, Itoh N, Shibasaki M (1995) Appl Organomet Chem 9:421

22. Iseki K, Oishi S, Sasai H, Shibasaki M (1996) Tetrahedron Lett 37:9081
23. Mimoto T, Imai J, Kisanuki S, Enomoto H, Hattori N, Akaji K, Kiso Y (1992) Chem Pharm Bull 40:2251
24. Kageyama S, Mitsuto T, Murakawa Y, Nomizu M, Ford H Jr, Shirasaka T, Gulnik S, Erickson J, Takada K, Hayashi H, Broder S, Kiso Y, Mitsuya H (1993) Antimicrob Agents Chemother 37:810
25. Sasai H, Kim WS, Suzuki T, Shibasaki M, Mitsuda M, Hasegawa J, Ohashi T (1994) Tetrahedron Lett 35:6123
26. Morisaki M, Koizumi N, Ikekawa N (1975) J Chem Soc Perkin Trans I 1421
27. Tanaka H, Abe E, Miyaura C, Kurihashi T, Konno K, Nishii Y, Suda T (1982) Biochem J 204:713
28. Kobayashi T, Okumura H, Azuma Y, Kiyoki M, Matsumoto K, Hashimoto K, Yoshikawa K (1990) J Dermatol 17:707
29. Sasai H, Tokunaga T, Watanabe S, Suzuki T, Itoh N, Shibasaki M (1995) J Org Chem 60:7388
30. Schwartz GK, Jiang J, Kelsen D, Albino AP (1993) J Natl Cancer Inst 85:402
31. Sasai H, Hiroi M, Yamada YMA, Shibasaki M (1997) Tetrahedron Lett 38:6031
32. Arai T, Yamada YMA, Yamamoto N, Sasai H, Shibasaki M (1996) Chem Eur J 2:1368
33. Young M, Pan W, Wiesner J, Bullough D, Browne G, Balow G, Potter S, Metzner K, Mullane K (1994) Drug Dev Res 32:19
34. Hammond HK, McKirnan MD (1994) J Am Coll Cardiol 23:475
35. Bach T (1994) Angew Chem 106:433; Angew Chem Int Ed Engl 33:417
36. Hollis YK, Bosnich B (1995) J Am Chem Soc 117:4570
37. Gröger H, Vogl EM, Shibasaki M (1998) Chem Eur J 4:1137
38. Yamada YMA, Yoshikawa N, Sasai H, Shibasaki M (1997) Angew Chem 109:1942; Angew Chem Int Ed Engl 36:1871
39. Rossiter BE, Swingle NM (1992) Chem Rev 92:771 and references cited therein
40. Sawamura M, Hamashima H, Shinoto H, Ito Y (1995) Tetrahedron Lett 36:6479 and references cited therein
41. Sasai H, Arai T, Satow Y, Houk KN, Shibasaki M (1995) J Am Chem Soc 117:6194
42. Sasai H, Arai T, Shibasaki M (1994) J Am Chem Soc 116:1571
43. Rappe AK, Casewit CJ, Colwell KS, Goddard III WA, Skiff WM (1992) J Am Chem Soc 114:10,024
44. Casewit CJ, Colwell KS, Rappe AK (1992) J Am Chem Soc 114:10,035
45. Rappe AK, Colwell KS, Casewitt CJ (1993) Inorg Chem 32:3438
46. Sasai H, Emori E, Arai T, Shibasaki M (1996) Tetrahedron Lett. 37:5561
47. Kobayashi S, Ishitani H (1994) J Am Chem Soc 116:4083 and references cited therein
48. Yamamoto H, Maruoka K, Furuta K (1989) In: Schnizer D (ed) Selectivities in Lewis acid promoted reactions. Kluwer, Dordrecht, p281
49. Morita T, Arai T, Sasai H, Shibasaki M (1998) Tetrahedron: Asymmetry 9:1445i
50. Katsuki K, Sharpless KB (1980) J Am Chem Soc 102:5974
51. Hanson R, Sharpless KB (1986) J Org Chem 51:1922
52. Zhang W, Loebach JL, Wilson SR, Jacobsen EN (1990) J Am Chem Soc 112:2801
53. Irie R, Noda K, Ito Y, Matsumoto N, Katsuki T (1990) Tetrahedron Lett 31:7345
54. Yamada T, Imagawa K, Nagata T, Mukaiyama T (1992) Chem Lett 2231
55. For other examples of asymmetric epoxidations, see: Tu Y, Wang ZK, Shi Y (1996) J Am Chem Soc 118:9806 and references therein
56. Bougauchi M, Watanabe S, Arai T, Sasai H, Shibasaki M (1997) J Am Chem Soc 119:2329
57. Budt KH, Jian-Qi L, Peyman A, Ruppert D (1992) Tetrahedron Lett 33:6625
58. Moore ML, Dreyer GB (1993) Perspect Drug Discovery Des 1:85
59. Burke TR Jr, Barchi JJ Jr, George C, Wolf G, Shoelson SE, Yan X (1995) J Med Chem 38:1386

60. Kafarski P, Lejczak B (1991) Phosphorus Sulfur Silicon Relat Elem 63:193
61. Powers JC, Boduszek B, Oleksyszyn J (1995) Georgia Tech Research Corp PCT Int Appl WO 95 29691; Chem Abstr (1996) 124:203,102m
62. Boduszek B, Oleksyszyn J, Kam CM, Selzer J, Smith RE, Powers JC (1994) J Med Chem 37:3969
63. Andrews KJM (1981) Hoffmann-LaRoche. Eur Pat 33,919; Chem Abstr (1982) 96:52,498r
64. Drauz K, Koban HG, Martens J, Schwarze W (1985) Degussa AG. US Pat 4,524,211; Chem Abstr (1984) 101:211,458x
65. Yokomatsu T, Yamagishi T, Shibuya S (1993) Tetrahedron:Asymmetry 4:1783
66. Rath NP, Spilling CD (1994) Tetrahadron Lett 35:227
67. Sasai H, Bougauchi M, Arai T, Shibasaki M (1997) Tetrahedron Lett 38:2717
68. Yokomatsu T, Yamagishi T, Shibuya S (1997) J Chem Soc, Perkin Trans I 1783
69. The $S_r^{(E)}$ values would be a measure of the susceptibility of the substrate to attack by the catalysts based on the distribution of electrons in the frontier orbital
70. Fukui K, Yonezawa T, Nagata C (1954) Bull Chem Soc Jpn 27:423
71. Kukhar VP, Soloshonok VA, Solodenko VA (1994) Phosphorus Sulfur Silicon Relat. Elem 92:239
72. Yager KM, Taylor CM, Smith III AB (1994) J Am Chem Soc 116:9377 and references cited therein
73. Sasai H, Arai S, Tahara Y, Shibasaki M (1995) J Org Chem 60:6656
74. Molander GA (1992) Chem Rev 92:292
75. Hoppe I, Schöllkopf K, Nieger M, Egert K (1985) Angew Chem 97:1066; Angew Chem Int Ed Engl 24:1036
76. Gröger H, Martens J (1996) Synth Commun 26:1903
77. Gröger H, Saida Y, Arai S, Martens J, Sasai H, Shibasaki M (1996) Tetrahedron Lett. 37:9291

Reactions of Ketones with Low-Valent Lanthanides: Isolation and Reactivity of Lanthanide Ketyl and Ketone Dianion Complexes

Zhaomin Hou* and Yasuo Wakatsuki

The Institute of Physical and Chemical Research (RIKEN), Hirosawa 2–1, Wako, Saitama 351–0198, JapanYasuo Wakatsuki
*E-mail: houz@postman.riken.go.jp

Recent progress in the chemistry of structurally well-defined lanthanide ketyl and ketone dianion complexes is reviewed, with particular emphasis on the ligand effects on the reactivity of these complexes. It has been demonstrated that the stability and reactivity of the ketyl radical and ketone dianion species strongly depend on the steric and electronic properties of the ancillary ligands, the structure of their parent ketones, as well as the nature of the metals to which they are bound. Fine-tuning these factors can control the reactivity of these species. Generation and reactions of dianionic thioketone and imine species are also briefly described.

Keywords: Ketyl radicals, Ketone dianions, Lanthanides, pinacol coupling, Ligand effects

1 **Introduction** 234

2 **One-Electron Reduction of Ketones** 234

2.1 Fluorenone Ketyl Complexes with an Aryloxide (ArO) Ligand . . . 235
2.2 Fluorenone Ketyl Complexes with a C_5Me_5 Ligand 237
2.3 Fluorenone Ketyl Complexes with an $N(SiMe_3)_2$ Ligand 238
2.4 Benzophenone Ketyl Complexes 240
2.5 Complexes Bearing Three Independent Ketyl Ligands 241
2.6 Structure-Reactivity Relation of Lanthanide Ketyl Complexes . . . 243

3 **Two-Electron Reduction of Ketones** 243

3.1 Isolation of a Ytterbium(II) Benzophenone Dianion Complex . . . 244
3.2 Reactions of Ketone Dianions with Organic Electrophiles 245
3.3 Reactions of Ketone Dianions with 2,6-Di-*tert*-Butylphenols 246
3.3.1 Benzophenone Dianion Species 246
3.3.2 Fluorenone Dianion Species 248

4 **Ketone-Reduction Paths Based on Isolated Intermediates** 248

5 **Dianionic Thioketone and Imine Species** 250

6 **Conclusions and Perspectives** 251

7 **References** 252

Topics in Organometallic Chemistry, Vol. 2
Volume Editor: S. Kobayashi

1 Introduction

Reactions of ketones with low-valent lanthanides such as SmI_2 and the Ln (Ln= Sm, Yb) metals constitute one of the most useful reactions in lanthanide-mediated organic synthesis, and occupy an important position in modern organic chemistry [1, 2]. It is well known that one-electron reduction of ketones by low-valent lanthanides easily produces the corresponding radical anion species, or ketyls. In the case of diaryl ketones, further one-electron transfer is also achievable to afford the corresponding ketone dianions. These highly reactive species play a very important role, as key intermediates, in a variety of useful reactions involving ketones, such as reductions, ketyl-olefin couplings, pinacol couplings, intermolecular cross couplings, and the Barbier-type reactions. Although other reducing metals such as alkali, alkaline earth, and low-valent titanium metals can also produce similar ketyls or ketone dianions upon reaction with ketones [3], the lanthanide species usually behave differently from those metal analogs and in many cases show higher reactivity and chemo-, regio- and stereo-selectivity.

Despite the extensive applications and importance of the lanthanide ketyl and ketone dianion species in organic synthesis, very little was known about their structures. Our understanding about the reactivity of this important class of species had relied solely on the analysis of the organic products obtained by hydrolysis work-up of the final reaction mixtures. Attempts to isolate these species were greatly hampered by their extremely high reactivity and air and moisture sensitivity [4].

It was not until very recently that significant progress has been made in this field. By using sterically demanding ancillary ligands under an extremely dry and oxygen-free inert atmosphere, several different types of lanthanide ketyl and ketone dianion complexes have been successfully isolated and many of them structurally characterized. Detailed studies on the reactivity of these well-defined complexes have offered unprecedented insights into the mechanistic aspects of the reactions of ketones with low-valent lanthanides, and also shed new lights on similar reactions with other reducing agents. This article is intended to highlight the recent progress in this area, with particular emphasis on the structure-reactivity relation of lanthanide ketyl and ketone dianion complexes.

2 One-Electron Reduction of Ketones

The most difficult problem in isolating a ketyl species generated in one-electron reduction of ketones is due to its rapid hydrogen abstraction and/or coupling reactions. To suppress these decomposition reactions, it is essential to use sterically demanding reducing agents and highly conjugated ketones [5]. It has been found that several types of lanthanide reducing agents are able to afford isolable and structurally characterizable ketyl complexes in the reactions with benzophenone and fluorenone. The structures and reactivity (stability) of the lanthanide ketyl complexes are strongly dependent on the ancillary ligands, solvents, and the structure of their parent ketones.

X-ray analyses have shown that the C-O bond distance of a ketyl unit is generally around 1.30 Å, which is longer than that of a free ketone (ca. 1.20 Å) and shorter than that of an alkoxide (ca. 1.40 Å). The radical carbon atom of a ketyl is still in a sp^2-hybrid state.

2.1 Fluorenone Ketyl Complexes with an Aryloxide (ArO) Ligand

Reactions of the divalent lanthanide aryloxide complexes $Ln(OAr)_2(L)_x$ (Ar= C_6H_2-*t*Bu_2-2,6-Me-4; **1a**: Ln=Sm, L=THF, x=3; **1b**: Ln=Yb, L=THF, x=3; **1c**: Ln=Sm, L=HMPA, x=2; **1d**: Ln=Yb, L=HMPA, x=2) with 1 equivalent of fluorenone in THF at room temperature give the corresponding ketyl complexes **2a-d** as deeply-colored crystals in high yields (Scheme 1) [6, 7]. Complexes **2a-d** all have a similar structure in which one fluorenone ketyl and two ArO ligands are placed at the equatorial and two L (L=THF or HMPA) ligands are located at the apical positions of a distorted trigonal bipyramid.

When the THF-coordinated complex **2a** is dissolved in hexane/ether, the two THF ligands are substituted by one molecule of OEt_2, and the ketyl radical dimerizes into a pinacolate (**3**) (Scheme 2) [6, 7]. More remarkably, the newly formed C-C bond in **3** (1.613(9) Å) can be easily broken to quantitatively regenerate the ketyl **2a** when **3** is dissolved in THF (Scheme 2), which unequivocally demonstrates that the pinacol coupling process of the ketyl radical is completely reversible. Addition of 2 equivalents of HMPA (per Sm) to a THF solution of **2a** or **3** gives almost quantitatively the corresponding HMPA-coordinated ketyl complex **2c** (Scheme 2). Since the coordination ability of HMPA is much stronger than that of THF [8], complex **2c** is much more stable than **2a**, and no change was observed when **2c** was treated similarly with hexane/ether. These results clearly demonstrate that the stability and reactivity of a ketyl species are strong-

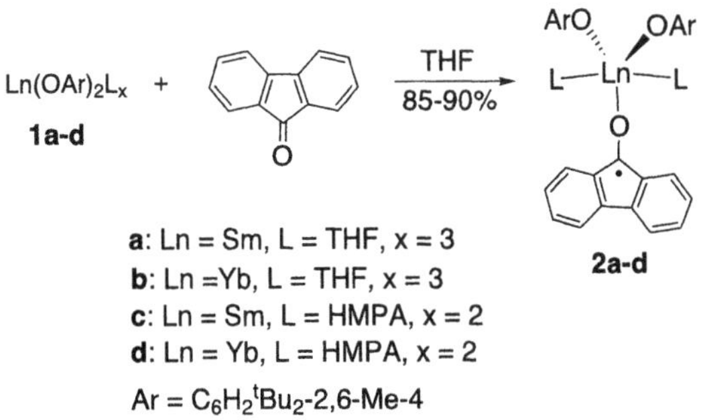

$Ln(OAr)_2L_x$ + fluorenone —THF, 85-90%→ **2a-d**

1a-d

a: Ln = Sm, L = THF, x = 3
b: Ln =Yb, L = THF, x = 3
c: Ln = Sm, L = HMPA, x = 2
d: Ln = Yb, L = HMPA, x = 2
Ar = $C_6H_2{}^tBu_2$-2,6-Me-4

Scheme 1

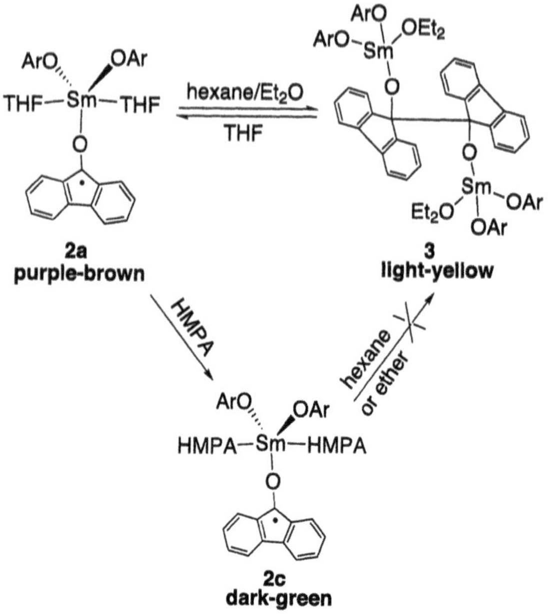

Scheme 2

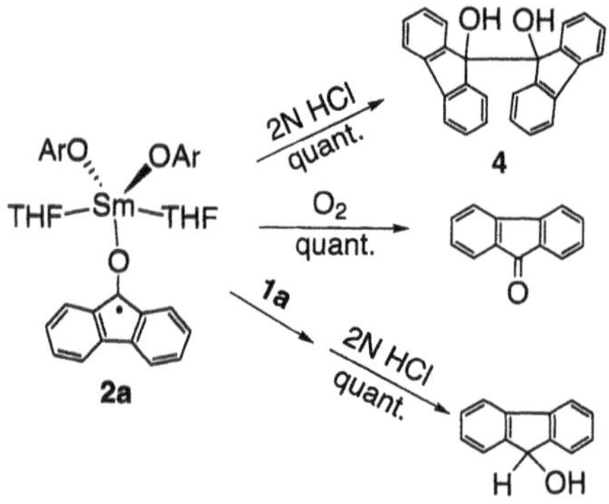

Scheme 3

ly influenced by the ancillary ligands bound to the central metal, and HMPA is a better ligand than THF for the stabilization of a ketyl species [5, 7, 9].

Reflecting the typical reactivity of a ketyl species, hydrolysis of **2a** gives the corresponding pinacol-coupling product **4**, while air oxidation of **2a** yields fluorenone almost quantitatively (Scheme 3). Reaction of **2a** with one equivalent of **1a** produced a THF-insoluble purple precipitate which upon hydrolysis afforded fluorenol quantitatively, suggesting that a fluorenone dianion intermediate was formed [7, 9].

2.2 Fluorenone Ketyl Complexes with a C_5Me_5 Ligand

The similar reactions of $(C_5Me_5)_2Ln(THF)_2$ (Ln=Sm, Yb) with 1 equivalent of fluorenone in THF give the corresponding ketyl complexes **5a,b** in 85–87% isolated yields (Scheme 4) [7]. Reactions of **5a,b** with 1 equivalent of HMPA afford the HMPA-coordinated ketyl complexes **5c,d**. In contrast with the ArO-ligated ketyl complexes **2a–d** which require two THF or HMPA ligands as an additional stabilizing moiety, the C_5Me_5-ligated complexes **5a–d** need only one THF or HMPA ligand for the stabilization of the ketyl species. This difference apparently results from the larger size of C_5Me_5 as compared to that of the ArO ligand. Addition of hexane/ether to the THF-coordinated ketyl complexes **5a,b** did not cause any further reaction (Scheme 4), showing that the C_5Me_5-supported **5a,b** are more stable and less reactive than the ArO-supported **2a,b** (see also Sect. 2.1).

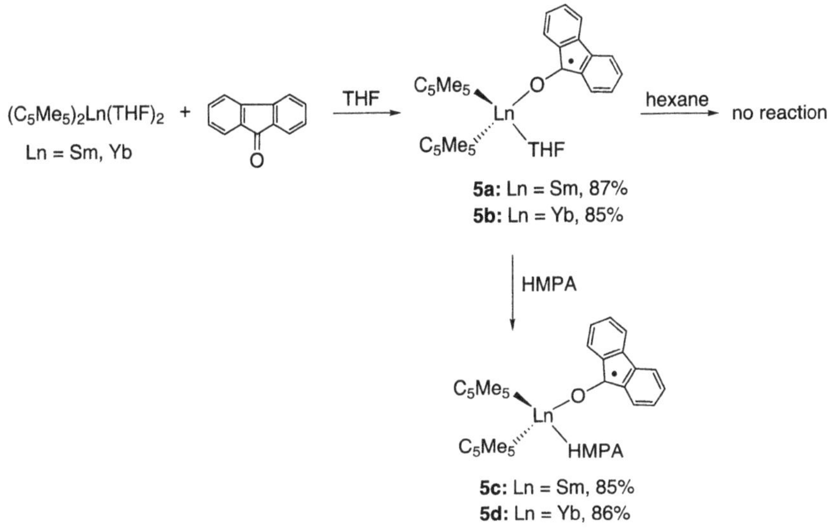

$(C_5Me_5)_2Ln(THF)_2$ + fluorenone —THF→ $(C_5Me_5)_2Ln(O\text{-fluorenyl ketyl})(THF)$ —hexane→ no reaction

Ln = Sm, Yb

5a: Ln = Sm, 87%
5b: Ln = Yb, 85%

↓ HMPA

$(C_5Me_5)_2Ln(O\text{-fluorenyl ketyl})(HMPA)$

5c: Ln = Sm, 85%
5d: Ln = Yb, 86%

Scheme 4

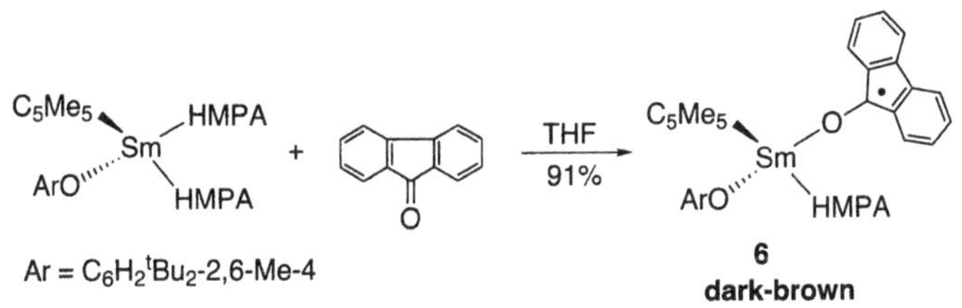

$(C_5Me_5)Sm(OAr)(HMPA)_2$ + fluorenone —THF, 91%→ $(C_5Me_5)Sm(OAr)(O\text{-fluorenyl ketyl})(HMPA)$

Ar = $C_6H_2{}^tBu_2$-2,6-Me-4

6
dark-brown

Scheme 5

Reaction of $(C_5Me_5)Sm(OAr)(HMPA)_2$ with 1 equivalent of fluorenone in THF also gives the corresponding ketyl complex **6** (Scheme 5) [7], demonstrating that the heteroleptic C_5Me_5/OAr ligand set [10] is also able to stabilize a ketyl species. It is also noteworthy that although the homoleptic analogs **2c** and **5c** are isolable, ligand redistribution of **6** to give either of these complexes was not observed. Complex **6** represents a rare example of a lanthanide complex which bears all different ligands.

2.3 Fluorenone Ketyl Complexes with an $N(SiMe_3)_2$ Ligand

The ketyl species stabilized by $N(SiMe_3)_2$ behaves a little differently from those by C_5Me_5 and the ArO ligand. Reaction of $Sm(N(SiMe_3)_2)_2(THF)_2$ with 1 equivalent of fluorenone in THF gave a brown solution whose UV-Vis spectrum was almost identical to that of the ArO-ligated ketyl complex **2a**, suggesting that the corresponding ketyl species **7a** was formed [7]. However, attempts to isolate the ketyl species **7a** from THF were unsuccessful and always resulted in the formation of the light yellow pinacolate **8a** (Scheme 6).

The central C-C bond in **8a** could also be cleaved by addition of a strongly coordinative solvent. Dissolving the light yellow **8a** in THF gave immediately a brown solution whose UV-Vis spectrum was the same as that of the solution originally obtained by the reaction of $Sm(N(SiMe_3)_2)_2(THF)_2$ with fluorenone in THF (Scheme 6). Addition of 4 equivalents of HMPA to this THF solution afforded the bis(HMPA)-coordinated ketyl complex **7b** together with the mono(HMPA)-coordinated pinacolate complex **8b** (Scheme 7) [7].

The isolation of the pinacolate **8a** rather than the ketyl **7a** from THF (Scheme 6) and the isolation of the mono(HMPA)-coordinated pinacolate **8b** together with the bis(HMPA)-coordinated ketyl **7b** in the reaction of **8a** or **7a** with HMPA (Scheme 7) contrast sharply with the similar reactions of the ArO-ligated complex **2a** or **3**, in which only the bis(THF or HMPA)-coordinated ketyl complexes were isolated (see also Sect. 2.1). These differences probably result from the difference in electron-donating ability between $N(SiMe_3)_2$ and the ArO ligand. Since $N(SiMe_3)_2$ is more electron-donating than the ArO ligand, dissociation of a neutral L (L=THF or HMPA) ligand from the central Sm atom in the

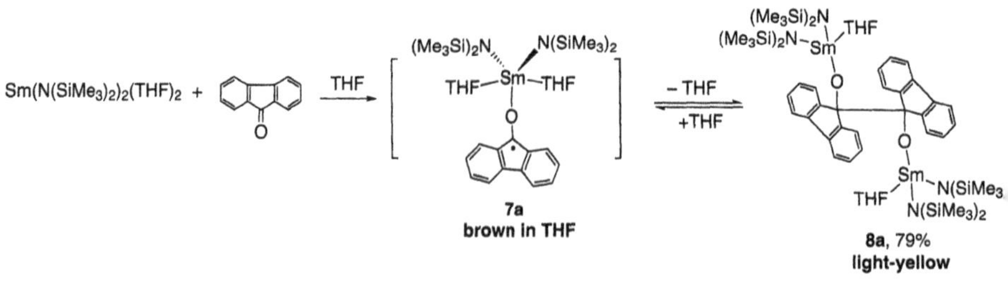

Scheme 6

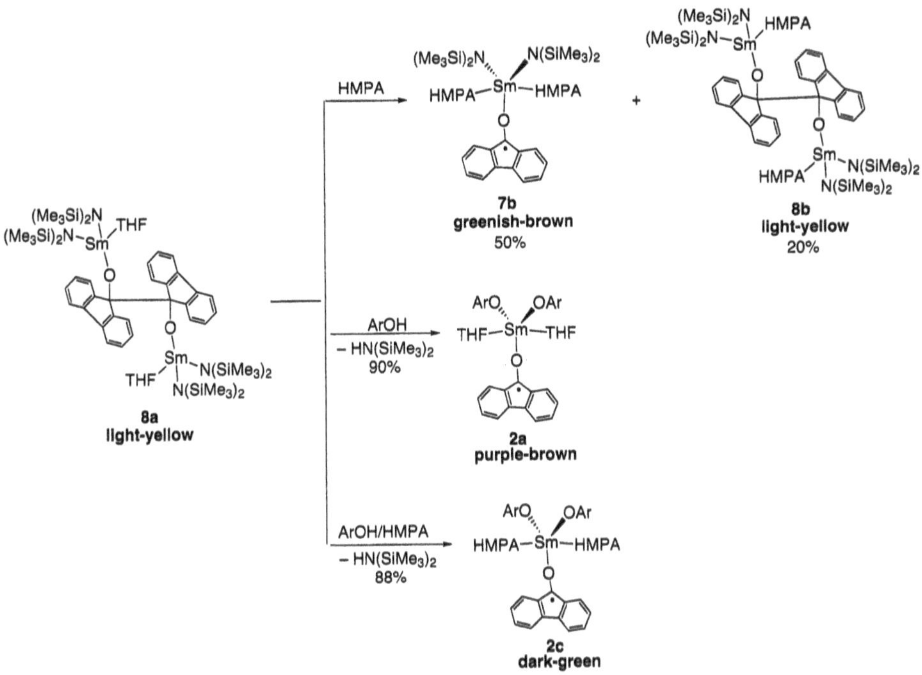

Scheme 7

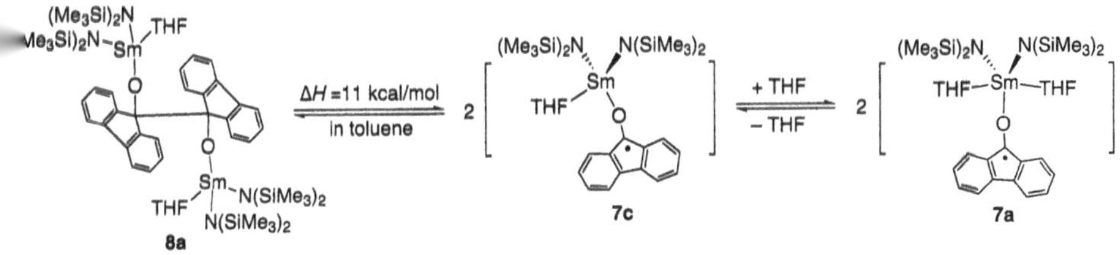

Scheme 8

$(Me_3Si)_2N$-ligated complexes **7a,b** would be easier than that in the ArO-ligated complexes **2a,c**, which thus causes pinacol-coupling of the ketyl radical in **7a,b** to occur more easily. In accord with this consideration, when 4 equivalents of ArOH were added to a THF solution of **8a**, the corresponding ArO-ligated ketyl complex **2a** was isolated in 90% yield with the release of $HN(SiMe_3)_2$ (Scheme 7). The similar reaction of **8a** with ArOH in THF/HMPA yielded selectively the corresponding bis(HMPA)-coordinated ketyl complex **2c** (Scheme 7) [7].

To gain more information concerning the C-C bond cleavage of the pinacolate **8a**, a variable-temperature UV-Vis spectroscopic study in toluene was carried out. The derived dissociation enthalpy ΔH_{diss} for **8a** to give **7c** is 11 kcal/mol (Scheme 8) [7]. This value is much smaller than that reported for silicon benzopinacolate, $Me_3SiOC(Ph)_2C(Ph)_2OSiMe_3$ (31 kcal/mol) [11], but comparable with that reported for the C_α-C_{para} bonds in the trityl dimer $[Ph_3C]_2$ (11–12 kcal/mol) [12] and that for the dimeric titanium(IV) alkoxide/enolate complex $[(^tBu_3SiO)_3Ti(OCPh_2)]_2$ (18 kcal/mol) [13]. Obviously the dissociation enthalpy for **8a** to give the bis(THF)-coordinated ketyl complex **7a** should be much smaller than this value, since this process is accompanied by the formation of two new Sm(III)-THF bonds (Scheme 8). Since the bond energy of a Sm(II)-THF bond in $(C_5Me_5)_2Sm(THF)_n$ (n=1 or 2) is about 5–7 kcal/mol [14], and a Sm(III)-THF bond is generally stronger than a Sm(II)-THF bond, the formation of two Sm(III)-THF bonds in the course of the transformation of **8a** to **7a** would release more than 10 kcal/mol, which could well compensate the enthalpy (11 kcal/mol) for the cleavage of the central C-C bond in **8a**. It is thus not difficult to understand that when a large excess of THF is present (e.g., in THF solution), the equilibria in Scheme 8 will be greatly shifted to the right, and complete dissociation of **8a** to **7a** will be easily achieved.

2.4 Benzophenone Ketyl Complexes

Compared to the planar fluorenone ketyl, benzophenone ketyl is less conjugated and thus more reactive [15]. Reactions of benzophenone with the lanthanide reducing agents which were successfully used for the isolation of fluorenone ketyl complexes did not give a structurally characterizable benzophenone ketyl species, and in many cases yielded the corresponding hydrogen abstraction products. For example, reaction of **1d** with 1 equivalent of benzophenone in THF gave after a few weeks the colorless ytterbium(III) aryloxide/diphenylmethoxide complex $Yb(OCHPh_2)_2(OAr)(HMPA)_2$ (**9**) via hydrogen radical abstraction from solvent by the initially generated ketyl radical species (Scheme 9) [7].

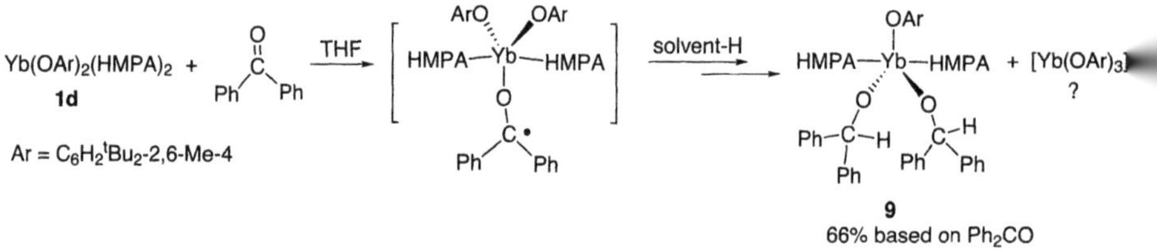

Scheme 9

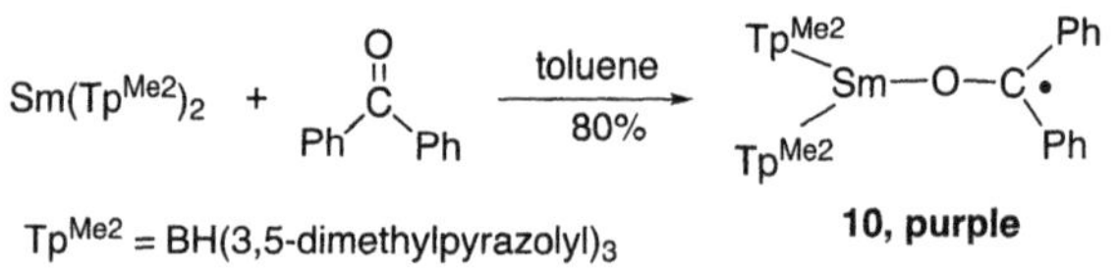

$Yb(OAr)_2(HMPA)_2$ + $Ph_2C{=}O$ (**1d**) —THF→ [ArO, OAr, HMPA–Yb–HMPA, O–C•(Ph)Ph] —solvent-H→ HMPA–Yb(OAr)–HMPA with two O–CH(Ph)Ph + $[Yb(OAr)_3]$?

Ar = $C_6H_2{}^tBu_2$-2,6-Me-4

9, 66% based on Ph_2CO

Scheme 10

$Sm(Tp^{Me2})_2$ + $Ph_2C{=}O$ —toluene, 80%→ $(Tp^{Me2})_2Sm$–O–C•(Ph)Ph

Tp^{Me2} = BH(3,5-dimethylpyrazolyl)$_3$

10, purple

Recently, Takats and coworkers have successfully isolated a structurally characterizable lanthanide benzophenone ketyl complex (**10**) by using a further sterically demanding reducing agent, $Sm(Tp^{Me2})_2$ (Tp^{Me2}=BH(3,5-dimethylpyrazolyl)$_3$) (Scheme 10) [16a]. Izod and coworkers have isolated and structurally characterized the corresponding benzophenone ketyl complex by using $Sm(C(SiMe3)_2(SiMe2OMe))_2(THF)$ as a rducing agent [16b]. These results again demonstrate that the behavior of a ketyl species is greatly influenced by the ancillary ligands.

2.5 Complexes Bearing Three Independent Ketyl Ligands

In contrast with the divalent lanthanide complexes, which are one-electron transfer agents and afford complexes consisting of one ketyl per metal ion upon reaction with ketones, the zero valent lanthanide metals are three-electron transfer agents, and their reactions with ketones, if straightforward, should yield complexes bearing three independent ketyl ligands. However, compared to mono(ketyl) complexes, the isolation of a multi(ketyl) metal species is generally more difficult and challenging, since intramolecular pinacol-coupling of ketyl radicals is also possible besides the common intermolecular reactions.

It has been found that the use of HMPA as a coordinating ligand can sufficiently suppress coupling reactions of the ketyl radicals in multi(ketyl) metal complexes to allow their isolation [5, 7, 15, 17]. Thus, reaction of Sm metal powder with 3 equivalents of fluorenone and 3 equivalents of HMPA in THF afforded the corresponding Sm(III) tris(fluorenone ketyl) complex **11a** as black-green blocks (Scheme 11) [7, 17]. An X-ray analysis has shown that this complex possesses an octahedral structure, in which the central Sm atom is coordinated in *mer*-type fashion by three ketyls and three HMPA ligands. The similar reaction of Yb metal with 3 equivalents of fluorenone gives the Yb(III) tris(ketyl) complex **11b** (Scheme 11), which is isostructural and isomorphous with **11a** [7].

Hydrolysis of **11a,b** gives almost quantitatively the corresponding pinacol **4** (Scheme 11). Interestingly, the reaction of **11a** with 0.5 equivalents of **4** yields the dimeric samarium(III) fluorenoxide/pinacolate complex **12** and fluorenone (Scheme 11) [7]. Further studies have confirmed that the formation of **12** is via hydrogen atom abstraction from the pinacol by one ketyl, followed by pinacol-coupling of the other two ketyls together with simultaneous release of two HMPA ligands in **11a** (Scheme 11). Homolytic C-C bond cleavage of the biradical species **13**, which is formed by dehydrogenation (–H·) of pinacol **4**, affords fluorenone. It is noteworthy that the small steric change, which is caused by the transformation of one of the three fluorenone ketyls to a fluorenoxy unit, has imposed crucial influence on the stability and reactivity of the other two ketyls.

In contrast with the case of fluorenone, similar reactions of Ln/HMPA (Ln= Sm, Yb) with benzophenone did not give an isolable ketyl species and yielded finally decomposition products, demonstrating again that benzophenone ketyl is more reactive than fluorenone ketyl [7]. These reactions are also in sharp con-

Scheme 11

trast with those of Na/HMPA [5] or Ca/HMPA [5, 15] with benzophenone, in which the corresponding benzophenone ketyl species have been successfully isolated and structurally characterized. These results show that the nature of the reducing metals also strongly influences on the reactivity and stability of a ketyl species and the lanthanide ketyl species are more reactive than those of alkali and alkaline earth metals.

2.6 Structure-Reactivity Relation of Lanthanide Ketyl Complexes

The results described above have demonstrated that the stability and reactivity of a ketyl species are extremely susceptible to the environment around the central metal ion. Both anionic and neutral (solvent) ligands play an important role in determining the behavior of a ketyl species. The sterically demanding bis(pentamethylcyclopentadienyl) ligand set $(C_5Me_5)_2$ together with a neutral L (L=THF or HMPA) ligand offers a good stabilizing environment for a lanthanide fluorenone ketyl species, while the less bulky $(ArO)_2$ or $((Me_3Si)_2N)_2$ ligand set requires two L (L=THF or HMPA) ligands for the stabilization of the same ketyl species. In the latter case, dissociation of one of the two L ligands from the central metal ion easily occurs to cause pinacol-coupling of the ketyl, and this takes place more easily for the $(Me_3Si)_2N$-ligated complexes than for the ArO-ligated ones, due to the stronger electron-donating ability of $N(SiMe_3)_2$. Re-coordination of an L ligand to the metal atom of the resulting pinacolate easily cleaves the central C-C bond and regenerates the ketyl, which thus makes the pinacol-coupling process reversible. The relatively small dissociation enthalpy for the pinacolate complexes (e.g., ΔH_{diss}=11 kcal/mol for **8a**) could well account for this reversibility. HMPA, as a neutral and strongly coordinating ligand, offers an excellent stabilizing moiety for both mono- and multi-ketyl complexes. The transformation of the tris(ketyl) complex **11a** to the fluorenoxide/pinacolate complex **12** provides unprecedented insights into the elementary steps of the hydrogenation and pinacol-coupling reactions of a ketyl species. Ketone-dependence of the stability of ketyls is also observed. In contrast to fluorenone ketyl complexes, the isolation of a structurally characterizable benzophenone ketyl complex is more difficult due to its less conjugated structure which makes it more reactive and less stable. However, by choosing an appropriate ligand set such as bis(hydrotris(3,5-dimethylpyrazolyl)borate), a structurally characterizable lanthanide benzophenone ketyl species can indeed be isolated. It is obvious from these results that the stability and reactivity of a ketyl radical species can be controlled through tuning the ancillary ligands bound to the central metal ion.

3 Two-Electron Reduction of Ketones

Among all low-valent lanthanides, the zero valent Sm and Yb metals are the most often used reducing agents for the study of two-electron reduction of ketones.

Although earlier studies showed that the formation of a samarium ketone dianion species was not as efficient as that of ytterbium [18–20], it was later found that samarium ketone dianion species could be generated similarly if the metal surface was sufficiently activated, e.g., by ICH_2CH_2I [21, 22]. Since the negatively charged carbon and oxygen atoms in a ketone dianion species are adjacent, the use of a sterically demanding lanthanide(II) reducing agent does not necessarily give a good result for the isolation of such species because of the steric repulsion between the resultant two closely located bulky metal moieties. Therefore, the control of the reactivity or stability of a ketone dianion species through tuning the ancillary ligands may not be as effective as in the case of ketyls. Moreover, compared to the mono-anionic ketyls, ketone dianions are more difficult to generate. They are more sensitive to air and moisture, and thus more difficult to handle. Probably due to these reasons, structurally characterized examples of ketone dianion species remain very scarce [23–25], and only one lanthanide ketone dianion complex, $[Yb(\mu\text{-}\eta^1,\eta^2\text{-}OCPh_2)(HMPA)_2]_2$, has been isolated and structurally characterized to date (see below) [24]. The known ketone dianions are limited solely to those bearing two aromatic groups on the carbonyl carbon atom.

Despite these limitations, it has been found that the lanthanide ketone dianions are an interesting class of species which show unique reactivities. As described below, in lanthanide ketone dianion species the carbonyl carbon atom is completely changed from electrophilic to nucleophilic, and shows strong nucleophilicity towards a variety of organic substrates. Due to the delocalization of the negative charges in the aromatic groups, some reactions can also take place at the aromatic ring.

3.1 Isolation of an Ytterbium(II) Benzophenone Dianion Complex

Reaction of Yb metal powder with 1 equivalent of benzophenone in THF/HMPA gives the Yb(II) benzophenone dianion complex $[Yb(\mu\text{-}\eta^1,\eta^2\text{-}OCPh_2)(HMPA)_2]_2$ (**14**) as purple crystals (Scheme 12) [21, 24]. X-ray analysis has revealed that complex **14** possesses a dimeric structure, in which the two Yb atoms are bridged by two benzophenone dianions. The bridge is built in such a manner

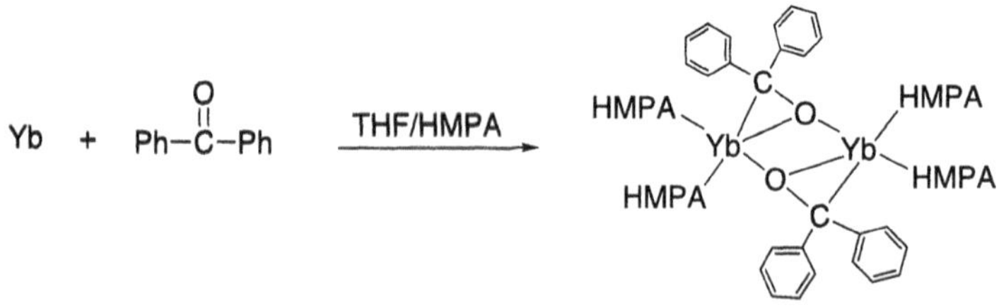

Yb + Ph–C(=O)–Ph → (THF/HMPA) → **14, purple**

Scheme 12

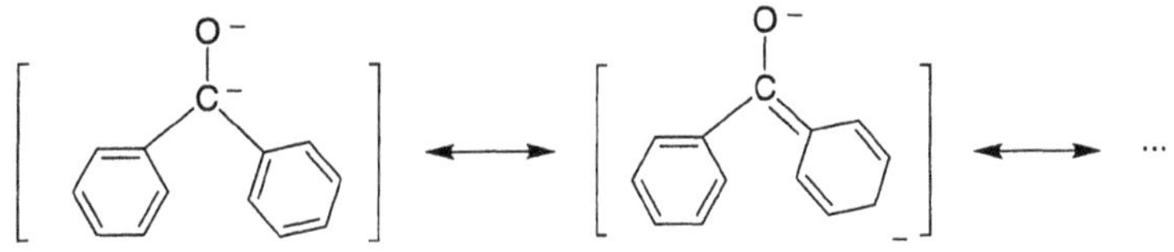

Scheme 13

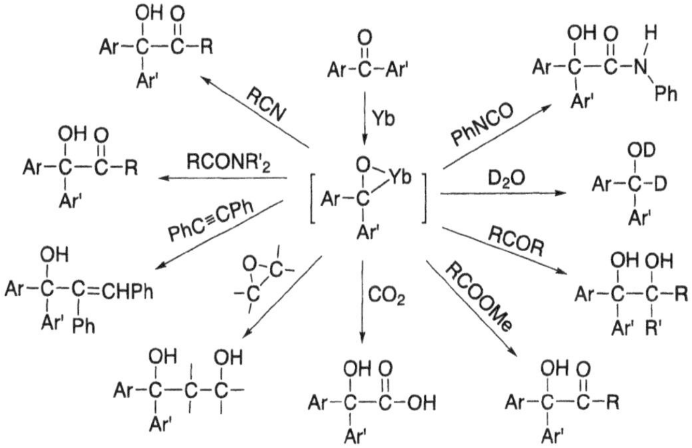

Scheme 14

that a benzophenone dianion uses its O atom to make a Yb-O bond with one Yb(II) ion and its C atom to make a Yb-C bond with the other Yb(II) ion to which the lone electron pair of the oxygen atom is also donated (Scheme 12). The C-O bond of the benzophenone dianion (1.39(6) Å) is longer than that of benzophenone ketyl (1.31(2) Å) [15, 16] and that of free benzophenone (1.23(1) Å) [26]. Similar to benzophenone ketyl, the carbonyl carbon atom in benzophenone dianion is still in an sp^2-hybrid state, which can thus allow good conjugation of the negative charges with the phenyl rings.

Consistent with the above structural observations, the ^{1}H NMR signals for the phenyl groups in **14** in THF-d_8 are greatly upfield shifted to as high as δ 5.63–7.04, demonstrating that the negative charges of the ketone dianion are indeed highly delocalized into the phenyl rings, especially to the para-positions (δ 5.63) (Scheme 13) [21,22]. As described in Sect. 3.3, this delocalization of electrons can have great influence on the reactivity of benzophenone dianion species.

3.2
Reactions of Ketone Dianions with Organic Electrophiles

It has been found that lanthanide ketone dianion species are excellent nucleophiles, which react smoothly with a variety of organic substrates such as ketones, nitriles, epoxides, CO_2, etc. to give after hydrolysis the corresponding unsymmetrical pinacols, α-hydroxy ketones, 1,3-diols, α-hydroxy carboxylic acids, etc., in good yields, respectively (Scheme 14) [18–20]. Compared to alkali metal ketone dianions, they are less basic and more nucleophilic, and selectively undergo addition reactions even with substrates bearing active α-protons, such as aliphatic ketones and nitriles.

3.3 Reactions of Ketone Dianions with 2,6-Di-*tert*-Butylphenols

Most of the reactions of ketone dianion species, including those of main group and titanium metals, are known to occur at the carbonyl unit, as demonstrated in Sect. 3.2. However, it was found that the protonation reactions of lanthanide benzophenone dianion species with 2,6-di-*tert*-butylphenols can take place not only at the carbonyl group but also at the phenyl ring, due to the delocalization of the negative charges onto the aromatic groups. The selectivity between reactions at the carbonyl group and those at the aromatic ring in lanthanide diaryl ketone dianion species depends greatly on the nature of both metals and ketones.

3.3.1 *Benzophenone Dianion Species*

Reaction of the Yb(II)-benzophenone dianion complex **14** with ArOH (Ar= C_6H_2-$t$$Bu_2$-2,6-R-4, R=H, Me) gave the corresponding ytterbium(II) bis(aryloxide) complex $Yb(OAr)_2(HMPA)_2$ (**1d**) as a major product, which was formed by the protonation of both the carbon and oxygen atoms of the carbonyl unit with ArOH (Scheme 15) [22, 24]. Besides **1d**, an ytterbium(III) enolate complex (**15**) was also obtained as a minor product. In contrast, when the analogous Sm(II) benzophenone dianion species was allowed to react with ArOH, the samarium(III) enolate complex **16**, which is analogous to **15**, was formed as a major product, while the samarium(II) bis(aryloxide) complex $Sm(OAr)_2(HMPA)_2$, an analog of **1d**, was not observed (Scheme 16) [22, 27]. The similar reaction of the Sm(II) benzophenone dianion species with ArOD yielded the corresponding

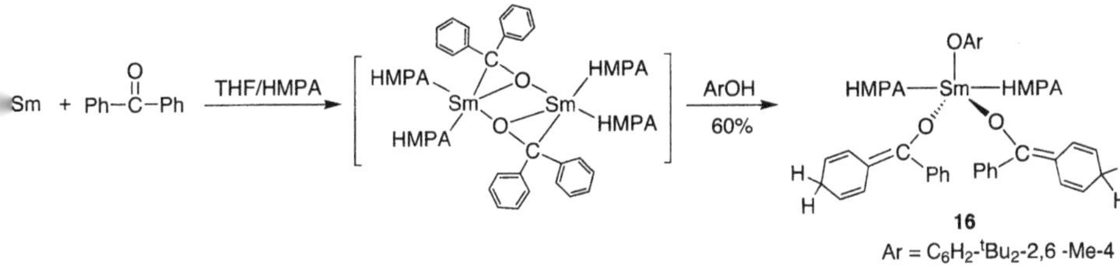

Scheme 16

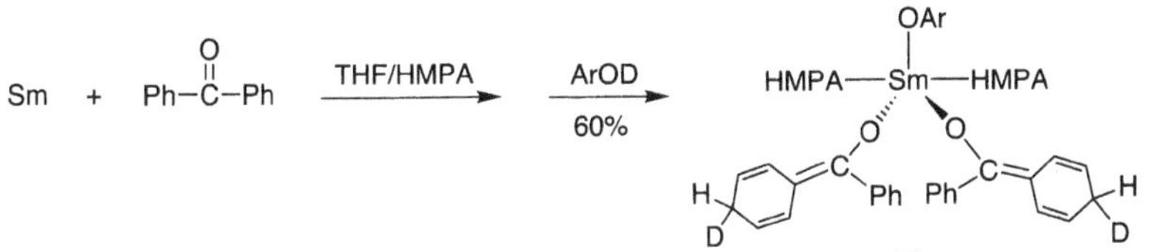

Scheme 17

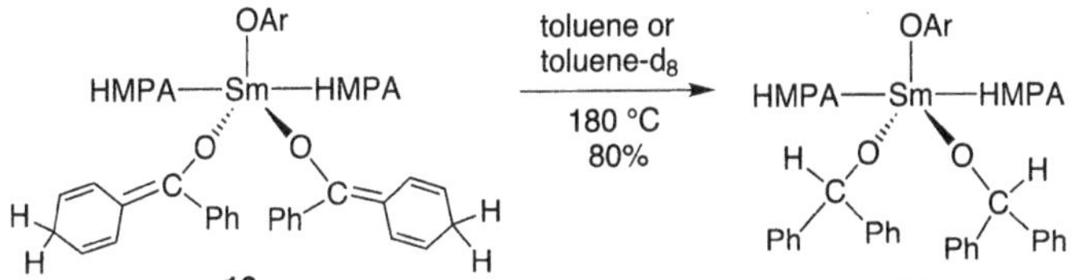

Scheme 18

deuterated enolate **17** (Scheme 17), suggesting that the formation of **15** and **16** is via the direct protonation of the para carbon atom of a phenyl group in the benzophenone dianion unit.

The formation of the enolate complexes **15** and **16** is apparently due to the delocalization of the negative charges in the phenyl groups of the benzophenone dianion species (see Scheme 13). The difference in reactivities observed between Sm(II) and Yb(II) benzophenone dianion species may result from differences in the two divalent metal ions. Since Sm(II) is more electron-donating [28] and bigger in size than Yb(II) [29], the negative charges in the Sm(II) benzophenone dianion species must be more delocalized into its phenyl rings, which are thus more easily protonated on the phenyl part and give a Sm(III) enolate species.

It is also interesting to note that the formation of the enolates **15** and **16** in the above reactions resembles well the Birch reduction of aromatic compounds bearing electron-withdrawing groups [30], in which metal enolate intermediates are believed to be formed via monoprotonation of dianionic species. However, such species have never been well characterized. Complexes **15** and **16** are

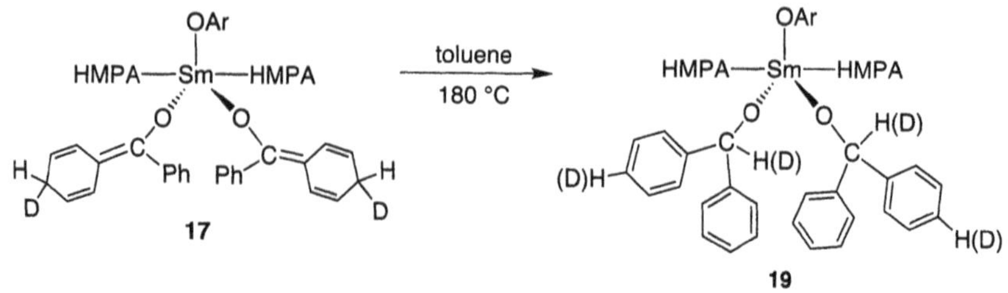

Scheme 19

rare examples of structurally characterized metal enolate compounds which model the enolate intermediates formed in the Birch reductions.

When the enolate complex **16** was heated in toluene or toluene-d_8 at 180 °C overnight, intramolecular hydrogen transfer reaction occurred to give the corresponding aryloxide/diphenylmethoxide complex $Sm(OC(H)Ph_2)_2(OAr)(HMPA)_2$ (**18**) (Scheme 18) [22, 24]. The similar reaction of the deuterated enolate **17** yielded **19** selectively (Scheme 19); scrambling of deuterium at the phenyl ring was not observed. These results suggest that the formation of **18** from **16** is a one-step 1,5-hydrogen shift process. It is also noteworthy that even at high temperatures (up to 180 °C) ligand redistribution in these heteroleptic lanthanide complexes was not observed.

3.3.2
Fluorenone Dianion Species

The protonation reaction of the Yb(II) fluorenone dianion species with ArOH is similar to the major reaction of the Yb(II) benzophenone dianion **14** with ArOH, occurring at both the C and O atoms of the carbonyl unit to afford **1d** and fluorenol (Scheme 20). However, the protonation reaction of the Sm(II) fluorenone dianion species with ArOH takes place only at the C atom and is accompanied by the oxidation of the Sm(II) ion to give the corresponding samarium(III) aryloxide/fluorenoxide complex **20** (Scheme 20), which again demonstrates the metal dependence of the reactivity of lanthanide ketone dianion species [22].

4
Ketone-Reduction Paths Based on Isolated Intermediates

As described above, with the combination of appropriate lanthanide reducing agents, solvents, and substrates, almost all types of the key intermediates involved in the reduction of ketones have been isolated and structurally characterized (Scheme 21), such as the one-electron reduction product (ketyl radical **21**), its reversible coupling product (pinacolate **24**), hydrogen abstraction product (alkoxide **23**) and further one-electron reduction product (ketone dianion **22**), and the selective protonation products of the ketone anion (alkoxide **23** and eno-

Ln +
THF/HMPA
Ln(HMPA)x
ArOH
ArO
Ln(HMPA)x
(Ln = Yb)
ArOH
– fluorenol
1d, 85%
(Ln = Sm)
ArOH
– H2
20, 64%

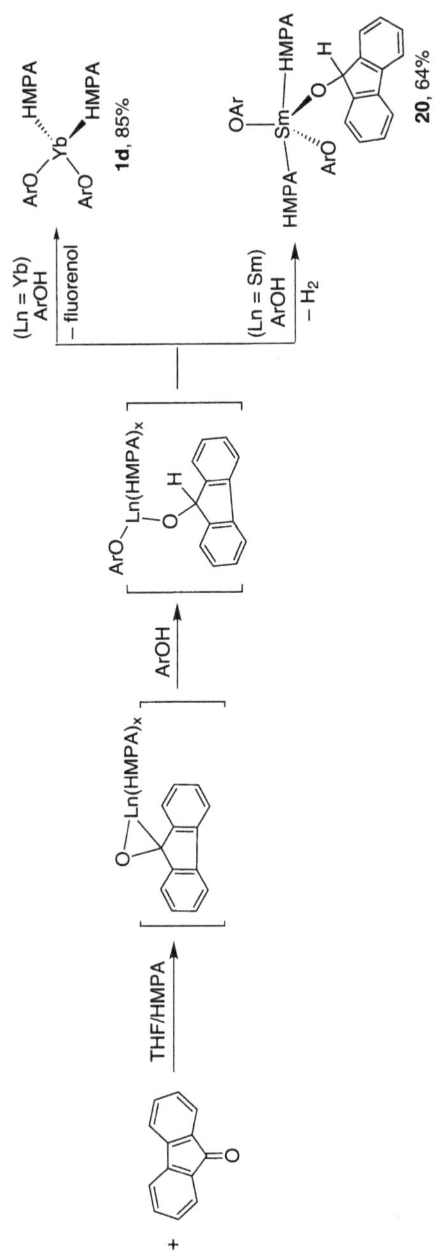

Scheme 20

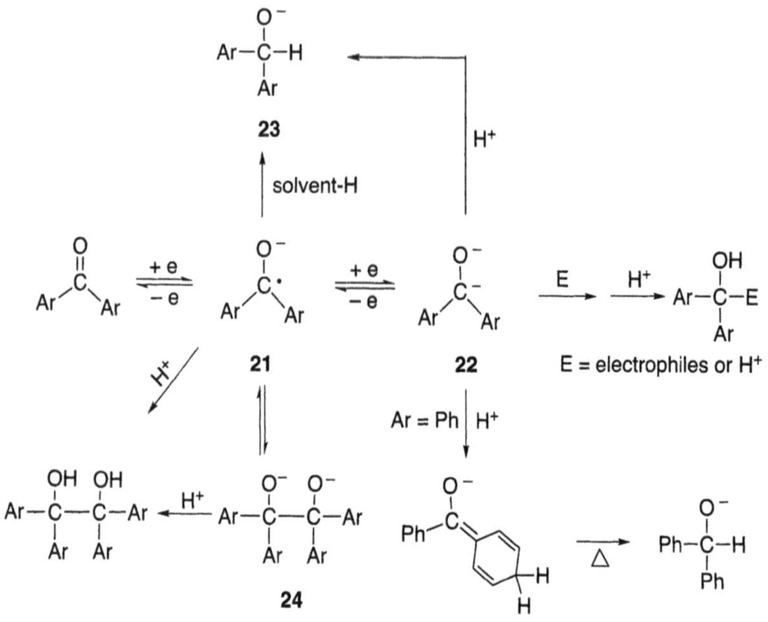

Scheme 21. Structurally characterized intermediates and their reactions

late **25**). Structural and reactivity studies of these isolated intermediates have given rise to an ever clearer picture of the reaction paths in the reaction of ketones with low-valent lanthanides (Scheme 21), and thus offered unprecedented insights into the mechanistic aspects of these and related reactions.

5 Dianionic Thioketone and Imine Species

Similar to diaryl ketones, aromatic thioketones and imines can also be reduced to the corresponding dianionic species. Fujiwara and coworkers have reported that the ytterbium diaryl thioketone dianion species, generated by reaction of Yb metal with the thioketones in THF/HMPA at low temperature, show good nucleophilicity towards organic substrates such as acetone and alkyl halides (Scheme 22) [31]. The thioketone dianion species seemed to be less stable than those of ketones. At room or higher temperatures, C-S bond cleavage reaction took place.

Reaction of *N*-(diphenylmethylidene)aniline with 1 equivalent of Yb metal in THF/HMPA gives the corresponding dianionic complex Yb(η^2-Ph_2CNPh)$(HMPA)_3$ (**26**) whose structure has been crystallographically determined (Scheme 23) [32]. The imine dianion complex **26** is more basic and less nucleophilic than ketone- and thioketone-dianion species. It reacts with CO_2 to give after hydrolysis the corresponding Yb(III) amino acid derivative [33], but with acetone to release

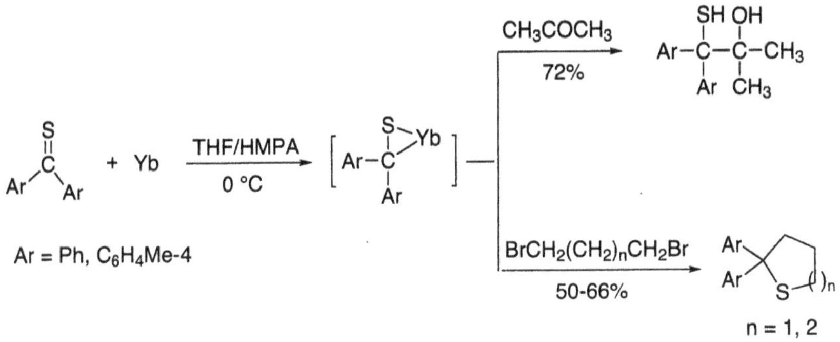

$Ar_2C{=}S$ + Yb →(THF/HMPA, 0 °C) [Ar–C(Ar)(S)–Yb]

→(CH_3COCH_3, 72%) Ar–C(SH)(Ar)–C(OH)(CH_3)–CH_3

→($BrCH_2(CH_2)_nCH_2Br$, 50-66%) Ar_2C(S)(CH_2)$_n$ ring, n = 1, 2

Ar = Ph, C_6H_4Me-4

Scheme 22

$Ph_2C{=}NPh$ + Yb →(THF/HMPA, room temp.) Ph_2C–N(Ph)–$Yb(HMPA)_3$

26, red-black 45%

→(CH_3COCH_3) $Ph_2CHNHPh$ > 99%

→(CO_2, H_2O) $Ph_2C(NHPh)COO)_3Yb \cdot 2HMPA \cdot 2H_2O$ 92%

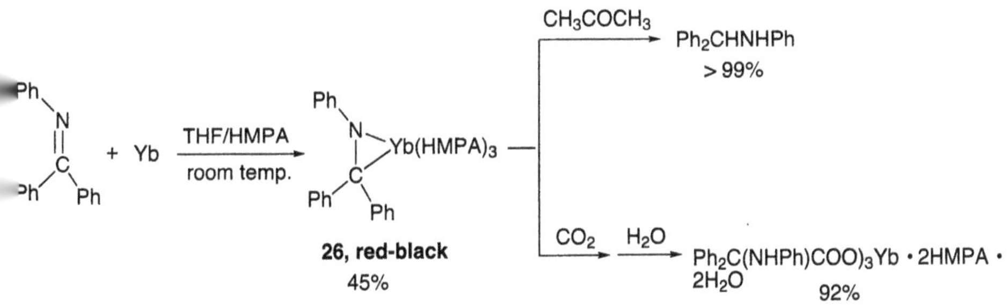

Scheme 23

$Ph_2CHNHPh$ almost quantitatively (Scheme 23) [32]. Isomerization of terminal alkynes to internal alkynes by **26** was also reported [34].

6 Conclusions and Perspectives

The results described in this article have clearly demonstrated that the highly reactive lanthanide ketyl and ketone dianion species can not only be isolated but also their reactivity be controlled through adjusting the environment around the central metals. For example, a ketyl species can be "deactivated" (stabilized) by binding to a bulky metal moiety and "reactivated" by reducing the bulkiness around the metal. Although the examples presented here are limited to those of aromatic ketones, these data should be conceptually useful for the understanding and design of reactions involving other organic carbonyl compounds. In view of the versatile uses of ketyls in organic synthesis and the high nucleophilicity of lanthanide ketone dianion species, if the "deactivation" and "reactivation" of alkyl ketone and/or aldehyde ketyl species can be achieved in a controllable way, and if alkyl ketone and/or aldehyde dianion species can also be generated, numerous classical C-C bond formation processes will be substantially improved and our capability to access a variety of new target compounds will be dramatically enhanced. These topics remain to be challenged.

Acknowledgments. We are grateful to all our coworkers who have so effectively participated in our work described here and whose names are shown in the list of references. Our studies were financially supported by grants from the Ministry of Education, Science, Sports, and Culture of Japan, and by the President's Special Research Grant of The Institute of Physical and Chemical Research (RIKEN).

7
References

1. For some reviews on lanthanide-mediated organic synthesis, see (a) Molander GA, Harris CR (1996) Chem Rev 96:307; (b) Molander GA (1992) Chem Rev 92:26; (c) Molander GA (1991) Samarium and ytterbium reagents. In: Trost BM, Fleming I, Schreiber S (eds) Comprehensive organic synthesis, vol 1. Pergamon, Oxford, p251; (d) Imamoto T (1991) Organocerium reagents. In: Trost BM, Fleming I, Schreiber S (eds) Comprehensive organic synthesis, vol 1. Pergamon, Oxford, p 231; (e) Kagan HB, Namy JL (1986) Tetrahedron 42:6573
2. Imamoto T (1994) Lanthanides in organic synthesis. Academic Press, London
3. For examples of the formation and reactions of other metal ketyl and ketone dianion speices, see: (a) Wirth T (1996) Angew Chem Int Ed Engl 35:61; (b) Fürstner A, Bogdanovic B (1996) Angew Chem Int Ed Engl 35:2442; (c) Huffman JW (1991) Reduction of C=X to CHXH by dissolving metals and related methods. In: Trost BM, Fleming I (eds) Comprehensive organic synthesis, vol 8. Pergamon, Oxford, p107; (d) Robertson GM (1991) Pinacol coupling reactions. In: Trost BM, Fleming I, Pattenden G (eds) Comprehensive organic synthesis, vol 3. Pergamon, Oxford, p563; (e) McMurry JE (1989) Chem Rev 89:1513; (f) Kahn BE, Riecke RT (1988) Chem Rev 88:733; (g) Pradhan SK (1986) Tetrahedron 42:6351
4. For a spectroscopic study of in situ generated lanthanide ketyl species, see Hirota N, Weissman SI (1964) J Am Chem Soc 86:2538
5. For a recent overview on metal ketyl complexes, see Hou Z, Wakatsuki Y (1997) Chem Eur J 3:105
6. Hou Z, Miyano T, Yamazaki H, Wakatsuki Y (1995) J Am Chem Soc 117:4421
7. Hou Z, Fujita A, Zhang Y, Miyano T, Yamazaki H, Wakatsuki Y (1998) J Am Chem Soc 120:754
8. (a) Hou Z, Zhang Y, Wakatsuki Y (1997) Bull Chem Soc Jpn 70:149; (b) Hou Z, Wakatsuki Y (1994) J Chem Soc Chem Commun 1205
9. Hou Z, Wakatsuki Y (1995) J Synth Org Chem Jpn 53:906
10. Hou Z, Zhang Y, Yoshimura T, Wakatsuki Y (1997) Organometallics 16:2963
11. (a) Neumann WP, Schroeder B, Ziebarth M (1975) Liebigs Ann Chem 2279; (b) Ziebarth M, Newmann WP (1978) Liebigs Ann Chem 1765
12. Neumann WP, Uzick W, Zarkadis AK (1986) J Am Chem Soc 108:3762
13. Covert KJ, Wolczanski PT, Hill SA, Krusic PJ (1992) Inorg Chem 31:66
14. Nolan SP, Stern D, Marks TJ (1989) J Am Chem Soc 111:7844
15. Hou Z, Jia X, Wakatsuki Y (1997) Angew Chem Int Ed Engl 36:1292
16. (a) Takats J (private communication). See also Takats J (1997) J Alloys Compd 249:51; (b) Clegg W, Eaborn C, Izod K, O'Shaughnessy P, Smith JD (1997) Angew Chem Int Ed Engl 36:2815.
17. Hou Z, Fujita A, Yamazaki H, Wakatsuki Y (1996) J Am Chem Soc 118:7843
18. Hou Z, Takamine K, Fujiwara Y, Taniguchi H (1987) Chem Lett 2601

19. Hou Z, Takamine K, Aoki O, Shiraishi H, Fujiwara Y, Taniguchi H (1988) J Chem Soc Chem Commun 668
20. Hou Z, Takamine K, Aoki O, Shiraishi H, Fujiwara Y, Taniguchi H (1988) J Org Chem 53:6077
21. Hou Z, Yamazaki H, Fujiwara Y, Taniguchi H (1992) Organometallics 11:2711
22. Yoshimura T, Hou Z, Wakatsuki Y (1995) Organometallics 14:5382
23. Bogdanovic B, Kruger C, Wermeckes B (1980) Angew Chem Int Ed Engl 19:817
24. Hou Z, Yamazaki H, Kobayashi K, Fujiwara Y, Taniguchi H (1992) J Chem Soc Chem Commun 722
25. Hou Z, Fujita A, Yamazaki H, Wakatsuki Y (1998) Chem Commun 669
26. Fleischer EB, Sung N, Hawkinson S (1968) J Phy Chem 72:4311
27. Hou Z, Yoshimura T, Wakatsuki Y (1994) J Am Chem Soc 116:11,169
28. cf. $E_0(Sm^{3+}/Sm^{2+})=-1.55$ V, $E_0(Yb^{3+}/Yb^{2+})=-1.15$ V in aqueous medium
29. Sm(II) is ca. 0.14 Å bigger than Yb(II) in radius when both have the same coordination number. See Shannon RD (1976) Acta Crystallogr Sect A 32:751
30. For examples of Birch reductions, see: Mander LN (1991) Partial reduction of aromatic rings by dissolving metals and by other methods. In: Trost BM, Fleming I (eds) Comprehensive organic synthesis, vol 8. Pergamon, Oxford, p489 and references cited therein
31. (a) Makioka Y, Uebori S. Tsuno M, Taniguchi Y, Takaki K, Fujiwara Y (1996) J Org Chem 61:372; (b) Makioka Y, Uebori S. Tsuno M, Taniguchi Y, Takaki K, Fujiwara Y (1994) Chem Lett 611
32. Makioka Y, Taniguchi Y, Fujiwara Y, Takaki K, Hou Z, Wakatsuki Y (1997) Organometallics 15:5476
33. Takaki K, Tanaka S, Fujiwara Y (1991) Chem Lett 493
34. Makioka Y, Saiki A, Takaki K, Taniguchi Y, Kitamura T, Fujiwara Y (1997) Chem Lett 27

Organo Rare Earth Metal Catalysis for the Living Polymerizations of Polar and Nonpolar Monomers

Hajime Yasuda

Department of Applied Chemistry, Faculty of Engineering, Hiroshima University, Higashi-Hiroshima 739-8527, Japan
e-mail: yasuda@ipc.hiroshima-u.ac.jp

This article deals with the rare earth metal initiated polymerization of polar and nonpolar monomers in a living fashion. For example, $[SmH(C_5Me_5)_2]_2$ or $LnMe(C_5Me_5)_2(THF)$ (Ln=Sm, Y and Lu) conducted the polymerization of methyl methacrylate(MMA) to give high molecular weight syndiotactic polymers (M_n>500,000, syndiotacticity>95%) quantitatively at low temperature (–95 °C). The initiation mechanism was discussed on the basis of X-ray analysis of the 1:2 adduct of $[SmH(C_5Me_5)_2]_2$ with MMA. Synthesis of high molecular weight isotactic poly(MMA) with very narrow molecular weight distribution was for the first time realized by the efficient catalytic function of $Yb[C(SiMe_3)_3]_2$. Living polymerizations of alkyl acrylates (methyl acrylate, ethyl acrylate, and butyl acrylate) were also possible by the excellent catalysis of $LnMe(C_5Me_5)_2(THF)$ (Ln=Sm, Y). By taking advantages of the living polymerization ability, we attempted ABA triblock copolymerization of MMA/butyl acrylate/MMA to obtain rubber-like elastic polymers. Organo rare earth metal complexes such as $LnOR(C_5R_5)_2$ or $LnR(C_5R_5)_2$ conducted the living polymerizations of various lactones such as β-propiolactone, δ-valerolactone and ε-caprolactone, and also conducted the block copolymerizations of MMA with various lactones. Lanthanum alkoxide(III) has good catalytic activity for the polymerization of alkyl isocyanates. Monodisperse polymerizations of lactide and various oxiranes were also achieved by the use of rare earth metal complexes. C_1 symmetric bulky organolanthanide(III) complexes such as $SiMe_2[2(3),4\text{-}(SiMe)_2C_5H_2]_2LnCH(SiMe_3)_2$ (Ln=La, Sm, and Y) show high catalytic activity towards linear polymerization of ethylene. Organolanthanide(II) complexes such as *racemic* $SiMe_2(2\text{-}SiMe_3\text{-}4\text{-}t\text{Bu-}C_5H_2)_2Sm(THF)_2$ as well as C_1 symmetric $SiMe_2[2(3),4\text{-}(SiMe_3)_2C_5H_2]_2Sm(THF)_2$ were found to have high activity for the polymerization of ethylene to give $M_n>10^6$ with M_w/M_n =1.6. Utilizing the high polymerization activity of rare earth metal complexes towards both polar and nonpolar monomers, block copolymerizations of ethylene with polar monomers such as methyl methacrylate and lactones were for the first time realized. 1,4-cis-Conjugated diene polymerization of 1,3-butadiene and isoprene became available by the efficient catalytic activity of $NdCl(C_5H_5)_2/AlR_3$ or Nd(octanoate)$_3$/AlR_3. The Ln(naphthenate)$_3$/$AliBu_3$ system allows selective polymerization of acetylene in *cis*-fashion.

Keywords: Living polymerization, Living copolymerization, Rare earth metal complexes, Alkyl methacrylate, Alkyl acrylates, Lactones, Ethylene, 1-Olefins, Conjugated dienes, Acetylene

1 **Introduction** 256
2 **Highly Stereospecific Living Polymerization of Alkyl Methacrylates** 257
3 **Living Polymerization of Alkyl Acrylates** 262

4 Block Copolymerization of Hydrophorbic and Hydrophilic Acrylates 264
5 Polymerization of Alkyl Isocyanates 265
6 Living Polymerization of Lactones 265
7 Stereospecific Polymerization of Oxiranes 267
8 Polymerization of Lactide 269
9 Stereospecific Polymerization of Olefins 269
10 Block Copolymerization of Ethylene with Polar Monomers 274
11 Polymerization of Styrene 276
12 Stereospecific Polymerization of Conjugated Dienes 276
13 Stereospecific Polymerization of Acetylene Derivatives 279
14 References 281

1 Introduction

Recent development of various living polymerizations of polar and nonpolar monomers allows the synthesis of high molecular weight polymers with very narrow molecular weight distribution. The indispensable conditions for the ideal living polymers are (1) number-average molecular weight M_n>100,000, (2) polydispersity index M_w/M_n<1.05, (3)stereoregularity >95%, and (4) high conversion in a short period of time >95%. Even the famous living polystyrene system [1] and the Group Transfer system [2] do not simultaneously satisfy requirements (1) and (2). Recently, to satisfy a long desire of many polymer chemists, all these four requirements were found to be met in the rare earth metal-initiated polymerization of methyl methacrylate (MMA), which gave M_n>500,000 with M_w/M_n=1.02–1.03 (syndiotacticity >95%) [3]. More recently, we have found isotactic polymerization of MMA by the efficient catalysis of non-metallocene type single component complex, $Yb[C(SiMe_3)_3]_2$, to give the poly(MMA) of M_n>200,000 with M_w/M_n=1.1 in high yield (isotacticity >97%).

Alkyl acrylates were for the first time polymerized in a living fashion with the aid of the unique catalytic action of rare earth metal complexes [4]. Since these monomers have an acidic α-H, termination and chain transfer reactions occur so frequently that their polymerizations generally do not proceed in a living manner. By taking advantages of the living polymerization ability of both MMA and alkyl acrylate, ABA or ABC type tri-block copolymerization was performed to obtain thermoplastic elastomers.

Living polymerization of lactones has been successfully performed by the catalysis of rare earth metal complexes producing M_w/M_n values of 1.07–1.08 [5]. Polymerizations of acrylonitrile and alkyl isocyanates have been successfully realized using $La[CH(SiMe_3)_2]_2(C_5Me_5)$ as initiator, and those of various oxiranes have been made using $Ln(acac)_3/AlR_3/H_2O$ system [6].

Organo rare earth metal initiators also show good activity towards non-polar monomers such as ethylene, 1-olefins, styrene, conjugated dienes, and acetylene derivatives. In fact, poly(ethylene) of M_n>1,000,000 with M_w/M_n=1.6 was made available by the use of a C_1 symmetric divalent complex, $Me_2Si[2(3),4-(SiMe_3)_2C_5H_2]_2Sm(THF)_2$. However, the magnitude of M_n for poly(olefin)s such as poly(1-pentene), polystyrene, and poly(1,5-hexadiene) were not made high enough even with the help of the characteristic catalytic action of rare earth metal complexes. The maximum M_n of 60,000 obtained for poly(1-hexene) was still far below the desired minimum of 400,000. Conjugated dienes and acetylene derivatives were polymerized by the use of rather complex rare earth metal catalysts, but the resulting M_w/M_n remained fairly large. The use of a mononuclear single-site catalyst is required for the polymerization of these derivatives.

2 Highly Stereospecific Living Polymerization of Alkyl Methacrylates

Although various living polymerization systems have been proposed, any of anionic [7], cationic [8], Group Transfer [2], and metal carbene initiated polymerizations [9] achieved no success in synthesizing living polymers of M_n>500,000 with M_w/M_n<1.05. On the other hand, high molecular weight poly(MMA) having an unusually low polydispersity has been synthesized by utilizing the unique initiating function of organolanthanide(III) complexes (Fig. 1) [3]. The relevant complexes include lanthanide hydrides, bulky alkyl lanthanide, bis-Me-bridged trimethylaluminum complexes of alkyllanthanides, and simple alkyl complexes (Fig. 2). The results of the polymerization of MMA with $[SmH(C_5Me_5)_2]_2$ initiator at different temperatures are summarized in Table 1. The most striking is very narrow molecular weight distributions, i.e. M_w/M_n=1.02–1.04 for M_n>60×10^3. Remarkably, $[SmH(C_5Me_5)_2]_2$ complexes give high conversion (polymer yield) in a relatively short period, and allow the polymerization to proceed over a wide range of reaction temperatures from –78 to 40 °C. Furthermore, syndiotacticity exceeding 95% is obtained when the polymerization temperature is lowered to –95 °C.

The typical initiator systems reported so far for the synthesis of highly syndiotactic poly(MMA) are bulky alkyllithium $CH_3(CH_2)_4CPh_2Li$ [10], Grignard reagent in THF [11], and some AlR_3 complexes [12]. Although the

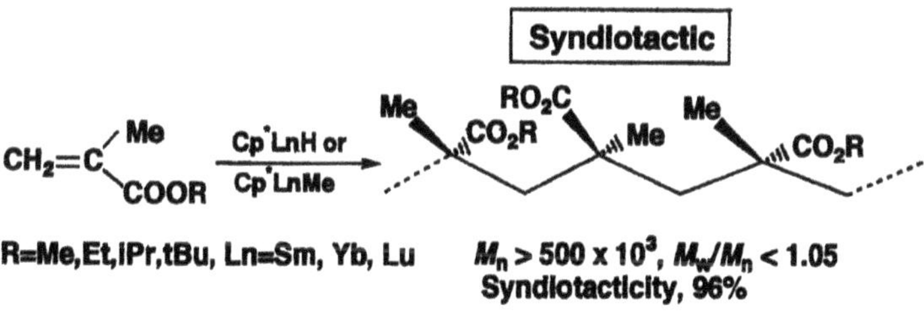

Fig. 1. Organolanthanide initiated syndiotactic polymerization of MMA

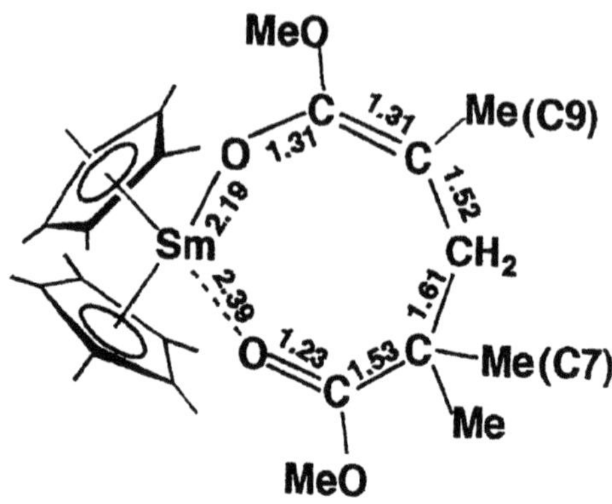

Fig. 2. X-ray analysis of $Sm(C_5Me_5)_2(MMA)_2H$

Table 1. Characterization of monodisperse poly(MMA) synthesized by $[SmH(C_5Me_5)_2]_2$ initiator

Polymerization temperature(°C)	MMA/initiator charged(mol/mol)	$M_n \times 10^3$	M_w/M_n	rr %	Conversion/% (react. time)
40	500	55	1.03	77.3	99 (1 h)
0	500	58	1.02	82.4	99 (1 h)
0	1500	215	1.03	82.9	93 (2 h)
0	3000	563	1.04	82.3	98 (3 h)
-78	500	82	1.04	93.1	97 (18 h)
-95	1000	187	1.05	95.3	82 (60 h)

M_n and M_w/M_n were determined by GPC using standard poly(MMA) with M_w measured by light scattering method. Solvent, toluene; solvent $/[M]_0=5$ (vol/vol).

$CH_3(CH_2)_4CPh_2Li$ initiator in THF conducted the polymerization of MMA at -78 °C, the M_n reached only 10,000 with $M_w/M_n=1.18$, while it gave isotactic polymers in toluene. iBuMgBr and $CH_2=CHC_6H_4CH_2MgBr$ in THF at lower temperature also gave high syndiotactic poly(MMA) but M_n remained as low as 14,000–18,000 and the yields were quite low. When iBuMgBr or *t*BuMgBr was used in toluene instead of THF, the resulting poly(MMA) had a high isotacticity of 96.7% with $M_n=19,900$ and $M_w/M_n=1.08$ [13]. $AlEt_3.PR_3$ complexes gave high syndiotacticity, but not a high molecular weight.

Ketene silyl acetal/nucleophilic agent systems initiate the polymerization of alkyl methacrylates. These well-known Group Transfer systems yielded living polymers with atactic sequences at relatively high temperature [2]. $Me_2C=C(OMe)OSiR_3$ and $R_2POSiMe_3$ can be used as initiators, and tris(dimethylamino)sulfonium bifluoride and Et_4CN are frequently used as catalysts. For example, the M_w/M_n of the resulting poly(MMA) was 1.06 for $M_n=3800$, and 1.15 for

M_n=6300. Thus, we conclude that organolanthanide-initiated polymerization is superior to other catalysts in obtaining monodisperse high molecular weight poly(MMA).

These findings motivated us to isolate the 1 : 1 or 1 : 2 adduct of $[SmH(C_5Me_5)_2]_2$ with MMA in order to elucidate the initiation mechanism. After expending much effort, we have obtained the desired 1:2 adduct as an air-sensitive orange crystal (mp 132 °C). The X-ray analysis of the adduct indicates that one of the MMA unit is linked to the metal in an enolate form while at the other end the penultimate MMA unit is coordinated to the metal through its C=O group. Thus an 8-membered cyclic intermediate was formed (Fig. 3).

Although similar cyclic intermediates have been proposed by Bawn and Ledwith [14] and Cram and Copecky [15] for the isotactic polymerization of MMA, no isolation of such active species has succeeded. On the basis of the X-ray structural data as well as the observed mode of polymerization, we can propose a coodination anionic mechanism involving an 8-membered transition state for the present organo rare-earth metal initiated polymerization of MMA. In the initiation step, the hydride attacks the CH_2 group of MMA, and a transient $SmOC(OCH_3)=C(CH_3)_2$ species should be formed. Then the incoming MMA molecule is supposed to participate in the 1,4-addition to produce an 8-membered cyclic intermediate. Further addition of MMA to the 1:2 addition compound results in liberation of the coordinated ester group and the 8-membered cyclic intermediate is again generated. The intermolecular repulsion between C(7) and C(9) (or the polymer chain) should be the essential factor in determining the syndiotacticity. The addition of 100 molar amounts of MMA to the 1:2 adduct yielded poly(MMA) of M_n=ca. 110×10^3 and M_w/M_n=1.03, proving that the 1:2 adduct is a real active species. Thus, initiation process occurs much faster than the propagation process.

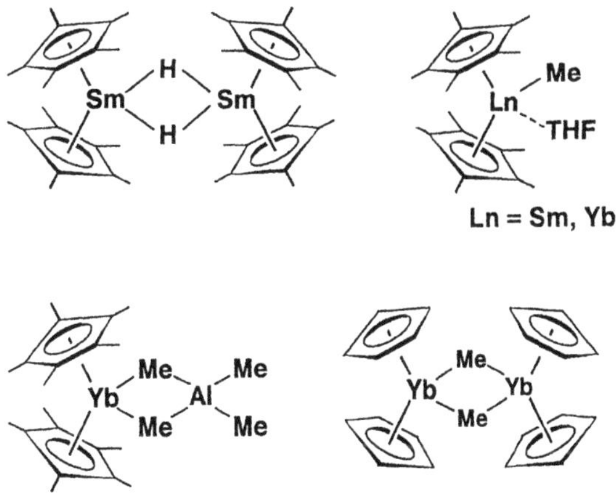

Fig. 3. Typical initiator used for the living polymerization of MMA

On the other hand, monodisperse isotactic poly(MMA) has been obtained with *t*BuMgBr in toluene at –40 °C (M_n=10.1×10^3, M_w/M_n=1.08, isotacticity =96.7%) [16] or *sec*BuMgBr in toluene at –78 °C (M_n=4.9×10^3, M_w/M_n=1.29, isotacticity =95.5%) [16]. Synthesis of isotactic poly(MMA) with much higher molecular weight is required. Recently, high molecular weight isotactic poly(MMA) (mm= 97%, M_n=500,000, M_w/M_n=1.12) was first obtained quantitatively by the use of a non-metallocene type rare earth complex $[(Me_3Si)_3C]_2Yb$ (Fig. 4) [17]. We propose here the initiation mechanism as noted below through noncyclic intermediate for this isotactic polymerization of MMA (Fig. 5). The polymerization should proceed by the enantiomorphic site control in toluene. In contrast, syndiotactic polymer was obtained when THF was used as solvent (rr=87%, M_n= 3.2×10^4, M_w/M_n=1.76) presumably due to chain end control through cyclic intermediate. By utilizing the resulting isotactic polymer, we have prepared a stereo-complex between isotactic and syndiotactic polymers by mixing these two polymers in acetone in a 1:2 ratio (Fig. 6). The resulting stereo-complex shows the intermediate physical property between the homo-isotactic and homo-syndotactic polymers (Table 2). Isotactic polymerization of MMA has also been achieved (mm=94%, M_n=134×10^3, M_w/M_n=6.7) by using $Me_2Si(C_5Me_4)[C_5H_3$-(1S),(2S),(5R)-neomenthyl]LaR (R=$CH(SiMe_3)_2$ or $N(SiMe_3)_2$) [18], while $Me_2Si(C_5Me_4)[C_5H_3$-(1S),(2S),(5R)-menthyl]LnR (Ln=Lu, Sm; R=$CH(SiMe_3)_2$ or $N(SiMe_3)_2$) complex produces syndiotactic poly(MMA) (rr=69%, M_n=177× 26010^3, M_w/M_n=15.7). Polydispersity is rather wide in these cases.

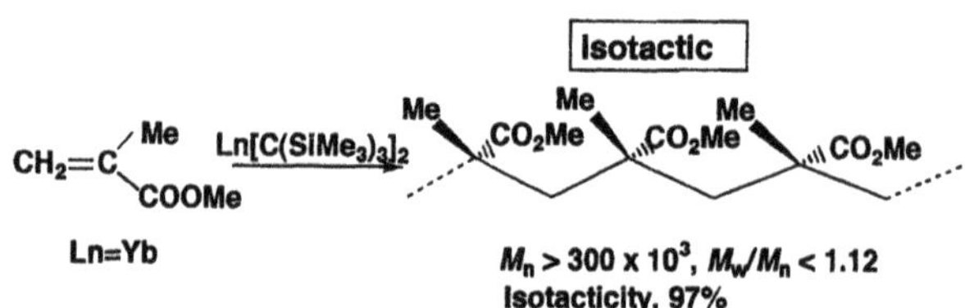

Fig. 4. Isotactic polymerization of MMA catalyzed by rare earth metal complex

Fig. 5. Proposed intermediates for syndiotactic and isotactic polymerization of MMA

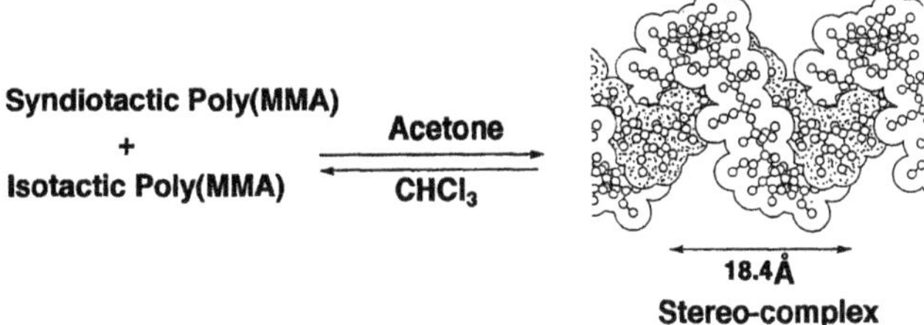

Fig. 6. Preparation of stereo-complex between isotactic and syndiotactic poly(MMA)

Table 2. Mechanical properties of isotactic, syndiotactic poly(MMA) and the stereo-complex

	Tensile modulus kgf/mm^2	Tensile strength kgf/mm^2	Elongation %
isotactic poly(MMA)	2.0	30.2	435
syndiotactic poly(MMA)	10.0	116.1	13
stereo-complex	4.6–5.8	86–121	20–30

Boffa and Novak found a divalent rare earth metal complex, $Sm(C_5Me_5)_2$, to be a good catalyst for polymerization of MMA [19]. The initiation started with the coupling of two coordinated MMA molecules to form Sm(III) species. The bis-allyl initiator, $Sm(\eta^3\text{-}CH_2\text{-}CH\text{-}CH)_2(C_5Me_5)_2$, was also effective for living polymerization of MMA. In this case, MMA must add to both ends of the hexadiene group. In the polymerization of MMA initiated with methylaluminum tetraphenylporphyrine, crowded Lewis acid such as MeAl(ortho-substituted phenolate)$_2$ serves as a very effective accelerator without damaging the living character of polymerization [20]. Thus, the polymer produced has narrow polydispersity (M_w/M_n=1.09) and sufficiently high molecular weight, M_n=25,500, but the stereoregularity is very poor. The use of a simple organoaluminum such as trimethylaluminum causes the occurrence of preferred termination.

The organolanthanide initiators also allowed stereospecific polymerization of ethyl, isopropyl, and *tert*-butyl methacrylates. The rate of polymerization and syndiotacticity decreased with increasing bulkiness of the alkyl group in the order Me>Et>iPr>>*t*Bu. Butyl methacrylate was also polymerized using Nd(octanoate)$_3$/$AliBu_3$ (Al/Nd=7–10), but the molecular weight distribution and stereoregularity were not reported [21].

In general, Ziegler-Natta catalysts such as $TiCl_4/MgCl_2/AlR_3$ and Kaminsky catalysts such as $Cp_2ZrCl_2/(AlMe_2\text{-}O\text{-})_n$ do not catalyze the polymerization of polar monomers. However, a mixture of cationic species $Cp_2ZrMe(THF)^+$ and Cp_2ZrMe_2 has been found to do so for MMA [17], allowing syndiotactic poly(MMA) (rr=80%, M_n=120,000, M_w/M_n=1.2–1.3) [22]. Recently, Soga et al. [23] reported the syndio rich polymerization of MMA catalyzed by $Cp_2ZrMe_2/Ph_3CB(C_6F_5)_4/ZnEt_2$ and also the isotactic polymerization of MMA catalyzed by *rac*-Et(ind)$_2$ZrMe$_2$/Ph$_3$CB(C$_6$F$_5$)$_4$/ZnEt$_2$ (mm=96.5%, M_n=39.3×10^4, M_w/M_n=1.43). The observed characteristic nature of rare earth metal initiator may originate from large ionic radii (1.0–1.17 Å) and low electronegativity (1.0–1.2). Their ionic radii are much larger than those of main group metals (Li 0.73; Mg 0.71; Al 0.68 Å) and the electronegativities resemble those of Li (1.0) and Mg (1.2).

3 Living Polymerization of Alkyl Acrylates

Living polymerization of alkyl acrylates is usually very difficult because the chain transfer or temination occurs preferentially, owing to a high sensitivity of the acidic α-proton to the nucleophilic attack. Exceptions are the living polymerization of a bulky acrylic ester catalyzed by alkyllithium/inorganic salt (LiCl) systems as well as the Group Transfer polymerizations of ethyl acrylate using ZnI$_2$ as the catalyst. Porphyrinatoaluminum initiator systems also induced the living polymerization of *tert*-butyl acrylate [24], but the upper limit of molecular weight attained was ca. 20, 000. We have found the efficient initiating properties of SmMe(C$_5$Me$_5$)$_2$(THF) and YMe(C$_5$Me$_5$)$_2$(THF) for living polymerization of acrylic esters (Table 3), i.e. methyl acrylate (MeA), ethyl acrylate (EtA), butyl acrylate (Bu), and *tert*-butyl acrylate (tBuA), although the reactions were non-stereospecific (Fig. 7) [25]. The initiator efficiency exceeded 90% except for tBuA. We therefore concluded that the reactions occur in living fashion. In fact, the M_n of poly(BuA) initiated by SmMe(C$_5$Me$_5$)$_2$(THF) increased linearly in pro-

Table 3. Polymerization of alkyl acrylates initiated by organolanthanide complexes

Initiator	Monomer	$M_n/10^3$	M_w/M_n	Tacticity/%			Conversion	
				rr	mr	mm	%	Efficiency %
[SmMe(C$_5$Me$_5$)$_2$(THF)]	MeA	48	1.04	30	50	20	99	89
	EtA	55	1.04	51		49	94	86
	nBuA	70	1.05	28	53	19	99	91
	*t*BuA	15	1.03	27	47	26	99	79

Polymerization conditions: 0 °C in toluene, initiator concentration 0.2 mol%.

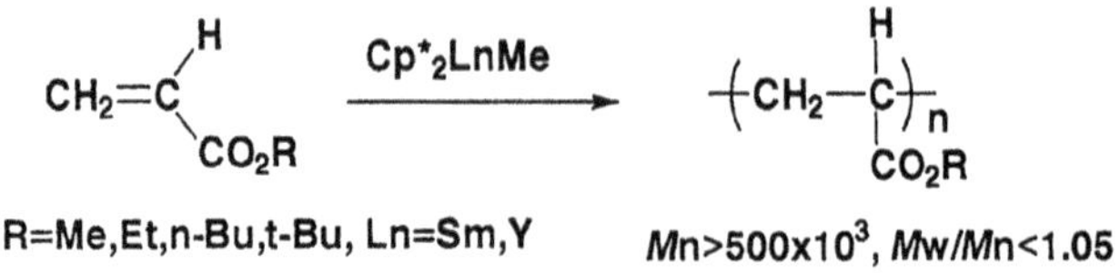

$$CH_2{=}C(H)(CO_2R) \xrightarrow{Cp^*_2LnMe} \left[CH_2-C(H)(CO_2R)\right]_n$$

R=Me,Et,n-Bu,t-Bu, Ln=Sm,Y

Mn>500x10^3, *Mw/Mn*<1.05

Fig. 7. Living polymerization of alkyl acrylates

portion to the conversion, while M_w/M_n remained unchanged, irrespective of the initiator concentration. In order to establish the characteristic nature of these initiation systems, the initiator concentration was decreased from 0.1 to 0.002 mol% of monomer and high molecular weight poly(EtA) of Mn=800,000 with narrow molecular weight distribution (Mw/Mn=1.05) was obtained.

ABA type triblock copolymerization of MMA/BuA/MMA should give rubber-like elastic polymers. The resulting copolymers should have two vitreous outer blocks, where the poly(MMA) moiety (hard segment) associates with the nodules, and the central soft poly(BuA) elastomeric block provides rubber elasticity. The first step polymerization of MMA gave M_n of 15,000 with M_w/M_n=1.04 and then a mixture of MMA and BuA was added to this growing end to result in the formation of desired ABA triblock copolymer (BuA polymerized more rapidly than MMA) (Fig. 8). Table 4 shows the typical mechanical properties of the ABA

Table 4. Mechanical properties of triblock copolymers

Copolymer	Tensile modulus (MPa)	Tensile strength (MPa)	Elongation (%)	Izod impact strength (J/m)	Compression set/% (70 °C, 22 h)
Poly(MMA)	610	80	21	18	100
poly(MMA/BuA/MMA) (20:47:33)	75	27	83	383 (N.B)	101
poly(MMA/BuA/MMA) (25:51:24)	46	22	8	390 (N.B)	103
poly(MMA/BuA/MMA) (8:72:20)	0.8	0.7	163	400 (N.B)	58
poly(MMA/BuA/MMA) (6:91:3)	0.2	0.1	246	410 (N.B)	97
poly(MMA/EtA/EtMA) (26:48:26)	119	22	276	34	62

N.B.; not break. EtA; ethylacrylate. EtMA; ethyl methacrylate

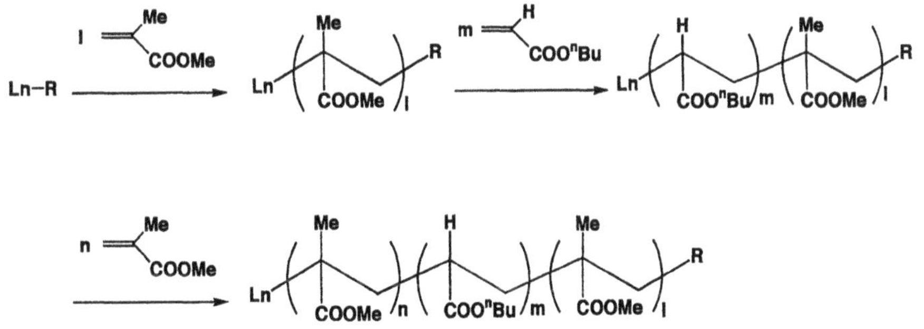

Fig. 8. ABA type block copolymerization of MMA/BuA/MMA

type copolymers thus obtained. Homo-poly(MMA) has large tensile modulus and large tensile strength, but is poor in elongation and shows very large compression set. In contrast, the triblock copolymer (8:72:20) shows 58% compression set and large elongation (163%), indicating that this polymer is a thermoplastic elastomer. The triblock copolymer of MMA/EtA/EtMA in the ratio of 26:48:26 also showed large elongation and small compression set. Thus, elastic copolymers of methacrylate with acrylate were for the first time obtained by using living polymerization ability of rare earth metal complexes.

4 Block Copolymerization of Hydrophobic and Hydrophilic Acrylates

Trimethylsilyl metacrylate (TMSMA) was found to proceed the living polymerization using $SmMe(C_5Me_5)_2(THF)$ as initiator to give high molecular weight syndiotactic polymer (M_n=56,000, rr=92%) with very low polydispersity (M_w/M_n=1.09). By utilizing this nature, we have performed the block copolymerization of TMSMA with MMA or butyl acrylate to obtain the adhesive materials upon hydrolysis of the resulting polymer [26]. The result is shown in Table 5. Thus, block copolymerization of TMSMA with MMA (or BuA) gave thermally stable adhesive materials for the first time (high thermal stability originates from high syndiotacticity) (Fig. 9).

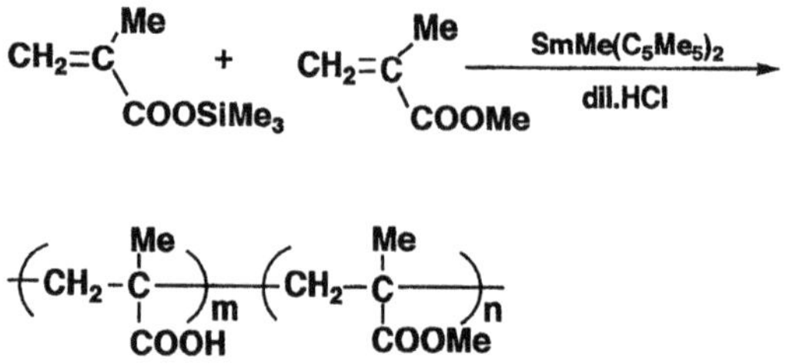

$$CH_2{=}C(Me)COOSiMe_3 + CH_2{=}C(Me)COOMe \xrightarrow[\text{dil.HCl}]{SmMe(C_5Me_5)_2} {-}\!(CH_2{-}C(Me)(COOH))_m{-}(CH_2{-}C(Me)(COOMe))_n{-}$$

Fig. 9. Block copolymerization of MMA with TMSMA

Table 5. Block copolymerization of TMSMA with MMA or BuA

Feed ratio		M_n (calcd)	poly(TMSMA)		poly(TMSMA/MMA or BuA)		Yield/%
TMSMA	MMA		M_n	Mw/Mn	M_n	Mw/M_n	
67	124	23,000	8,900	1.13	28,100	1.17	99
TMSMA	BuA						
30	42	10,100	52,000	1.11	10,500	1.16	98
67	93	22, 500	81,000	1.10	22,600	1.26	99
30	304	43,700	51,000	1.11	77,300	1.15	99

Solvent, toluene; Polymerization of TMSMA 2 min, Polymerization temp. 0 °C.

5 Polymerization of Alkyl Isocyanates

Polyisocyanates have attracted much attention owing to their liquid crystalline properties, stiff-chain solution characteristics, and induced optical activities associated with the helical chain conformation. Pattern and Novak discovered that such titanium complexes as $TiCl_3(OCH_2CF_3)$ and $TiCl_2(C_5H_5)(OCH_2CF_3)$ initiate the living polymerization of isocyanates at ambient temperature, giving polymers with narrow molecular weight distribution [27]. When hexyl isocyanate was added to $TiCl_3(OCH_2CF_3)$ the polymerization took place, to give M_n (M_w/M_n= 1.1–1.3) increasing linearly with the initial monomer-to-initiator mole ratio or the monomer conversion over a wide range. Recently, Fukuwatari et al. found lanthanum isopropoxide to serve as a novel anionic initiator for the polymerization of hexyl isocyanate at low temperature (–78 °C), which led to very high molecular weight (M_n>10^6) and rather narrow molecular weight distribution (M_w/M_n=2.08–3.16) [28]. Other lanthanide alkoxides such as $Sm(OiPr)_3$, $Yb(OiPr)_3$, and $Y(OiPr)_3$ also induced the polymerization of hexyl isocyanate. Furthermore, it was shown that butyl, isobutyl, octyl and *m*-tolyl isocyanates were polymerized using lanthanum isopropoxide as initiator. However, *tert*-butyl and cyclohexyl isocyanates failed to polymerize with these initiators under the same reaction conditions. When the reaction temperature was raised to ambient temperature, only cyclic trimers were produced at high yields. More recently we have found that $La(C_5Me_5)_2[CH(SiMe_3)]_2$ also initiates the polymerization of hexyl isocyanate in 90–96% yields in THF at –20 to –40 °C (M_n=59×10^4, M_w/M_n=1.57–1.90) [29] (Fig. 10).

6 Living Polymerization of Lactones

$AlEt_3$-H_2O or $AlEt_3$-catalyzed polymerizations of 3-methyl-β-propiolactone and ε-caprolactone have been reported [30, 31] but this type of polymerization generally gives a broad molecular weight distribution. We have explored the polymerization of various lactones including β–propiolactone (PL), 3-methyl-β-propiolactone (MePL), δ-valerolactone (VL) and ε-caprolactone initiated by single organolanthanides, and found that VL and CL led to the living polymerization (Fig. 11) , yielding polymers with M_w/M_n=1.05–1.10 at quantitative yields (Table 6). For ε-caprolactone, M_n obtained with the $SmMe(C_5Me_5)_2(THF)$ or

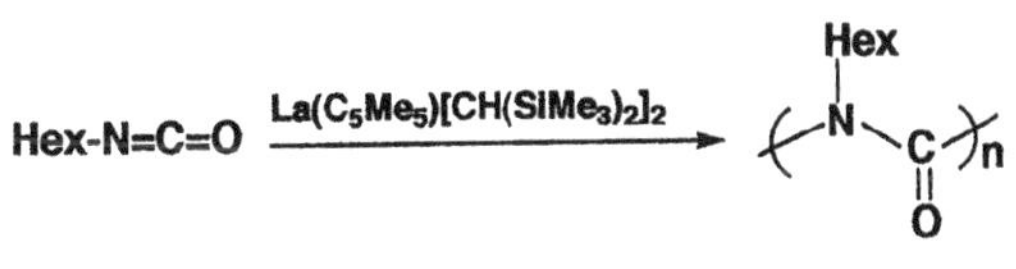

Hex-N=C=O $\xrightarrow{La(C_5Me_5)[CH(SiMe_3)_2]_2}$ $\left(N(Hex)-C(=O)\right)_n$

Fig. 10. Living polymerization of alkyl isocyanate

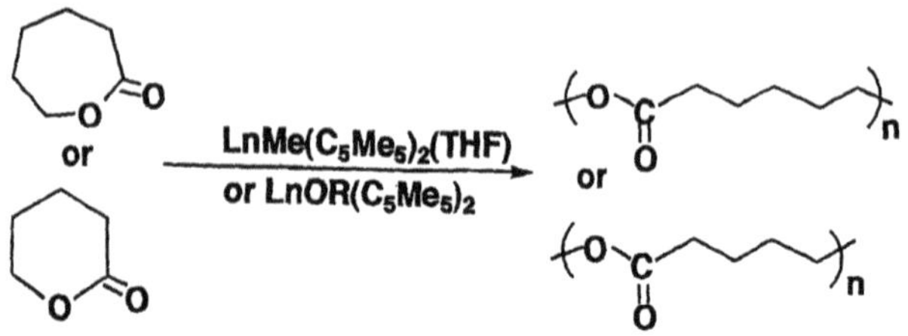

Fig. 11. Living polymerization of lactones

Table 6. Living polymerization of lactones with organolanthanide complexes

Initiator	Monomer	$M_n/10^3$	M_w/M_n	Conversion/%
$SmMe(C_5Me_5)_2(THF)$	VL	75.2	1.07	80.1 (7 h)
	CL	109.4	1.09	92.0 (7 h)
$[SmH(C_5Me_5)_2]_2$	VL	65.7	1.08	90.5 (7 h)
	CL	71.1	1.19	88.7 (5 h)
$[YOMe(C_5H_5)_2]_2$	CL	162.2	1.10	87.5 (5 h)
	PL	60.5	3	94.5 (5 h)

PL β-propiolactone, VL δ-varelolactone, CL ε-caprolactone. Polymerization, 0 °C in toluene.

$[SmH(C_5Me_5)_2]_2$ system increased with increasing conversion, but M_w/M_n remained constant, irrespective of the conversion. For β-propiolactone, the use of $YOR(C_5Me_5)_2$ was more favorable. On the other hand, divalent organolanthanide complexes can initiate the polymerization of lactones, but the resulting polymers had rather broad molecular weight distributions.

At the early stage of the polymerization of lactone with $Ln(OR)(C_5Me_5)_2$, one mole of ε-caprolactone coordinates to the metal, as is the case for the reaction of YCl_3 with ε-caprolactone, giving the six-coordinate mer complex, YCl_3.(ε-caprolactone)$_3$, in which each caprolactone molecule is coordinated as a monodentate ligand through its carbonyl oxygen [32]. The polymerization starts with the coordination of ε-caprolactone to form the 1:1 complex, $Ln(OR)(C_5Me_5)_2$ (ε-caprolactone) (Fig. 12), and in its propagation step the alkoxide attacks the C=O group to produce $Ln[O(CH_2)_5C(O)OR](C_5Me_5)_2$ (ε-caprolactone) [5]. In the $SmMe(C_5Me_5)_2$ initiator system, the reaction is initiated by the attack of ε-caprolactone or δ-valerolactone to result in the formation of an acetal without ring opening. This process has been confirmed by ^{13}C NMR studies on the stoichiometric reaction products.

Anionic block copolymerizations of MMA with lactones proceeded smoothly to give copolymers with M_w/M_n=1.11–1.23 when the monomers were added in this order (Fig. 13).

However, when the order of addition was reversed, no copolymerization took place, i.e. no addition of MMA to the polylactone active end occurred [5].

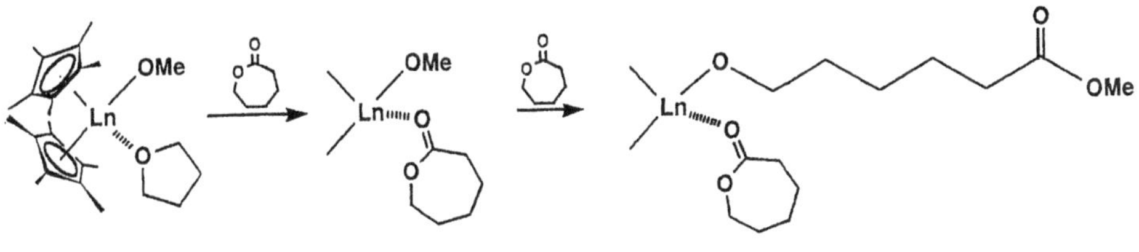

Fig. 12. Initiation mechanism for polymerization of lactones

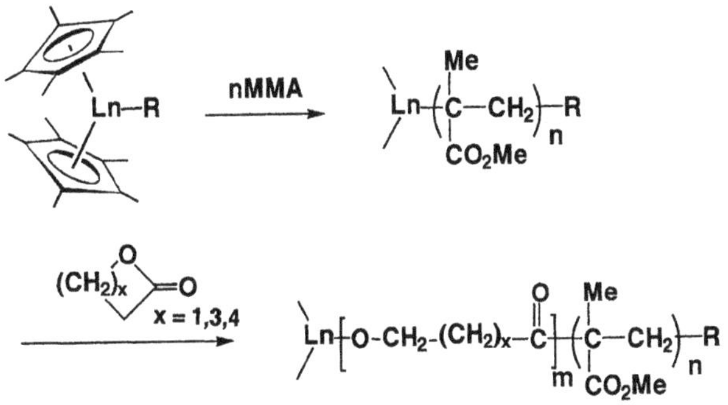

Fig. 13. Copolymerization of MMA with lactones

7
Stereospecific Polymerization of Oxiranes

Organolanthanide(III) complexes such as $LnMe(C_5Me_5)_2$ or $LnH(C_5Me_5)_2$ do not initiate the polymerization of oxirans, but more complex systems like $Ln(acac)_3/AlR_3/H_2O$ and $Ln(2\text{-}EP)_3/AlR_3/H_2O$ (2-EP; di-2-ethylhexylphosphate) polymerize with good initiation activity [33]. High molecular weight poly(ethylene oxide) is one of the common water-soluble polymers useful as adhesives, surfactants, plasticizers, and dispersants as well as for sizing material. Poly(ethylene oxide) of M_n=2.85×10^6 was obtained with the $Y(2\text{-}EP)_3/AliBu_3/H_2O$ system at the ratio of $Y/Al/H_2O$=1/6/3. The initiator activity varies with the molar ratio of the components. Polymerization of propylene oxide was reported to proceed with the $Ln(acac)_3/AlEt_3/H_2O$ system, and light rare earth elements (Y, La, Pr, Nd, Sm) produced very high molecular weight poly(propylene oxide) at Al/Ln=6 in a short period of time (2 h) in toluene [34]. The $Nd(acac\text{-}F_3)_3/AlR_3/H_2O$ (Ln=Y, Nd) systems [35] gave isotactic poly(propylene oxide), while the $Cp_2LnCl/AlR_3/H_2O$, $Sm(OiPr)_3/AlR_3/H_2O$ or Y(2-ethylhexanoate)$_3/AlR_3/H_2O$ system produced relatively low molecular weight isotactic species of this polymer (Fig. 14).

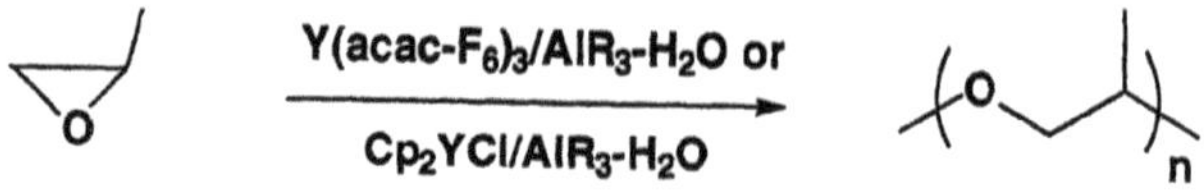

Y(acac-F$_6$)$_3$/AlR$_3$-H$_2$O or Cp$_2$YCl/AlR$_3$-H$_2$O

Fig. 14. Rare earth metal initiated polymerization of oxirans

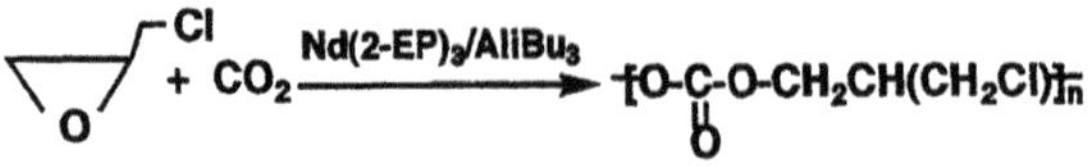

+ CO_2 Nd(2-EP)$_3$/AliBu$_3$ → $-[O-C(=O)-O-CH_2CH(CH_2Cl)]_n-$

Fig. 15. Copolymerization of epichlorohydrine with CO_2

Random copolymerization of propylene oxide with ethylene oxide proceeded smoothly with the Nd(2-EP)$_3$/AlEt$_3$/H$_2$O system at 80 °C [36]. From the copolymerization composition curve, the monomer reactivity ratios were evaluated to be r_1(EO)=1.60 and r_2(PO)=0.45. The conversion increased with an increase in the Al/Nd ratio and saturated at a molar ratio of 16. Since the Nd(2-EP)$_3$/AlEt$_3$/H$_2$O system generates a growing poly(propylene oxide) chain having a very long life, block copolymerization with ethylene oxide can be achieved successfully. The Ln(acac)$_3$/AliBu$_3$/H$_2$O (1:8:4) systems, especially in the case of Nd derivative, also initiated the polymerization of epichlorohydrin (EPH) to yield a polymer of $M_v=16.5\times10^5$ with 21% crystallinity. Toluene is used preferably but aliphatic hydrocarbons is not suitable because poly(epichlorohydrin) precipitates from the solvent during the polymerization. The relative monomer reactivities evaluated for the propylene oxide (PO)-allyl glycidylether(AGE) system were r_1(PO)=2.0 and r_2(AGE)=0.5, and those for the epichlorohydrin-AGE system were r_1(EPH)=0.5 and r_2(AGE)=0.4 [37]. This combination of monomer reactivity ratios indicates that the polymerization with Ln(acac)$_3$/AliBu$_3$/H$_2$O follows a coordination anionic mechanism, but that with the AliBu$_3$/H$_2$O system does cationic polymerization mechanism.

Although the copolymerization of propylene oxide with CO_2 takes place effectively with the organozinc-additives or the (tetraphenyl)porphyrin-AlCl system [38], the copolymerization of epichlorohydrine with CO_2 hardly occurs with these catalysts. Shen et al. [39] showed that a rare earth metal catalyst such as the Nd(2-EP)$_3$/AliBu$_3$ (Al/Nd=8) system was very effective for the copolymerization of epichlorohydrine with CO_2 (30–40 atm) at 60 °C (Fig. 15). The content of CO_2 in the copolymer reached 23–24 mol% when 1,4-dioxane was used as solvent.

The ZnEt$_2$/H$_2$O [40], AlEt$_3$/H$_2$O [41], and Cd salt [42] systems are well-known initiators for the polymerization of propylene sulfide. Shen et al. [43] examined this polymerization with the Ln(2-EP)$_3$/AliBu$_3$/H$_2$O system and found that high molecular weight polymers were produced at a low concentration of Nd (6.04×10^{-3} mol/l) at the ratio of Nd/Al/H$_2$O=1/8/4. The polymerization activity decreased in the order Yb=La>Pr>Nd=Eu>Lu>Gd>Dy>Ho>Er. The ^{13}C-NMR spectrum indicated that β-cleavage occurs preferentially over the α-cleavage and the ratio of these ring openings changes little with the initiator system and the polymerization temperature. The polymers obtained were amorphous according to DSC and XRD analyses.

8 Polymerization of Lactide

Shen et al. [44] succeeded ring opening polymerization of D,L-lactide (racemic species) using $Nd(naphthenate)_3/AliBu_3/H_2O$ (1:5:2.5), $Nd(P_{204})_3/AliBu_3/H_2O$ ($P_{204}=[CH_3(CH_2)_3CH(CH_2CH_3)CH_2O]_2P(O)OH]$, and $Nd(P_{507})_3/AliBu_3$ $/H_2O$ ($P_{507}=(iC_8H_{17}O)_2P(O)OH$) systems, obtaining the polymers whose molecular weights were M_n=3.1–3.6×10^4 and the conversions larger than 94%. When the $Ln(naphthenate)_3/AliBu_3/H_2O$ system was used, nearly the same results were obtained irrespective of the metals used (La, Pr, Sm, Gd, Ho, Tm). Divalent (2,6-tBu$_2$-4-Me-phenyl)$_2$Sm(THF)$_4$ was also found to be active for the polymerization of D,L-lactide at 80 °C in toluene, giving M_n of 1.5–3.5×10^4 [45]. A more recent finding is that the Ln(O-2,6-tBu$_2$-C$_6$H$_3$)$_3$/iPrOH (1:1–1:3) system initiates a smooth homo-polymerization of L-lactide, CL (caprolactone) and VL (valerolactone) to give relatively high molecular weights (M_n>24×10^3) with low polydispersity indices (M_w/M_n=1.2–1.3) (Fig. 16) [46]. Ring-opening polymerization of D,L-lactide was also carried out by using Ln(OiPr$_3$)$_3$ as the catalyst at 90 °C in toluene. The catalytic activity increased in the order La>Nd>Dy>Y and the molecular weight reaches 4.27×10^4(conversion 80%) [47]. However, the molecular weight distribution is not clear at present. The block copolymerization of CL with L-lactide proceeded effectively and gave a polymer of very narrow molecular weight distribution (M_w/M_n=1.16). On the other hand the addition of CL to the living poly(L-lactide) end led to no success.

9 Stereospecific Polymerization of Olefins

Bulky organolanthanide(III) complexes such as $LnH(C_5Me_5)_2$ (Ln=La, Nd) were found to catalyze with high efficiency the polymerization of ethylene. These hydrides are, however, thermally unstable and cannot be isolated as crystals. Therefore, thermally more stable bulky organolanthanides were synthesized by

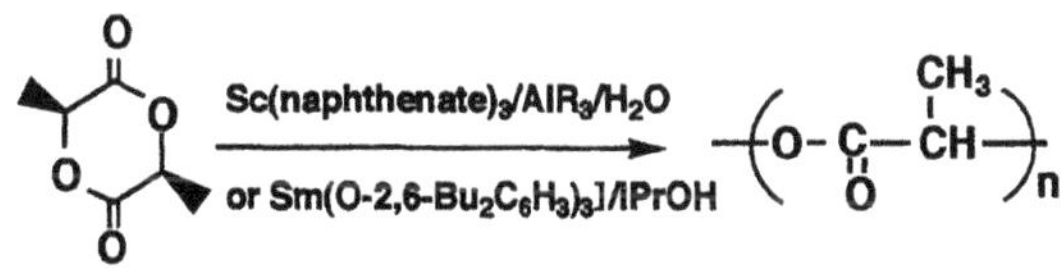

Sc(naphthenate)$_3$/AlR$_3$/H$_2$O or Sm(O-2,6-Bu$_2$C$_6$H$_3$)$_3$]/iPrOH

Fig. 16. Polymerization of D,L-lactide

introducing four trimethylsilyl groups into the Me_2Si bridged Cp ligand, as shown in Fig. 17. The reaction of the dilithium salt of this ligand with anhydrous $SmCl_3$ produced a mixture of two stereo-isomeric complexes. The respective isomers were isolated by utilizing their different solubilities in hexane, and their structures were determined by X-ray crystallography. One of them has a C_2 symmetric (racemic) structure in which two trimethylsilyl groups are located at the 2,4-positions of the Cp rings, while the other has a C_1 symmetric structure in which two trimethylsilyl groups are located at 2,4- and 2,3-positions of each Cp ring. Both were converted to alkyl derivatives when they were allowed to react with bis(trimethylsilyl)methyllithium [48]. The Cp'-Sm-Cp' angle of the *racemic* precursor is 107°, which is about 15° smaller than that of non-bridged $SmMe(C_5Me_5)_2(THF)$.

Meso type ligands were synthesized by forcing two trimethylsilyl groups to be located at the 4-position of the ligand with introduction of two bridges. Actually, the complexation of this ligand with YCl_3 yielded a *meso* type complex, and the structure of the complex was determined by X-ray analysis. The *meso* type alkyl complex was synthesized in a similar manner [49]. Table 7 summarizes the results of ethylene polymerization with these organolanthanide(III) complexes. Interestingly, only C_1 type complexes can initiate the polymerization, implying that the catalytic activity varies with the structure of the complex. The X-ray structure of the C_1 symmetric complex is given in Fig. 18, where the Cp'-Sm-Cp' angle is seen to be 108°, a very small dihedral angle. The polymerization of ethylene with $SmH(C_5Me_5)_2$ in the presence of $PhSiH_3$ formed PhH_2Si capped polyethylene(M_n=9.8 x10^4, Mw/M_n=1.8), and the copolymerization of ethylene with 1-hexene or styrene gave PhH_2Si capped copolymer (comonomer content 60 and 26 mol%, respectively; M_n=3.7×10^3, Mw/M_n=2.9 for ethylene-1-hexene copolymer, M_n=3.3×10^3 for ethylene-styrene copolymer) [50].

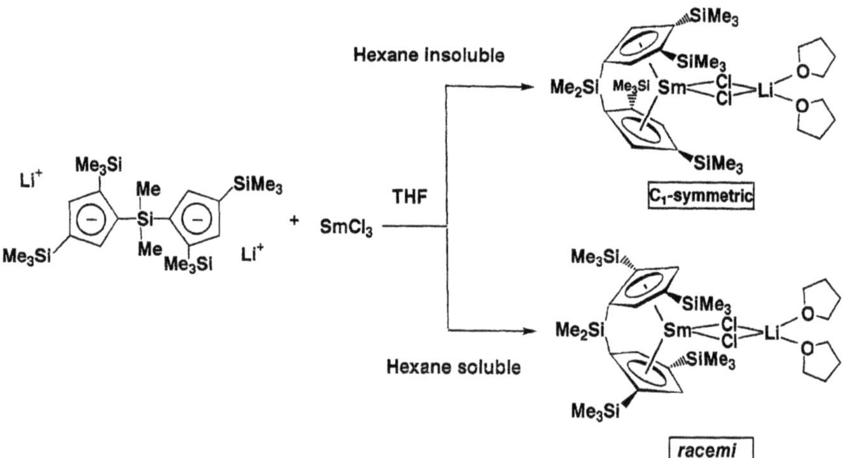

Fig. 17. Synthesis of organolanthanide(III) for polymerization of ethylene

Table 7. Ethylene polymerization by organolanthanide(III) complexes

Initiator	$M_n/10^4$	M_w/M_n	Activity (g/mol•h)
$(C_5Me_5)_2SmCH(SiMe_3)_2$			no polymerization
$SiMe_2[2,4\text{-}(SiMe_3)_2C_5H_2]_2SmCH(SiMe_3)_2$ (*racemic*)			no polymerization
$SiMe_2(Me_2SiOSiMe_2)(3\text{-}SiMe_3\text{-}C_5H_2)YCH(SiMe_3)_2$ (*meso*)			no polymerization
$SiMe_2[2(3),4\text{-}(SiMe_3)_2C_5H_2]_2SmCH(SiMe_3)_2$ (C_1)	41.3	2.19	3.3×10^4
$SiMe_2[2(3),4\text{-}(SiMe_3)_2C_5H_2]_2YCH(SiM_3)_2$ (C_1)	33.1	1.65	18.8×10^4

Initiator concentration, 0.2 mol%. Ethylene was introduced by bubbling at atmospheric pressure.

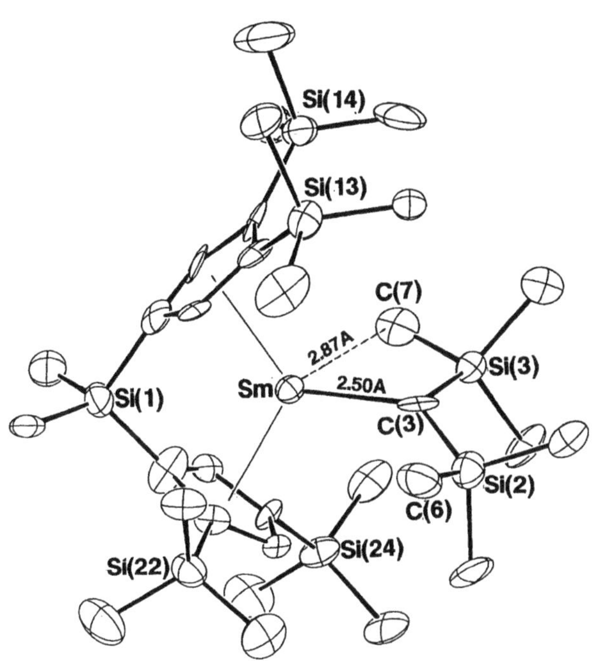

Fig. 18. X-ray structure of C_1 symmetric $SiMe_2[2(3),4\text{-}(SiMe_3)_2C_5H_2]_2SmCH(SiMe_3)_2$

Racemic, *meso*, and C_1 symmetric divalent organolanthanide complexes can be synthesized by allowing the dipotassium salt of the corresponding ligand to react with SmI_2 [51]. Figure 20 shows their structures determined by ^1H NMR and X-ray analyses. Table 8 shows the results of the ethylene polymerization with divalent samarium complexes. It is seen that the *meso* type complex has the

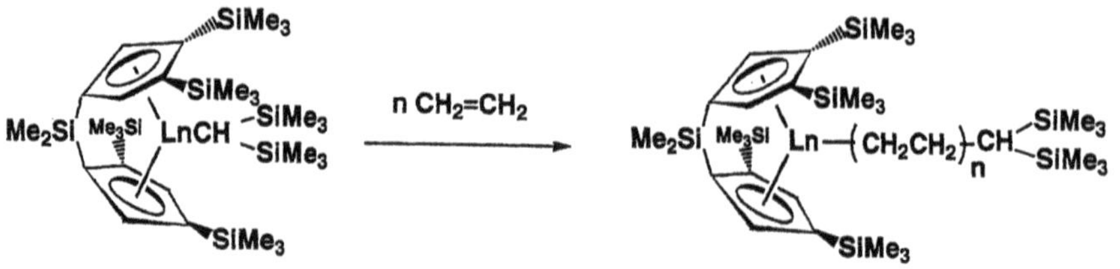

Fig. 19. Initiation mechanism for polymerization of ethylene catalyzed by rare earth metal (III) complexes

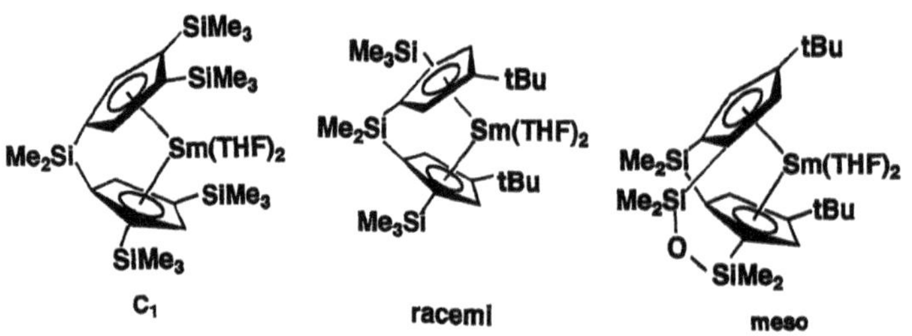

Fig. 20. X-ray and NMR analysis of racemic, meso and C_1 symmetric complexes(II)

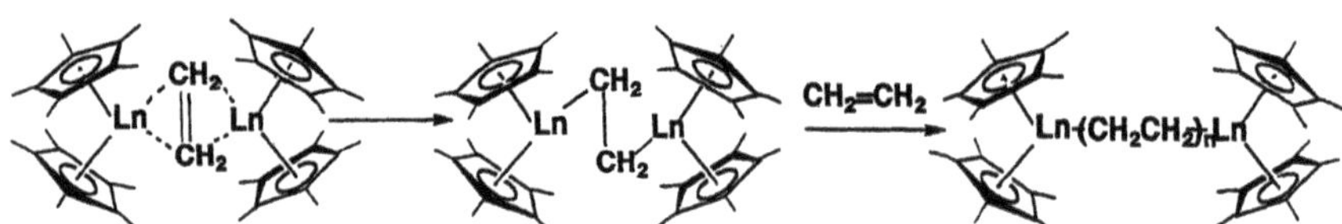

Fig. 21. Initiation mechanism for polymerization of ethylene catalyzed by rare earth meta (II) complexes

highest activity for the polymerization of ethylene, but the molecular weights of the resulting polymers are the lowest. On the other hand, the *racemic* and C_1 symmetric complexes produce much higher molecular weight polyethylene but the activity is rather low.

Particularly, the very high molecular weight polyethylene (M_n>100×10^4) obtained with C_1 complex deserves attention (Fig. 21). For the polymerization of α-olefins, only the *racemic* divalent complex showed good activity at 0 °C in toluene: poly(1-hexene) M_n=24,600, M_w/M_n=1.85; poly(1-pentene) M_n=8,700, M_w/M_n=1.58. Thus, we see that the reactivity of divalent organolanthanide complexes varies depending on their structure. The poly(1-alkene) obtained revealed highly isotactic structure (>95%) when examined by ^{13}C NMR (Fig. 22). The dihedral angles of Cp'-Ln-Cp' of racemic and meso type divalent complexes were 117 and 116.7°, respectively. These values are much smaller than those of

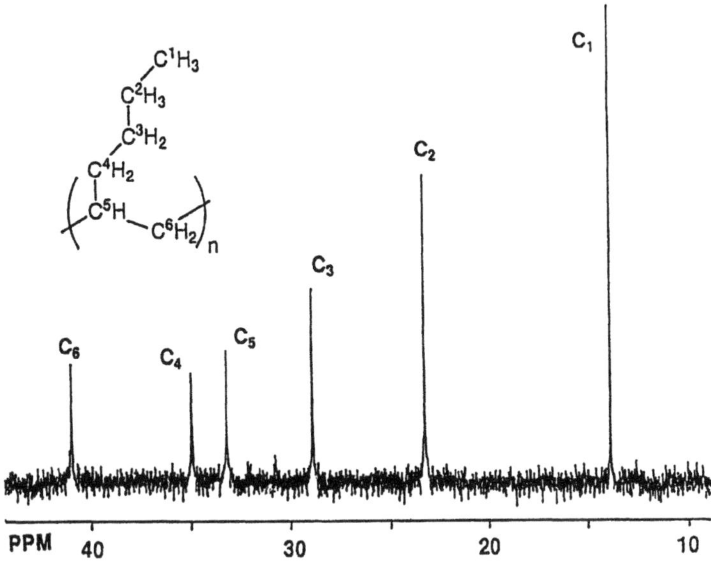

Fig. 22. ^{13}C NMR spectrum of isotactic poly(1-hexene)

Table 8. Ethylene polymerization by divalent samarium complexes

Initiator	Time min	$M_{n}/10^4$	M_w/M_n	Activity (g/mol•h)
$(C_5Me_5)_2Sm(THF)_2$	1	2.28	1.25	43
		2.46	2.28	41
Racemic	1	11.59	1.43	6
		35.63	1.60	14
Meso	5	1.94	3.29	14
		4.73	3.49	47
C_1 Symmetry	5	100.8	1.60	1.6
		>100.0	1.64	1.1

$Sm(C_5Me_5)_2(THF)$ (136.7°) [52] and $Sm(C_5Me_5)_2$ (140.1°) [53]. Therefore, it can be concluded that the complexes having smaller Cp'-Ln-Cp' angles are more active for the polymerization of ethylene and α–olefins.

More recently, YR_2[tris(pyrazoyl)borate] of trivalent state was found to have excellent activity for the polymerization of ethylene (Fig. 23). When R is C_6H_5, resulting M_w reaches 1.2×10^6 (M_w/M_n=4.10). However, the present initiator is inactive for the polymerization of 1-olefins such as propylene and 1-hexene [54].

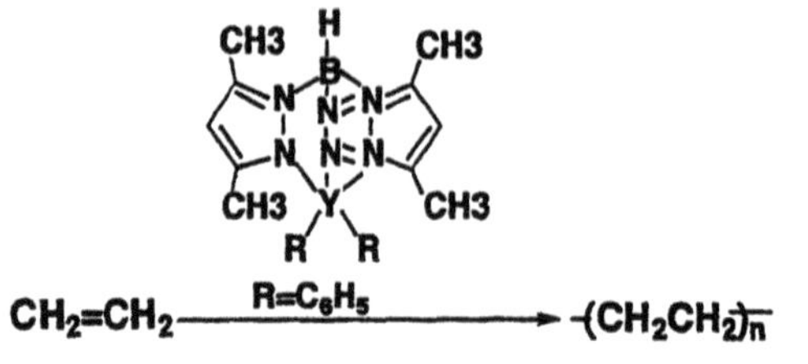

$$CH_2{=}CH_2 \xrightarrow{R=C_6H_5} -(CH_2CH_2)_n-$$

Fig. 23. YR_2[tris(pyrazoyl)borate] initiated polymerization of ethylene

$$(C_5Me_5)_2Sm(CH_2CH_2)_nR \xrightarrow{mMMA}$$

$$(C_5Me_5)_2Sm{-}O{-}C(OMe){=}C(Me){-}CH_2{-}(C(Me)(COOMe){-}CH_2)_{m-1}{-}(CH_2CH_2)_n\,R$$

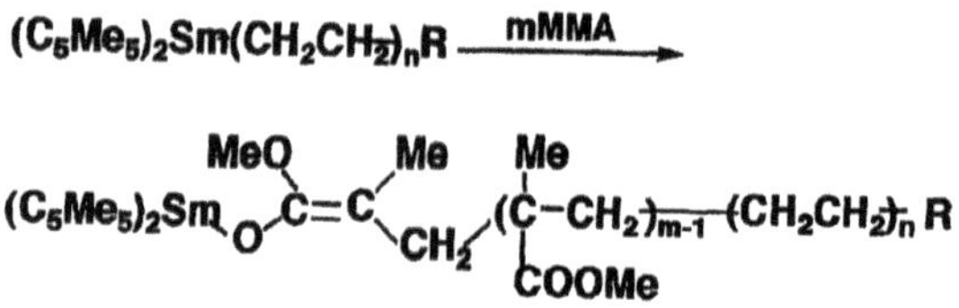

Fig. 24. Block copolymerizations of ethylene with MMA

10 Block Copolymerization of Ethylene with Polar Monomers

Block copolymerization of ethylene or propylene with polar monomers is yet to be attained in polyolefin engineering. The success of this type of block copolymerization should give hydrophobic polymeric materials having remarkably high adhesive, dyeing, and moisture adsorbing properties. The following is the first example of a well-controlled block copolymerization using the unique dual catalytic function of $LnR(C_5Me_5)_2$ (Ln=Sm, Yb, and Lu; R=H, Me) complexes toward polar and nonpolar olefins [55]. Ethylene was copolymerized with MMA first by the homopolymerization of ethylene (17–20 mmol) with $SmMe(C_5Me_5)_2$(THF) or $[SmH(C_5Me_5)_2]_2$(0.05 mmol) at 20 °C in toluene under atmospheric pressure, and then sequential addition of MMA (10 mmol) (Table 9). The initial step proceeded very rapidly, completed in 2 min, and gave a polymer of M_n=ca. 10,100 and M_w/M_n=1.42–1.44. However, the second step was rather slow, with the reaction taking 2 h at 20 °C (Fig. 24). The polymer obtained was soluble in 1,2-dichlorobenzene and 1,2,4-trichlorobenzene at 100 °C but insoluble in THF and $CHCl_3$. This fact indicates quantitative conversion to the desired linear block copolymer. Repeated fractionation in hot THF did not change the molar ratio of the polyethylene and poly(MMA) blocks, though poly(MMA) blended with polyethylene can easily be extracted with THF. With the copolymerization, the elution maximum in GPC shifted to a higher molecular

Table 9. Block copolymerization of ethylene with polar monomers

Polar monomer	Polyethylene block[a]		Polar polymer block[b]		Unit ratio
	$M_n/10^3$	M_w/M_n	$M_n/10^3$	M_w/M_n	
MMA	10.3	1.42	24.2	1.37	100:103
		1.39	12.8	1.37	100:13
MeA	6.6	1.40	15.0	1.36	100:71
		2.01	3.0	1.66	100:4
EtA	10.1	1.44	30.8	2.74	100:85
		1.97	18.2	3.84	100:21
VL	10.1	1.44	7.4	1.45	100:20
CL	6.6	1.40	23.9	1.76	100:89

[a] Determined by GPC using standard polystyrene
[b] Determined by ^{1}H NMR
Polymerization was carried out at 0 °C.

weight region, with its initial unimodal pattern unchanged. The relative molar ratio of the polyethylene and poly(MMA) blocks was controllable at will in the range of 100:1 to 100:103 if the M_n of the initial polyethylene was fixed to ca. 10,300. ^{1}H and ^{13}C NMR spectra for the copolymers as well as their IR absorption spectra were superimposable onto those of the physical mixtures of the respective homopolymers. The molar ratio of the poly(MMA) and polyethylene blocks, however, decreased as the M_n of the prepolymer increased, especially when M_n exceeds over ca. 12,000 at which value polyethylene began precipitating as fine colorless particles. It is noteworthy that smooth block copolymerization of ethyl acrylate or methyl acrylate to the polyethylene growing chain (M_n=6600–24,800) can be realized by the sequential addition of the two monomers.

We have extended the above work to the block copolymerization of ethylene with lactones. δ-Valerolactone and ε-caprolactone were incorporated to the growing polyethylene end at ambient temperature and the expected AB type copolymers (100:1 to 100:89) were obtained in high yield.

The treatment of the resulting block copoly(ethylene/MMA) (100:3, M_n= 35,000) and block copoly(ethylene/ε–caprolactone) (100:11, M_n=12,000) with dispersed dyes (Dianix AC-E) made them deeply dye with three primary colors, though polyethylene itself was inert to these dyes. Reversed addition of the monomers (first MMA or lactones and then ethylene) induced no block copolymerization at all, even in the presence of excess ethylene, and only homo-poly(MMA) and homo-poly(lactone) were produced. Hence, these copolymers can be said to have a very desirable chemical reactivity.

More recently, Yang et al. [56] have examined a new approach in which a reactive functional group was introduced into polyolefins using methylenecyclopropane. Thus, ethylene (1.0 atm) was copolymerized with methylenecyclopropane (0.25–2.5 ml) using $[LnH(C_5Me_5)_2]_2$ (Ln=Sm, Lu) in toluene at 25 °C, and it was shown that 10–65 of exo-methylenes were incorporated per 1000 $-CH_2-$ units. The resulting polymer had a M_w of 66–92×10^3. Yet its M_w/M_n was larger than 4.

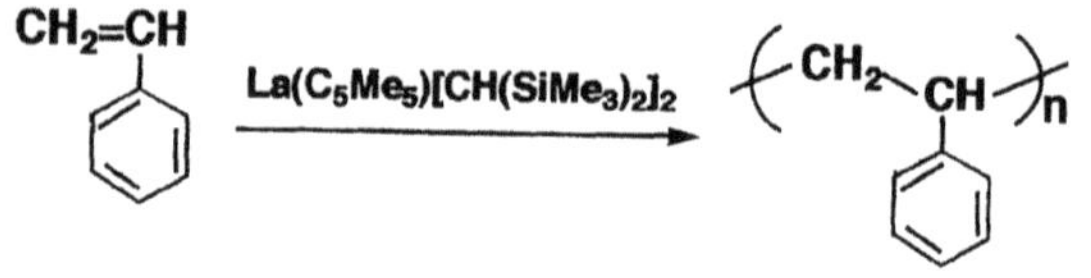

Fig. 25. Styrene polymerization by $La(C_5Me_5)[CH(SiMe_3)_2]_2$

11 Polymerization of Styrene

Styrene polymerization was performed by using binary initiator systems such as $Nd(acac)_3/AlR_3$ or $Nd(P_{507})_3/AlR_3$. *Syndio*-rich polystyrene was obtained at a ratio of Al/Nd=10–12 [57]. More recently, it was shown that the $Gd(OCOR)_3/iBu_3Al/Et_2AlCl$ catalytic system initiates the copolymerization of styrene with butadiene, but gives only atactic polystyrene [58]. The $Sm(OiPr)_3/AlR_3$ or $Sm(OiPr)_3/AlR_2Cl$ (Sm/Al=1–15) catalytic system also initiates the polymerization of styrene to give a high molecular weight polymer (M_n= 300,000), low in polydispersity but atactic in stereoregularity [59]. The cationic polymerization of styrene using $Ln(CH_3CN)_9(AlCl_4)_3(CH_3CN)$ was also examined [60], with the finding that the activity increased in the order La (conversion 73%)>Tb=Ho>Pr=Gd>Nd>Sm=Yb>Eu (conversion 54%), while M_n decreased with increasing the polymerization temperature from 0 (20×10^3) to 60 °C (13×10^3). A more recent study showed that the single component initiator $[(tBuCp)_2LnCH_3]_2$ (Ln=Pr, Nd, Gd) initiated the polymerization of styrene at relatively high temperature, 70 °C, with a conversion of 96% and the M_n of 3.3×10^4 for $[(tBuCp)_2NdCH_3]_2$ [61], though stereoregularity was very poor. The activity varied greatly with the lanthanide element; and catalytic activity increased in the order Nd>Pr>Gd>>Sm=Y (the Sm and Y complexes showed practically no activity). Therefore, the reaction is supposed to follow the radical initiation mechanism. Styrene polymerization was also performed successfully using the single component initiators, $[(Me_3Si)_2N]_2Sm(THF)_2$, $[(Me_3Si)_2CH]_3Sm$, and $La(C_5Me_5)[CH(SiMe_3)_2]_2(THF)$ at 50 °C in toluene without addition of any cocatalyst (Fig. 25). The resulting polymers had M_n=1.5–1.8×10^4 and M_w/M_n=1.5–1.8, and show atactic properties [62].

Thus no success has yet been achieved in synthesizing syndiotactic polystyrene with rare earth metal complexes, in contrast to the synthesis of highly syndiotactic polystyrene with $TiCl_3(C_5Me_5)/(AlMe\text{-}O\text{-})_n$ (syndiotacticity >95%) [63,64].

12 Stereospecific Polymerization of Conjugated Dienes

Organolanthanide(III) based binary initiator systems were used by Yu et al. [65] for stereospecific polymerization of butadiene and isoprene. Typically, the polymerization of butadiene catalyzed by $C_5H_5LnCl_2$.THF/AlR_3 yielded polymers with a *cis*-1,4-content as high as 98% (Fig. 26). The polymerization activity decreased in the order Nd>Pr>Y>Ce>Gd and iBu_2AlH>iBu_3Al>Et_3Al>Me_3Al, while the viscosity of the polymer decreased in the order Et_3Al>iBu_3Al> iBu_2AlH. Although the $NdCl_3$/iBu_3Al system exhibited practically no initiating activity, the use of $NdCl_3$/PrOH instead of the solvent-free metal chloride brought about high polymerization activity and high stereoregularity in the *cis*-1,4-polymerization of butadiene [66]. The M_n of the polymer increased linearly with increasing conversion and reached 1530×10^3 at 85% conversion when $NdCl_3$•iPrOH/$AlEt_3$ (1:10) was used in heptane at –70 °C, but M_w/M_n (1.8–2.5) showed no change with conversion. The number N of polymer chains per metal atom was 1.09–1.43 at –70 °C, and increased to 2.0–3.0 when the polymerization temperature was raised to 0 °C. Most noteworthy is a very high *cis*-content realized at –70 °C, which amounted to 99.4%.

This indicates the existence of the anti-π-allyl-Nd species rather than the syn-π-allyl-Nd species in the polymerization system. The *cis*-1,4-content of poly(butadiene) increased as the $AlEt_3$ concentration was lowered [66].

The arene organolanthanide system, $Nd(C_6H_6)(AlCl_4)_3$/$AliBu_3$ (Al/Nd=30), also induces the catalytic polymerization of isoprene to give *cis*–1,4 polymers in 92–93% selectivity at low conversion (17–36%). Neither of the $Nd(C_6Me_6)(AlCl_4)_3$/$AliBu_3$ (1:30) nor $NdCl_3$/$AlCl_3$/$AliBu_3$ (1:3:30) systems showed catalytic activity for the polymerization of isoprene [67]. The random copolymerization of isoprene with butadiene went smoothly with the use of the $Nd(C_6H_6)(AlCl_4)_3$/$AliBu_3$ system and gave an isoprene/butadiene(1:4) copolymer at high yield, but no data for *Mw*/*Mn* and *Mn* were reported [68]. The (β-CH_3-π-allyl)$_2LnCl_5Mg_2(TMEDA)_2$/AlR_3 and (allyl)$_4$LnLi systems also initiate the 1,4-polymerization of isoprene in 50% stereoregularity at high conversion. Highly selective *cis*-1,4-polymerizations of conjugated dienes were obtained by the use of homogeneous $(CF_3COO)_2NdC1$•EtOH/$AlEt_3$ (1:7) initiator

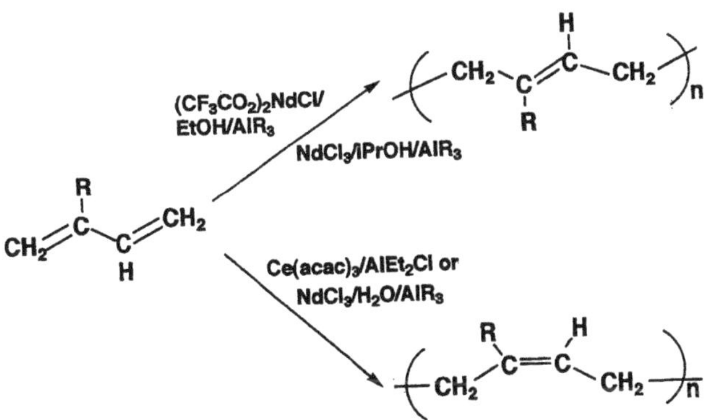

Fig. 26. Rare earth metal initiated polymerizations of conjugated dienes

system, i.e. 97.5% *cis*-selectivity for butadiene and 96.7% for isoprene. Although bimetallic species like $(CF_3COO)EtNd(\mu\text{-}Cl)(\mu\text{-}H)AlEt_2$ are proposed as active one, their exact structures are still unknown [69]. The molecular structure of dimer complexes, $[(CF_3C)(\mu^2\text{-}O)_2(\mu^3\text{-}O)_4YAlEt_2(THF_2)]_2$ and $[(CF_3)(\mu^2\text{-}O)_3(\mu^3\text{-}O)_3NdAlEt_2(THF_2)]_2$, generated during the reaction between $(CF_3COO)_2NdC1$ and $AlEt_3$, has recently been elucidated by X-ray analysis. However, these complexes are considered to be byproducts because they give polymers of low stereoregularity only at low yield [70].

A remarkable solvent effect was observed on the activity of the $Nd(OCOC_7H_{15})_3/Et_2AlCl/iBu_3Al$ system, which initiates the *cis*-1,4-polymerization of butadiene (>98%) and isoprene (>95%). Aliphatic compounds such as pentane, 1-pentene, and 2-pentene behave as good solvents, while aromatic compounds such as toluene and mesitylene act as inhibitors [71]. The coordination of aromatic compounds to the metal center may be responsible for the remarkable suppression of the polymerization, as was the case in the polymerization using cobalt catalysts [72]. A binary initiator system, $LnCl_3[(BuO)_3PO]_3/AliBu_3$, and a ternary system, $Ln(naphthenate)_3/AliBu_3/Al_2Et_3Cl_3$ (naphthenate; $C_{10}H_7COO$), also behave as good initiators for the *cis*-1,4-polymerization of isoprene [73]. In both cases, the polymerization activity varies with the nature of the metal in the order Nd>Pr>Ce>Gd>La>>Sm>Eu. Thus, the activity increases with increasing electronegativity and does so with decreasing M^{3+} ionic radius, except for Sm^{3+} and Eu^{3+} which can be easily converted to the M^{2+} species in the presence of a reducing agent. No relationship was observed between the initiating activity and the Ln-O or Ln-Cl bond energy determined by IR and laser Raman spectroscopy.

Block copolymerization of butadiene with isoprene (32:68–67:33) giving high *cis*-1,4-polymers have also been successfully made with the $Ln(naphthenate)_3/AliBu_3/Al_2Et_3Cl_3$ system at temperature –78 to 33 °C. Noteworthy is the relatively long life time of this initiator. Thus, it was possible to copolymerize isoprene 1752 h after the polymerization of butadiene. The $(iPrO)_2HLn_2Cl_3HAlEt_2$ species (Ln=Gd, Dy, Er, Tm) [74] prepared from either $Ln(iPrO)_3/Et_2AlCl/Et_3Al$ or $(iPrO)_2LnCl/Et_3Al$ also can initiate the *cis*-1,4-polymerization of butadiene and isoprene. The most probable structure of this complex as evidenced by X-ray analysisis is (iPrO)HLnEt(Cl)AlHEt(Cl)LnCl(OiPr). The polymerization activity decreased in the order Gd>Dy>Er>Tm and the *cis*-content ranged from 92 to 95% in the case of Gd derivatives. Random copolymerization of butadiene with isoprene was also performed using $Nd(C_6H_6)(AlCl_4)_3/AlR_3$ (Al/Nd=30) in benzene. Both monomers were incorporated in the copolymer selectively with *cis*–1,4-butadiene 96.1–96.4% and *cis*–1,4-isoprene 97.5–98.3%. The conversion increased with increase in the polymerization temperature from 0 (10%) to 80 °C (80–100%) [75].

Some rare earth metal based initiators induce the *trans*-1,4-polymerization of conjugated dienes at high yield. The $Ce(acac)_3/AlEt_2Cl$ system as well as the $CeCl_3/AlEt_2Cl$ and $GdCl_3/AlEt_3$ systems were most effective catalyst system to carry out this type of polymerization for isoprene with high selectivity (91–

97%) [76]. The marked difference in the selectivity between the $Ce(acac)_3/AlEt_3$ and the $C_5H_5LnCl_2/AlR_3$ or $(iPrO)_2LnCl/AlR_3$ initiator systems may be due to a specific action of small amounts of water present in the system. Actually, metal compound hydrates like $NdCl_3{\cdot}6H_2O$, $PrCl_3{\cdot}6H_2O$ and $UO_2(C_2H_3O_2)_2{\cdot}6H_2O$ can initiate the *trans*-polymerization in the presence of AlR_3 [77]. Prolonged aging of the initiator system decreased the activity significantly, presumably owing to an irreversible self-reaction of the intermediate generated from the organolanthanide and water. A precise recent study on the effect of water has revealed that a maximum conversion was attainable at an $H_2O/AlEt_2Cl$ ratio of 1.1–1.2, which produces $(AlEt\text{-}O\text{-})_n(AlCl\text{-}O)_m$ species. However, the molecular weight was independent of the amount of water added to $AlEt_2Cl$.

Butadiene-styrene copolymerization was attempted using the $L_3Ln\text{-}RX\text{-}AlR_3$ system [78]. Especially, $(CF_3COO)_3Nd/C_5H_{11}Br/AliBu_3$ (1:3:15) was found to be active for this type of copolymerization, with the *cis*-content of butadiene unit amounting 97.8% and the styrene content to ca. 32%. However, for the system of isoprene/styrene, the *trans*-1,4-polyisoprene copolymer was produced exclusively.

13 Stereospecific Polymerization of Acetylene Derivatives

Polyacetylene (PA) is one of the simplest conjugated polymers, useful for manufacturing lightweight high energy density plastics for storage batteries, solar energy cells, etc. Acetylene can be polymerized to give high *cis* PA film with $Ti(OiBu)_4/AlEt_3$ [79] or $Co(NO_3)_2/NaBH_4$ [80] at temperatures lower than –78 °C. Recently, it has been reported that $Ln(naphthenate)_3/AlR_3/Donor$ (1:10:2–3) (Donor=acetone, ether, ethyl acetate) systems also can initiate stereoregular *cis* polymerization of acetylene at 30 °C, which leads to silvery metallic film [81, 82]. The polymer yield increased with increasing polymerization temperature from –15 to 45 °C. A *cis*-polyacetylene with 95% selectivity was obtained when the Al/Ln ratio was adjusted to ca. 5. The polymerization activity decreased in the order Y=Ce>Nd=Tb>Pr>La>Lu>Gd>Tm=Er>Ho=Yb=Eu>Sm>Dy. The *trans* content of the film amounted to 100% when the temperature was raised to 180 °C. The elements leading to PA film with a *cis* content exceeding 95% are La, Pr, Nd, Sm, Gd, Tb, Dy, Ho, Er, Tm, and Y. The electrical conductivity of the films was $294{\times}10^{-8}\,Scm^{-1}$ for La, $181{\times}10^{-8}\,Scm^{-1}$ for Nd, $194{\times}10^{-8}\,Scm^{-1}$ for Gd, $490{\times}10^{-8}\,Scm^{-1}$ for Tb, and $184{\times}10^{-8}\,Scm^{-1}$ for Tm. Differential scanning calorimetry revealed two exothermic peaks at 200 and 380 °C and an endothermic peak at 460 °C. These peaks were attributed to *cis-trans* isomerization, hydrogen migration, and chain decomposition, respectively [82].

$Sc(naphthenate)_3/ROH/AlR_3$ (1/2/7) has been found to exhibit an activity similar to the lanthanide series catalyst [83] (Fig. 27). The *cis* PA film obtained with it showed an electrical conductivity of $14.4\,Scm^{-1}$ when the polymer was doped with I_2 at a ratio of $(CHI_{0.04})_n$, and the TEM measurement suggested the formation of fibrils of about 20–30 nm in size.

The $(P_{204})_3Ln/AlR_3$ system also exhibited good activity for the polymerization of acetylene when Al/Ln ratio was 5 [83]. The polymerization was conducted by the conventional method, and polymers with silver metallic appearance were obtained. The addition of an oxygen-containing donor was effective for enhancing the polymerization rate and the *cis*-content. The effects were especially marked for P_{204} (PO/Nd=1.1). The activity decreased in the order Nd=Tb>Ce>Pr=Y>La>Er>Ho>Sm=Eu>Yb=Lu>Gd>Tm>Dy and the *cis*-content decreased in the order Pr (95%)>Lu=Tb=Dy (92%)>Er=Y=Sm=Gd (87–89%). The polymerization activities of $Nd(P_{507})_3$, $Nd(P_{204})_3$, and $Nd(P_{215})$ were compared and found to increase in this order, the result is consistent with the basicities of the ligands ($P_{507}H$=pKa 4.10, $P_{204}H$=pKa 3.32, $P_{215}H$=pKa 3.22). The M-C bond is supposed to weaken as the electron donating ability of the ligand increases. The $Nd(iPrO)_3/AlEt_3$ (Al/Nd=10) system [84] was also shown to be a good initiator for the polymerization of acetylene. The soluble fraction obtained was considered to be *trans*-polyacetylene, which was shown to have a molecular weight of 277–540. Its ^1H NMR spectrum revealed methyl groups at δ= 0.826 ppm and terminal vinyl groups at 4.95 ppm.

Phenylacetylene was polymerized to give a polymer of high *cis* configuration by the use of the $Ln(naphthenate)_3/AlEt_3$ system [85, 86], with the activity decreasing in the order Gd>Lu>Nd=Ce>Ho>Sm>Dy=Eu>Er>Pr>La>Y=Tm>Yb, and the *cis*-content exceeding 90%. It had M_n and M_w of 2×10^5 and 4×10^5, respectively, and was crystalline according to XRD and SEM measurement. Its softening point was in the range 215–230 °C. Other terminal alkynes such as 1-hexyne, 1-pentyne, 3-methyl-1-pentyne, 4-methyl-1-pentyne, 3-methyl-1-butyne, and phenylacetylene were found to polymerize quantitatively in the *cis*-fashion with the $Ln(naphthenate)_3/AlR_3/C_2H_5OH$ (Ln=Sc, Nd) or $Ln(P_{204})_3/AlR_3/C_2H_5OH$ (1:7:3) system. The highest molecular M_n obtained was 16.8×10^4 for poly(1-pentyne). Trimethylsilylacetylene was oligomerized to $H(Me_3SiC{=}CH)_nCH_2CHMe_2$ (n=2–3) by the use of $LnX_3(Donor)/AliBu_3$ (Ln=Gd, Pr, Nd, Tb, Dy, Lu; X=Cl, Br) [87, 88]. The catalytic dimerization of terminal alkynes using $(C_5Me_5)_2LnCH(SiMe_3)_2$ (Ln=Y, La, Ce) has been reported recently. Here the dimer was a mixture of 2,4-disubstituted 1-buten-3-yne and 1,4-disubstituted 1-

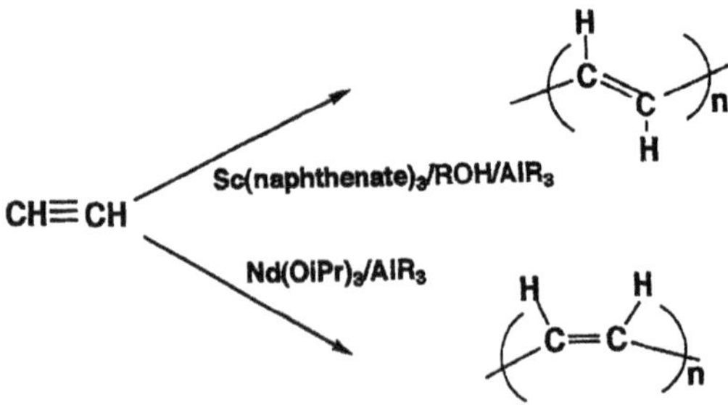

Fig. 27. Rare earth metal initiated polymerizations of acetylene derivatives

buten-3-yne for phenylacetylene and (trimethylsilyl)acetylene, but it was a 2,4-disubstituted dimer for alkylacetylene [89]. Selective formation of 2,4-disubstituted 1-buten-3-ynes has already been achieved with the $(C_5Me_5)_2TiCl_2$/RMgX catalyst [90].

14 References

1. Klein JW, Lamps JP, Gnaonou Y, Remp P (1991) Polymer 32:2278
2. (a) Webster OW, Hertler WR, Sogah DY, Farnham WB, RajanBabu TV (1983) J Am Chem Soc 105:5706; (b) Sogah DY, Hertler WR, Webster OW, GM Cohen (1987) Macromolecules 20:1473
3. (a)Yasuda H, Yamamoto H, Yokota K, Miyake S, Nakamura A (1992) J Am Chem Soc 114:4908; (b)Yasuda H, Yamamoto H, TakemotoY, Yamashita M, Yokota K, Miyake S, Nakamura A (1993)Makromol Chem Macromol Symp 67:187; (c) Yasuda H, Yamamoto H, Yamashita M, Yokota K, Nakamura A, Miyake S, Kai Y, Kanehisa N (1993) Macromolecules 22:7134; (d) Yasuda H, Tamai H (1993) Prog Polym Sci 18:1097; (e) Yasuda H, Ihara E (1997) Adv Polym Sci 133:53
4. Ihara E, Morimoto M, Yasuda H (1995) Macromolecules 28:7886
5. Yamashita M, Takemoto Y, Ihara E, Yasuda H (1996) Macromolecules 29:1798
6. Tanaka K, Ihara E, Yasuda H (1996) (unpublished results)
7. (a)Szwarc M (1983) Adv Polym Sci 49:1; (b) Nakahama S, Hirao A (1990) Prog Polym Sci (1990) 13:299; (c)Inoue S (1988) Macromolecules 21:1195
8. (a) Sawamoto M, Okamoto O, Higashimura T (1987) Macromolecules 20:2693; (b) Kojima K, Sawamoto M, Higashimura T (1988) Macromolecules 21:1552
9. (a) Gillon LR, Grubbs RH (1986) J Am Chem Soc (1986) 108:733; (b)Grubbs RH, Tumas W (1989) Science 243:907; (c) Schrock RR, Feldman J, Canizzo LF, Grubbs RH (1989) Macromolecules 20:1169
10. Cao ZK, Okamoto Y, Hatada K (1986)Kobunshi Ronbunshu 43:857
11. (a) Joh Y, Kotake Y (1976) Macromolecules 3:337; (b) Hatada K, Nakanishi H, Ute K, Kitayama T (1986) Polym J 18:581
12. Abe H, Imai K, Matsumoto M (1968) J Polym Sci C23:469
13. Nakano N, Ute K, Okamoto Y, Matsuura Y, Hatada K (1989) Polym J 21:935
14. Bawn CEH, Ledwith A (1962) Quart Rev Chem Soc 16:361
15. Cram DJ, Kopecky KR (1959) J Am Chem Soc 81:2748
16. Hatada K, Ute K, Tanaka Y, Kitayama T, Okamoto (1986) Polym J 18:1037
17. Nitto Y, Hayakawa T, Ihara E, Yasuda H (1998) (unpublished result)
18. Yamamoto Y, Giardello MA, Brard L, Marks TJ (1995) J Am Chem Soc 117:3726
19. Boffa LS, Novak BM (1994) Macromolecules 27:6993
20. (a)Kuroki M, Watanabe T, Aida T, Inoue S (1991) J Am Chem Soc 113: 5903; (b) Aida T, Inoue S (1996) Acc Chem Res 29:39
21. Sun J, Wang G, Shen Z (1993)Yingyong Huaxue 10:1
22. Collins S, Ward DG, Suddaby KH (1994) Macromolecules 27:7222
23. Soga K, Deng H, Yano T, Shiono T (1994) Macromolecules 27:7938
24. Hosokawa Y, Kuroki M, Aida T, Inoue S (1991) Macromolecules 24:8243
25. (a) Ihara E, Morimoto M, Yasuda H (1996) Proc Japan Acad 71:126; (b) Ihara E, Morimoto M, Yasuda H (1995) Macromolecules 28:7886; (c)Suchoparek M, Spvacek J (1993) Macromolecules 26:102; (d)Madruga EL, Roman JS, Rodriguez MJ (1983) J Polym Sci Polym Chem Ed 21: 2739
26. Kakehi T, Yasuda H (1997) (unpublished result)

27. (a)Patten TE, Novak BM (1991) J Am Chem Soc 113:5065; (b)Patten TE, Novak BM (1993) Makromol Chem Macromol Symp 67:203
28. Fukuwatari N, Sugimoto H, Inoue S (1996) Macromol Rapid Commun 17:1
29. Tanaka K, Ihara E, Yasuda H (1998) (unpublished result)
30. (a) Hofman A, Szymanski R, Skomkowski S, Penczek S (1980) Makromol Chem 185:655; (b) Agostini DE, Lado JB, Sheeton JR (1971) J Polym Sci A1:2775
31. Cherdran H, Ohse H, Korte F (1962) Makromol Chem 56:187
32. Evans JE, Shreeve JL, Doedens RJ (1993) Inorg Chem 32:245
33. Zhang Y, Chen X, Shen Z (1989) Inorg Chim Acta 155:263
34. Wu J, Shen Z (1990) J Polym Sci Polym Chem Ed 28:1995
35. Itoh H, Shirahama H, Yasuda H (1998) (unpublished results)
36. Shen Z, Wu J, Wang G (1990) J Polym Sci Polym Chem Ed 28:1965
37. Wu J, Shen Z (1990) Polymer J 22:326
38. (a) Inoue S, Tsuruta T (1969) J Polym Sci Polym Lett Ed 7:287; (b) Kobayashi M., Inoue S, Tsurtuta T (1973) J Polym Sci Polym Chem Ed 11:2383
39. (a) Shen X, Zhang Y, Shen Z (1994) Chinese J Polym Sci 12:28; (b)Shen Z, Chen X, Zhang Y (1994) Macromol Chem Phys 195:2003
40. (a) Machoon JP, Sigwalt P (1965) Compt Rend 260:549; (b)Belonovskaja GP (1979) Europ Polym J 15:185
41. Dumas P, Spassky N, Sigwalt P (1972) Makromol Chem 156:65
42. Guerin PB, Sigwalt P (1974) Eur Polym J 10:13
43. Shen Z, Zhang Y, Pebg J, Ling L (1990) Sci China 33:553
44. Shen Z, Sun J, Wu Li, Wu L (1990) Acta Chim Sinica 48:686
45. Yao Y, Shen Q (private communication)
46. Stevels WM, Ankone JK, Dijkstra PJ, Feijen J (1996) Macromolecules 29:3332
47. Shen Y, Zhang F, Zhang Y, Shen Z (1995) Acta Polym Sinica 5:222
48. (a) Yasuda H, Ihara E (1993) J Synth Org Chem Jpn 51:931; (b)Yasuda H, Ihara E, Yoshioka S, Nodono M, Morimoto M, Yamashita M (1994) Catalyst design for tailor-made polyolefins. Kodansha-Elsevierp, p237
49. Nodono M, Ihara E, Yasuda H (unpublished result)
50. Fu P, Marks TJ (1995) J Am Chem Soc 117:10,747
51. Ihara E, Nodono M, Yasuda H, Kanehisa N, Kai Y (1996) Macromol Chem Phys 197:1909
52. Evans WJ, Ulibbari TA, Ziller JW (1988) J Am Chem Soc 110:6877
53. Evans WJ, Ulibbari TA, Ziller JW (1990) J Am Chem Soc 112:219
54. Lang DP, Bianconi PA (1996) J Am Chem Soc 118:52,453
55. Yasuda H, Furo M, Yamamoto H, Nakamra A, Miyake S, Kibino N (1992) Macromolecules 25:5115
56. Yang Y, Seyam AM, Fuad PF Marks TJ (1994) Macromolecules 27:4625
57. Yang M, Chan C, Shen Z (1990) Polym J 22:919
58. Kobayashi E, Aida S, Aoshima S, Furukawa J (1994) J Polym Sci Polym Chem Ed 32:1195
59. Hayakawa T, Ihara E, Yasuda H (1995) 69th National Meeting of Chem Soc Jpn 2B530
60. Cheng XC, Shen Q (1993) Chinese Chem Lett 4:743
61. Hu J, Shen Q (1993) Cuihua Xuebao 11:16
62. Tanaka K, Yasuda H (1997) (unpublished result)
63. Ishihara N, Seimiya T, Kuramoto M, Uoi M (1986) Macromolecules 19:2464
64. Ishihara N, Kuramoto M, Uoi M (1988) Macromolecules 21: 3356
65. Yu G, Chen W, Wang Y (1981) Kexue Tongabo 29:412
66. Ji X, Png S, Li Y, Ouyang J (1986) Scientia Sinica 29: 8
67. Jin S, Guan J, Liang H, Shen Q (1993) J Catalysis (Cuihua Xuebao) 159:14
68. Hu J, Liang H, Shen Q (1993) J Rare Earths 11:304

69. (a) Jin Y, Li F, Pei F, Wang F, Sun Y (1994) Macromolecules 27:4397; (b) Li F, Jin Y, Pei K, Wang F (1994) J Macromol Sci Pure Appl Chem A31:273
70. (a)Jin Y, Li X, Sun Y, Ouyang J (1982) Kexue Tongbau 27:1189; (b) Jin Y, Li X, Lin Y, Jin S, Shi E, Wang M (1989) Chinese Sci Bull 34:390
71. Ricci G, Boffa G, Porri L (1986) Makromol Chem Rapid Commun 7:355
72. (a) Natta G, Porri L (1964) Polym Prepr 5:11,630; (b)Natta G, Polli L (1966) Adv Chem Ser 52:24
73. Wong F (1994) J Macromol Sci Pure Appl Chem A31:273
74. Li X, Sun Y, Jin Y (1986) Acta Chim Sinica 44:1163
75. Hu Y, Ze L, Shen Q (1993) J Rare Earths 11:304
76. Lee DH, Wang JK, Ahn TO (1987) J Polym Sci Polym Chem Ed 25:1407
77. Lee DH, Ahn TO (1988) Polymer 29:71
78. Wang P, Jin Y, Pei F, Jing F, Sun Y (1994) Acta Polym Sinica 4:392
79. Ito T, Shirakawa H, Ikeda S (1974) J Polym Sci Polym Chem Ed 12:11
80. Luttinger LB (1962) J Org Chem 27:159
81. Shen Z, Wang Z, Can Y (1985) Inorg Chim Acta 110:55
82. Shen Z, Yang M, Shi M, Cai Y (1982) J Polym Sci Polym Chem Ed 20:411
83. Shen Z, Yu L, Yang M (1984) Inorg Chim Acta 109: 55
84. Hu X, Wang F, Zhao X, Yan D (1987) Chin J Polym Sci 5: 221
85. ZhaoJ, Yang M, Yuan Y, Shen Z (1988) Zhonggono Xitu Xuebao 6:17
86. Shen Z, Farona MF (1984) J Polym Sci Polym Chem Ed 22:1009
87. Shen Z, Farona MF (1983) Polym Bull 10:298
88. MullagalievI R, Mudarisova RK (1988) Izv Akad Nauk SSSR Ser Khim 7:1687
89. HeersH J, Teuben JH (1991) Organometallics 10:1980
90. Akita M, Yasuda H, Nakamura A (1984) Bull Chem Soc Jpn 57: 480

Polymer-Supported Rare Earth Catalysts Used in Organic Synthesis

Shū Kobayashi

Graduate School of Pharmaceutical Sciences, The University of Tokyo, Hongo, Bunkyo-ku, Tokyo 113- 0033, Japan
E-mail: skobayas@mol.f.u-tokyo.ac.jp

Abstract. Three types of polymer-supported rare earth catalysts, Nafion-based rare earth catalysts, polyacrylonitrile-based rare earth catalysts, and microencapsulated Lewis acids, are discussed. Use of polymer-supported catalysts offers several advantages in preparative procedures such as simplification of product work-up, separation, and isolation, as well as the reuse of the catalyst including flow reaction systems leading to economical automation processes. Although the use of immobilized homogeneous catalysts is of continuing interest, few successful examples are known for polymer-supported Lewis acids. The unique characteristics of rare earth Lewis acids have been utilized, and efficient polymer-supported Lewis acids, which combine the advantages of immobilized catalysis and Lewis acid-mediated reactions, have been developed.

Keywords: Lewis acids, Polymer-supported catalysts, Rare earth triflate, Combinatorial synthesis, Carbon–carbon bond-forming reactions

1 **Introduction** 285
2 **Nafion-Based Rare Earth Catalysts** 286
3 **Polyacrylonitrile-Based Rare Earth Catalysts** 289
4 **A Microencapsulated Rare Earth Lewis Acid. (A New Type of Polymer-Supported Catalyst)** 295
5 **References and Notes** 303

1 Introduction

Use of polymer-supported catalysts offers several advantages in preparative procedures. Simplification of product work-up, separation, and isolation as well as reuse of the catalyst including use of flow reaction systems could lead to economical automation processes. Although the use of immobilized homogeneous catalysts is of continuing interest [1], few successful examples are known for polymer-supported Lewis acids [2,3]. This is probably due to the instability of most Lewis acids to air (moisture) and water. During preparation of polymer-supported catalysts, many manipulations have to be carried out in air or in the presence of water. On the other hand, it was recently found that some rare earth Lewis acids are stable in water [4]. Utilizing this very unique characteristic, several efforts to develop efficient polymer-supported Lewis acids combining the advantages of immobilized catalysis and Lewis acid-mediated reactions, have been made. In this chapter, three types of polymer-supported rare earth catalysts, Nafion-based rare earth catalysts, polyacrylonitrile-based rare earth catalysts, and microencapsulated Lewis acids are discussed [5].

2
Nafion-Based Rare Earth Catalysts

Recently, scandium triflate [$Sc(OTf)_3$] was found to be stable in water and successful Lewis acid catalysis was carried out in both water and organic solvents [6–8]. $Sc(OTf)_3$ coordinates to Lewis bases under equilibrium conditions, and thus activation of carbonyl compounds using a catalytic amount of the acid has been achieved [6,7]. In addition, effective activation of nitrogen-containing compounds such as imines, amino aldehydes, etc. has been performed successfully [8]. Encouraged by the characteristics and the usefulness of $Sc(OTf)_3$ as a Lewis acid catalyst, a polymer-supported scandium catalyst was prepared.

Nafion (NR-50, Du Pont) was chosen as the supporting framework [9]. Three equivalents of Nafion were treated with $ScCl_3 \cdot 6H_2O$ in acetonitrile under reflux conditions [10]. After 40 h, 96% of the $ScCl_3 \cdot 6H_2O$ was consumed and the polymer thus prepared (Nafion-Sc) contained 1.3% Sc, according to ICP analysis. Choice of solvent is important at this stage; only 27% of the $ScCl_3 \cdot 6H_2O$ was consumed when 1,2-dichloroethane was used as the solvent. This Nafion-Sc catalyst was then tested in several synthetic reactions [11]. First, allylation reactions of carbonyl compounds were investigated. Allylation reactions of carbonyl compounds are among the most important carbon–carbon bond-forming reactions, and the products, homoallylic alcohols having hydroxyl and double bond groups, are synthetically useful intermediates [12]. Nafion-Sc was also found to be effective in the allylation reactions of carbonyl compounds with tetraallyltin, and selected examples are shown in Table 1 [13]. In all cases, the reactions proceeded smoothly in both organic and aqueous solvents to afford the desired homoallylic alcohols in high yields. Not only aldehydes, but also ketones worked well. Moreover, use of aqueous solvents enabled the reactions of non-protected carbohydrates [14]. Non-protected sugars reacted directly with tetraallyltin to give the adducts, which are useful intermediates for the synthesis of higher sugars. Salicylaldehyde and 2-pyridinecarboxaldehyde reacted with tetraallyltin to afford the corresponding homoallylic alcohols in good yields. These compounds are known to react with the Lewis acids rather than the nucleophile under general Lewis acidic conditions.

Nafion-Sc was also found to be effective in some other reactions (Schemes 1–3). In typical Lewis acid-mediated reactions, such as Diels–Alder, Friedel–Crafts acylation, and imino Diels–Alder reactions, Nafion-Sc worked efficiently to afford the corresponding adducts in high yields.

It was also found that Nafion-Sc could be easily recovered and reused. The catalyst was recovered simply by filtration and washing with a suitable solvent, and the activity of the recovered Nafion-Sc was comparable to the fresh catalyst; in the reaction of benzaldehyde with tetraallyltin in H_2O/MeOH/toluene(1:7:4)

Table 1. Sc-Nafion-catalyzed allylation reactions of carbonyl compounds with tetraallyltin[a]

Carbonyl Compound	Product	Solvent[b]	Yield (%)
PhCHO	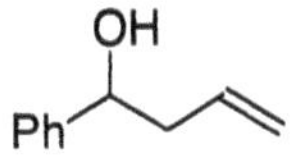 OH, Ph	A B C	91 90 91
 Ph, CHO	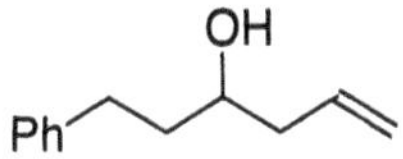 OH, Ph	A C	83 83
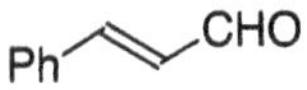 Ph, CHO	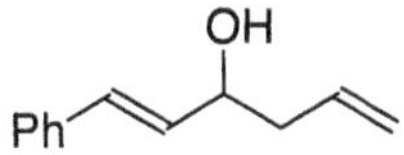 OH, Ph	A C	93 98
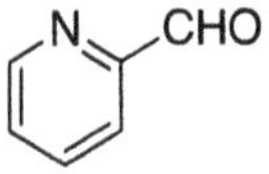 N, CHO	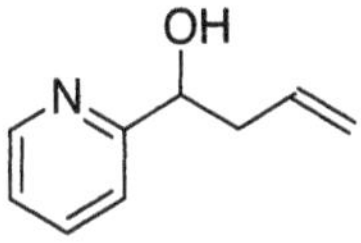 OH, N	A	78
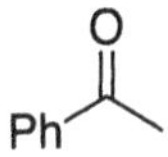 O, Ph	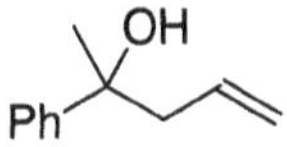 OH, Ph	A	57[c]
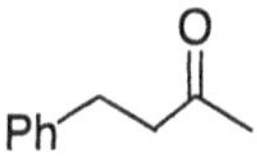 O, Ph	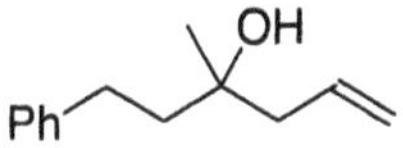 OH, Ph	A C	64[c] 87[c]
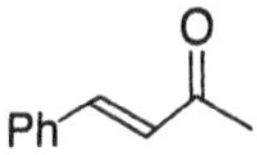 O, Ph	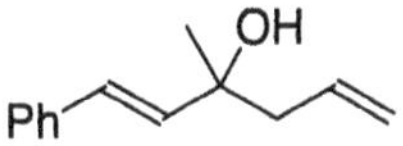 OH, Ph	A C	95[c] 85[c]
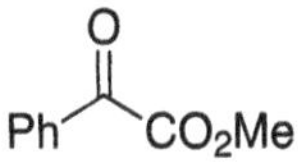 O, Ph, CO_2Me	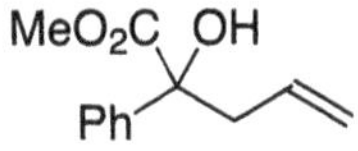 MeO_2C, OH, Ph	A C	81[c] 84[c]
D-arabinose	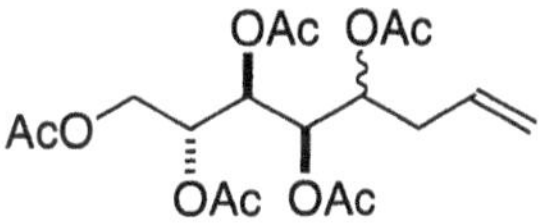 OAc, OAc, AcO, OAc, OAc	A	76[c,d,e]
D-ribose	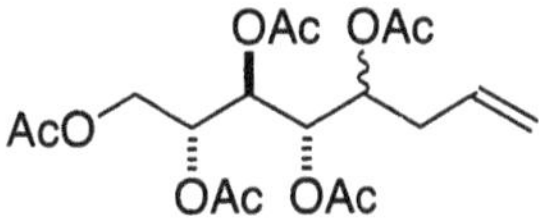 OAc, OAc, AcO, OAc, OAc	A	91[c,d,f]
D-glucose	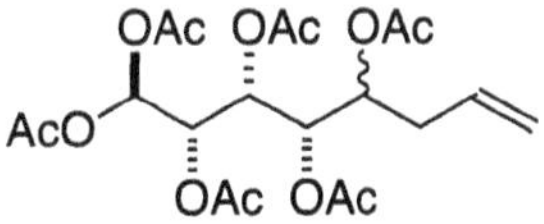 OAc, OAc, OAc, AcO, OAc, OAc	A	64[c,d,g]

[a] The reaction was carried out at rt unless otherwise noted.
[b] A: H_2O/THF (1:9); B: H_2O/MeOH/toluene (1:7:4); C: CH_3CN.
[c] The reaction was carried out at 60 °C.
[d] The products were isolated after acetylation.
[e] *syn*/*anti*=59:41.
[f] *syn*/*anti*=65:35.
[g] Diastereomeric ratio=68:32. Relative configuration assignment was not made.

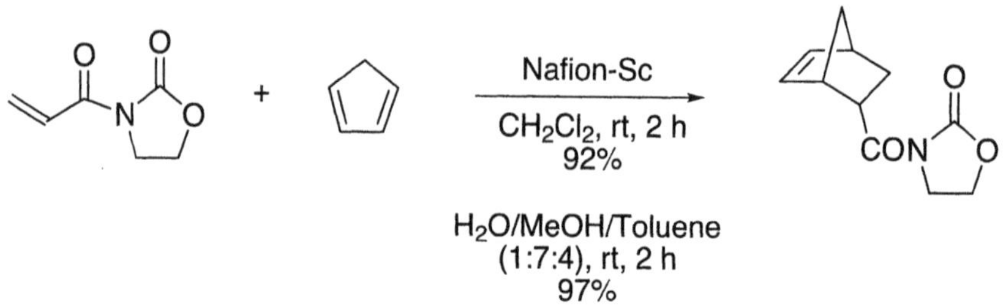

Nafion-Sc

CH_2Cl_2, rt, 2 h
92%

H_2O/MeOH/Toluene
(1:7:4), rt, 2 h
97%

Scheme 1

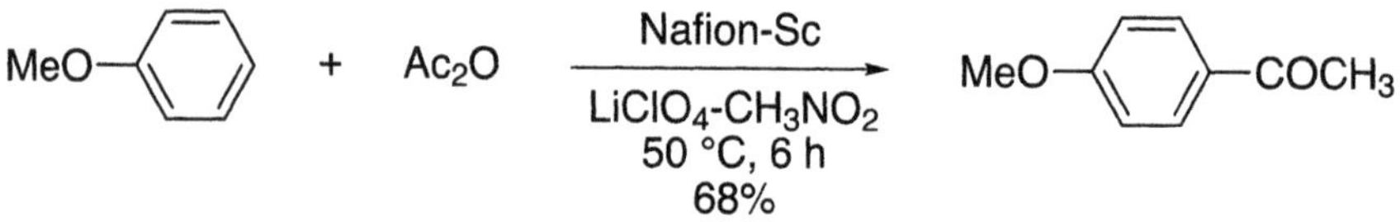

MeO + Ac_2O → MeO–C_6H_4–$COCH_3$

Nafion-Sc

$LiClO_4$-CH_3NO_2
50 °C, 6 h
68%

Scheme 2

$PhCOCHO{\cdot}H_2O$ + NH_2 MeO +

Nafion-Sc

H_2O/MeOH/Toluene
(1:7:4), rt, 2 h
92%

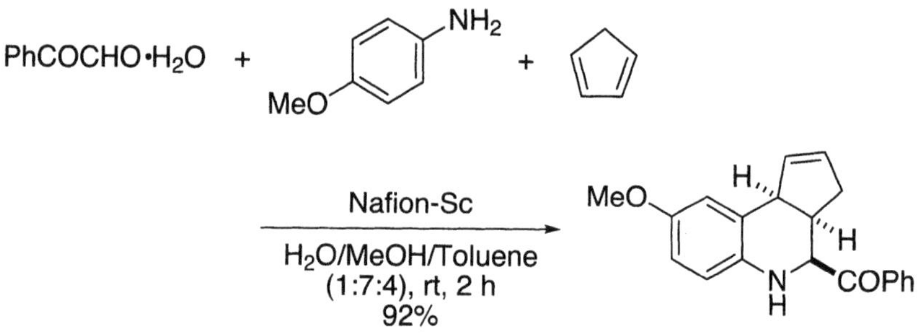

Scheme 3

[15] at rt; lst, 91% (15 h); 2nd, 81% (48 h); 3rd, 93% (40 h); in the reaction of 3-acryloyl-1,3-oxazolidin-2-one with cyclopentadiene at rt; 1st, 92% (2 h); 2nd, 87% (2 h); 3rd, 91% (2 h).

Nafion-Sc was also used in a flow system. Nafion-Sc was packed in a glass tube and substrates were passed through the column using a motor pump (Fig. 1) [16]. Results of the reaction of benzaldehyde with tetraallyltin in aqueous solution are shown in Scheme 4. It is noted that even higher yields were obtained in the 2nd, 3rd, and 4th runs than in the 1st run.

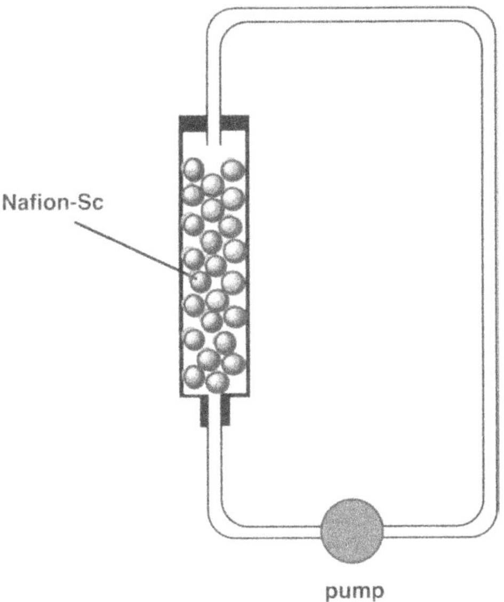

Fig. 1. A flow system using Nafion-Sc

PhCHO + $(CH_2=CHCH_2)_4Sn$ $\xrightarrow[\text{CH}_3\text{CN, rt}]{\text{Nafion-Sc}}$ PhCH(OH)CH$_2$CH=CH$_2$

Run	Yield (%)	Run	Yield (%)
1	75	3	84
2	93	4	84

Scheme 4. Nafion-Sc-catalyzed allylation using a flow system

3
Polyacrylonitrile-Based Rare Earth Catalysts

One of the drawbacks of polymer-supported catalysts is their low reactivity. Bearing in mind that the low reactivity might be ascribed to insolubility of the catalysts, a new polymer-supported catalyst, which was partially soluble in an appropriate solvent and is precipitated after completion of the reaction and recovered quantitatively by filtration, was searched. After several trials, a new scandium catalyst, polyallylscandium trifylamide ditriflate (PA-Sc-TAD), was finally developed. PA-Sc-TAD was readily prepared according to Scheme 5. Polyacrylonitrile was treated with $BH_3{\cdot}SMe$ in diglyme for 36 h at 150 °C. The resulting amine (1) was reacted with Tf_2O in the presence of Et_3N in 1,2-dichlo-

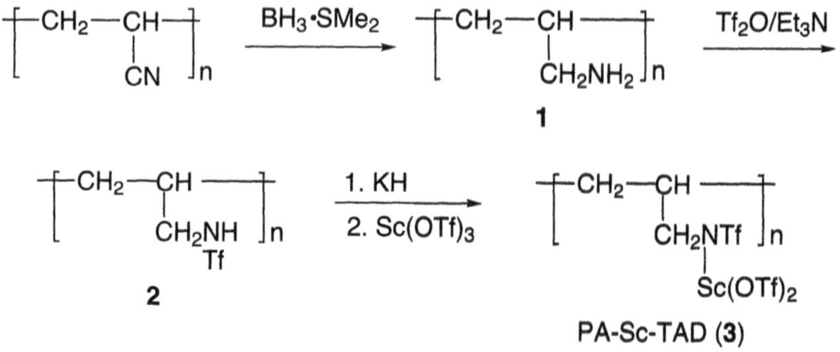

$[CH_2-CH(CN)]_n$ —$BH_3 \cdot SMe_2$→ $[CH_2-CH(CH_2NH_2)]_n$ **1** —Tf_2O/Et_3N→

$[CH_2-CH(CH_2NHTf)]_n$ **2** —1. KH, 2. $Sc(OTf)_3$→ $[CH_2-CH(CH_2NTf\text{-}Sc(OTf)_2)]_n$ PA-Sc-TAD (**3**)

Scheme 5. Preparation of PA-Sc-TAD (3)

roethane for 10 h at 60 °C to afford sulfonamide **2** [17]. After compound **2** and KH were combined, $Sc(OTf)_3$ was added and the mixture was stirred in THF for 48 h at rt to give **3** [18]. PA-Sc-TAD (**3**) is gummy, but is dispersed and partially soluble in a mixture of CH_2Cl_2/CH_3CN. The dispersed catalyst assembles again when hexane is added.

In the presence of PA-Sc-TAD (**3**), quinoline synthesis based on a three-component coupling reaction [8b,c] was performed. An aldehyde, an aromatic amine, and an alkene were mixed in CH_2Cl_2/CH_3CN (2:1) at 60 °C for 12 h. Hexane was then added and the catalyst was filtered off. The filtrate was concentrated *in vacuo* to afford a crude adduct. After purification by column chromatography, the desired quinoline derivative was obtained in a high yield.

This method is especially useful for construction of a quinoline library (Scheme 6) [19]. The procedure is very simple; it involves just mixing the catalyst (PA-Sc-TAD), an aldehyde, an aromatic amine, and an alkene (alkyne). After filtration, the filtrates are concentrated to give almost pure quinoline derivatives in most cases. Dehydrating agents such as MS 4A, $MgSO_4$, etc. are not necessary. It is noted that PA-Sc-TAD is water-tolerant and that substrates having water of crystallization can be used directly. PA-Sc-TAD can be easily recovered and continuous use is possible without any loss of activity (Scheme 7).

Combinatorial synthesis, a synthetic strategy which leads to large chemical libraries, is beginning to make a significant impact especially on the drug discovery process. Although polymer-supported substrates (reagents) have often been employed for library construction [20], there are some disadvantages to this method. First, the reactions of polymer-supported reagents are sometimes slow, and differences in reactivity between the substrates lead to lack of diversity of the library in some cases. Secondly, the loading level of polymer-supported substrates is generally low (<0.8 mmol/g), and large-scale preparation is difficult. The present reactions using the polymer-supported catalysts provide a new method for synthesizing a large number of structurally distinct compounds. In addition, a characteristic feature of the present method compared with conventional combinatorial synthetic technology using polymer-supported reagents is

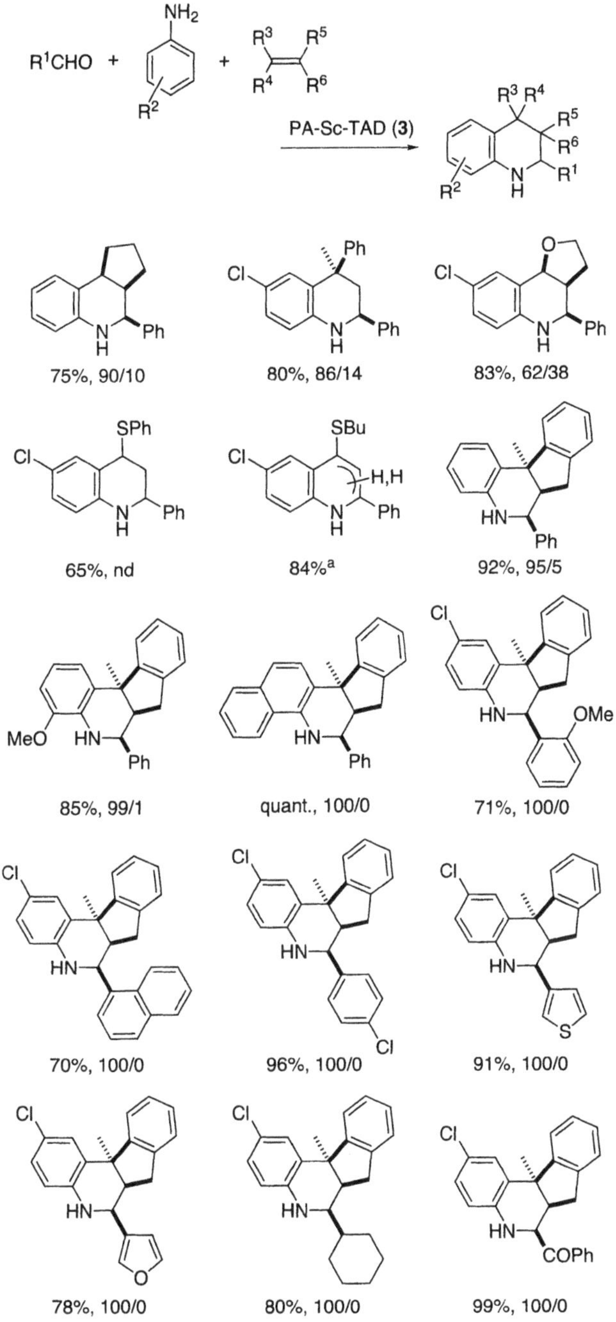

R^1CHO + (aniline, R^2) + ($R^3R^4C{=}CR^5R^6$) → PA-Sc-TAD (3) → tetrahydroquinoline (R^1–R^6)

75%, 90/10

80%, 86/14

83%, 62/38

65%, nd

84%[a]

92%, 95/5

85%, 99/1

quant., 100/0

71%, 100/0

70%, 100/0

96%, 100/0

91%, 100/0

78%, 100/0

80%, 100/0

99%, 100/0

[a]Butyl ethynyl sulfide was used as a dienophile.

Scheme 6. Synthesis of quinoline derivatives

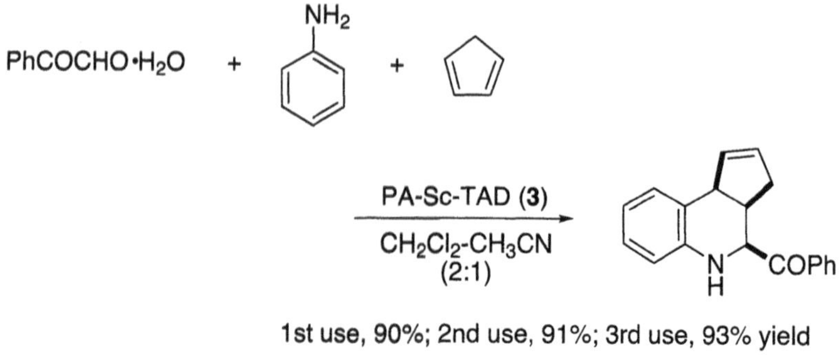

PhCOCHO·H_2O + NH_2 +

PA-Sc-TAD (**3**)

CH_2Cl_2-CH_3CN (2:1)

N H COPh

1st use, 90%; 2nd use, 91%; 3rd use, 93% yield

Scheme 7

that more than 100-mg scale syntheses with a large array of diverse molecular entries have been achieved with high purities (high yields and high selectivities). The number of commercially available aromatic aldehydes, aliphatic aldehydes, heterocyclic aldehydes, and glyoxals and glyoxylates that could be applicable to this synthesis is more than 200, and more than 200 aromatic amines and 50 alkenes (and alkynes) [21] are commercially available. Therefore, a quinoline library of more than a million compounds with high quantity and quality could be prepared by using an automation system based on this method. Moreover, the tetrahydroquinoline derivatives thus obtained are easily oxidized to dihydroquinoline or quinoline derivatives, which could double or triple the size of the library. Similarly, diverse amino ketone, amino ester, and amino nitrile derivatives were synthesized using PA-Sc-TAD as a catalyst (Schemes 8 and 9) [22].

In addition to several advantages of polymer-supported catalysts in synthesis, such as simplification of reaction procedures, easy separation of catalysts and products, application to automation systems, etc., it is desirable to achieve high selectivities by utilizing the bulkiness of polymers, the stability of polymer-coordinated complexes, etc. [23]. Such examples have recently been reported using a polymer-supported scandium catalyst (PA-Sc-TAD) [24]. It is well recognized that aldimines are less reactive than aldehydes towards nucleophilic addition because of the difference in electronegativity between oxygen and nitrogen, the steric hindrance of aldimines, etc. [25]. In fact, in the model competition reaction of benzaldehyde and *N*-benzylideneaniline with propiophenone lithium enolate, the enolate attacks the aldehyde exclusively in the coexistence of the aldimine. On the other hand, it was recently reported that in the presence of a catalytic amount of $Yb(OTf)_3$, aldimines react with silyl enolates selectively to afford the corresponding adducts in high yields, even in the coexistence of aldehydes [26]. Scandium salts were then used in the same competition reactions. Although both scandium and ytterbium elements belong to group 3, the competition reaction of benzaldehyde and *N*-benzylideneaniline with the silyl enol ether of propiophenone in the presence of 20 mol% of $Sc(OTf)_3$ gave a mixture of aldehyde-adduct **4a** (13%) and aldimine-adduct **5a** (58%). Interestingly, when

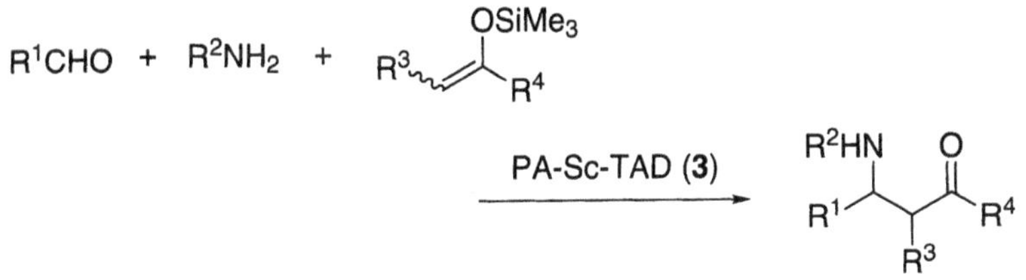

Scheme 8

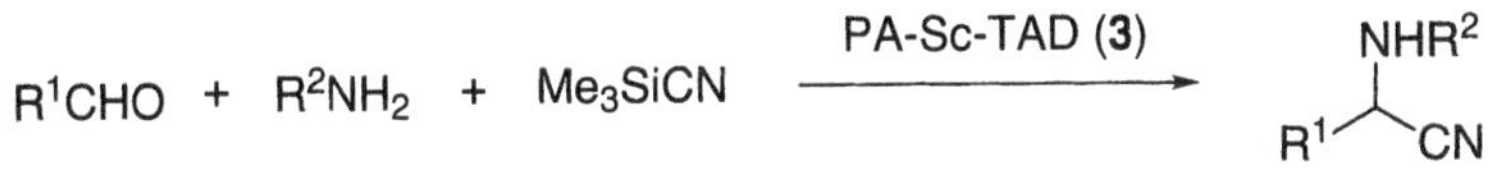

Scheme 9

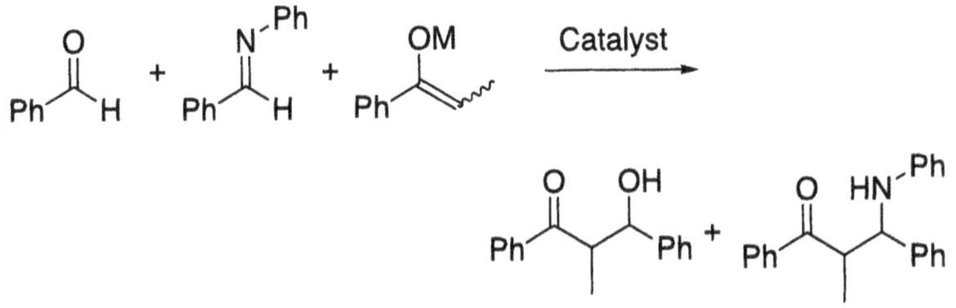

M	Catalyst /equiv	Solvent /temp(°C)	Yield (%) 4a	Yield (%) 5a	4a/5a
Li	—	THF/-78	97	trace	>99/1
TMS	$Sc(OTf)_3$/0.2	CH_2Cl_2/-23	13	58	18/82
TMS	PA-Sc-TAD/0.06	CH_2Cl_2:CH_3CN (2:1)/rt	1	95	1/99

Scheme 10. A competition reaction between an aldehyde and an aldimine

a polymer-supported scandium catalyst (PA-Sc-TAD) was used, the aldimine selectivity was improved dramatically, and the aldimine reacted with the silyl enolate to give the corresponding adduct in excellent selectivity (Scheme 10). Other combinations of aldehydes, aldimines, and silyl enolates were then tested, and the results are summarized in Scheme 11. In all cases, aldimines reacted with silyl enolates selectively in the coexistence of aldehydes to afford the corresponding adducts in high yields.

The unique selectivities obtained using the polymer-supported catalyst can be explained by the following equations. A catalytic amount of a scandium Lewis

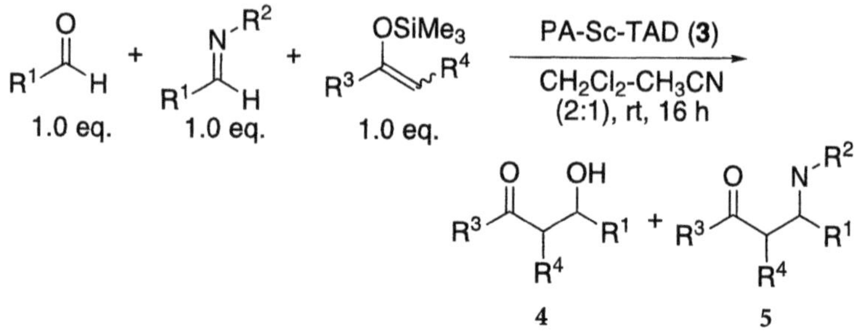

R^1	R^2	R^3	R^4	Yield (%) 4	5	4/5
Ph	Ph	Ph	Me	1	97	1/99
c-C_6H_{11}	Ph	Ph	Me	trace	91	<1/>99
2-furyl	Ph	Ph	Me	trace	>99	<1/>99
c-C_6H_{11}	(p-Cl)Ph	Ph	Me	trace	97	<1/>99
Ph	Ph	SEt	H	trace	58	<1/>99

Scheme 11. Aldimine-selective reaction using a polymer-supported catalyst

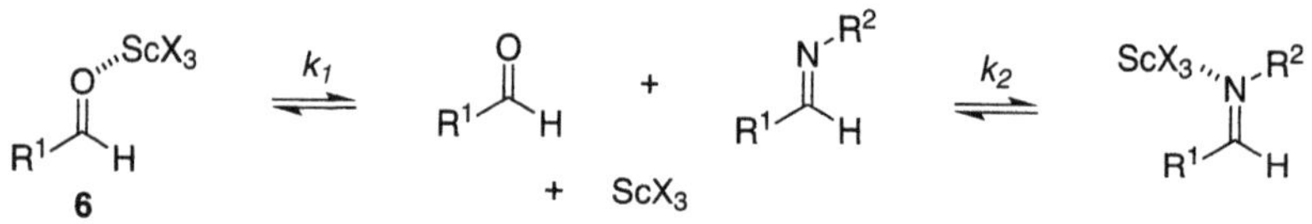

Scheme 12

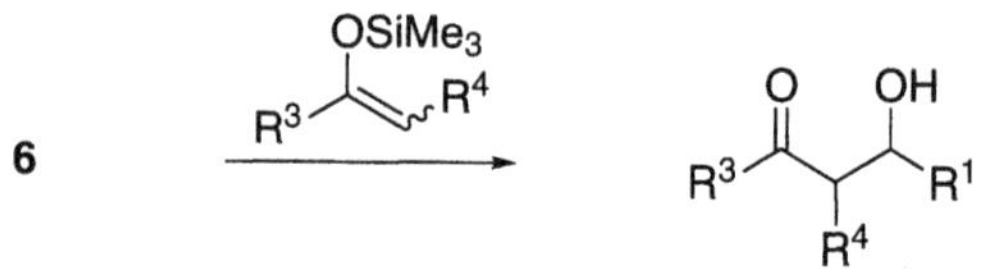

Scheme 13

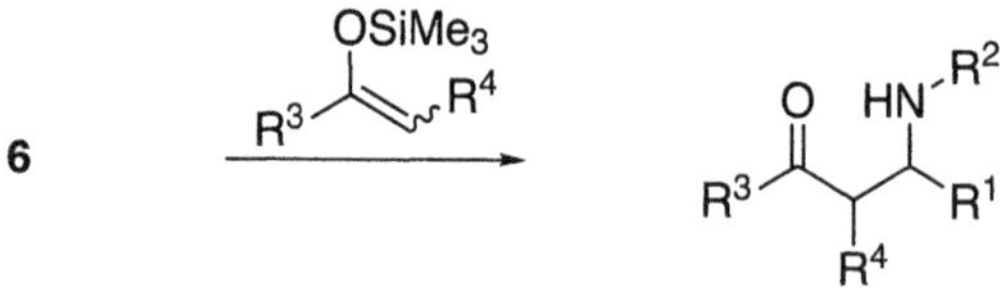

Scheme 14

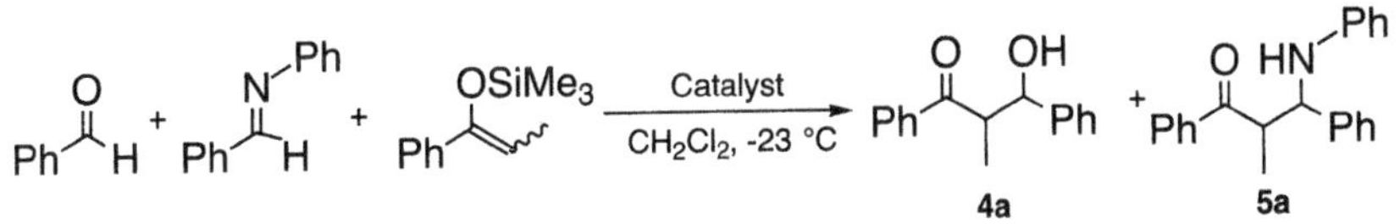

Catalyst	Yield (%) 4a	5a	4a/5a
$Sc(OTf)_3$	13	58	18/82
$Sc(OTf)_2(NTf_2)$	16	77	17/83
$Sc(NTf_2)_3$	3	77	4/96
$ScCl_3$	trace	43	<1/>99
$Sc(PF_6)_3$	2	76	3/97
$Sc(SbF_6)_3$	3	64	5/95
$Sc(AsF_6)_3$	26	49	35/65
$Sc(ClO_4)_3$	trace	87	<1/>99
$Sc(BPh_4)_3$	18	51	16/84
$\{[(3,5\text{-}CF_3)_2Ph]_4B\}_3Sc$	10	66	13/87

Scheme 15

acid first coordinates an aldimine, but the coordination occurs under equilibrium conditions (Schemes 12–14). Aldehyde/Lewis acid or aldimine/Lewis acid complexes should be stabilized by counter anions of scandium compounds (Scheme 15), and the ratios of **4**/**5** shown in Scheme 15 could depend on the stability of the complexes influenced by the counter anions. On the other hand, when the polymer-supported scandium catalyst (PA-Sc-TAD) was used, the catalyst also coordinated aldimines first. In this case, however, the aldimine/polymer-supported catalyst complexes are more stable than aldimine/*nonpolymer* Lewis acid complexes due to the *polymer effect*. Hence, the aldimines activated by the polymer-supported catalyst reacted with silyl enolates to afford aldimine adducts exclusively.

These aldimine-selective reactions using the polymer-supported catalyst will be applied to other nucleophilic addition reactions in organic chemistry.

4 A Microencapsulated Rare Earth Lewis Acid (A New Type of Polymer-Supported Catalyst)

In Sects. 1 and 2, polymer-supported scandium Lewis acids based on Nafion and a polyacrylonitrile derivative were discussed. While several useful reactions

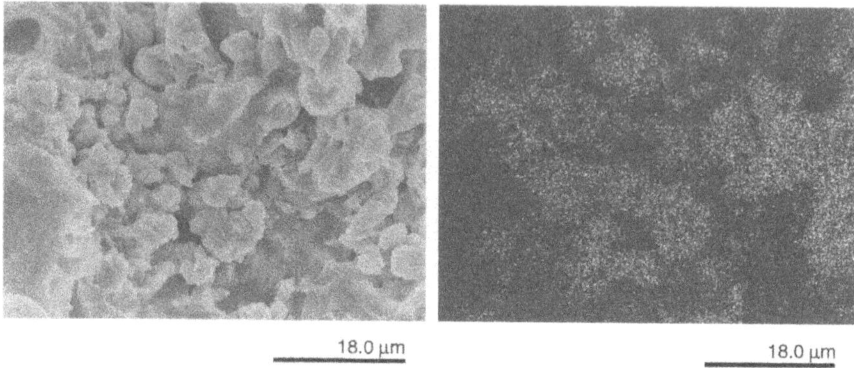

Fig. 2. Scanning electron microscopy (SEM) micrograph (*left*) and scandium energy dispersive X-ray (EDX) map of MC $Sc(Otf)_3$

have been developed, it was found that their reactivity, especially towards aldehydes and ketones, was lower than that of the nonpolymer Lewis acids. In general, catalysts are immobilized on polymers via coordinate bonds or covalent bonds. While the stability of polymer catalysts having coordinate bonds is often low, preparation of polymer catalysts having covalent bonds can be troublesome and their activities are generally lower than those of the nonpolymer catalysts [27]. As an unprecedented polymer-supported Lewis acid, attention is focused on a *microencapsulated* Lewis acid.

Microcapsules have been used for coating and isolating substances until such time as their activity is needed, and their application to medicine and pharmacy has been extensively studied [28]. Recently, much progress has been made in this field, and, for example, the size of the microcapsules that can be achieved has been reduced from a few μm to nanometers [28,29]. The idea is to apply this microencapsulation technique to the immobilization of Lewis acid catalysts onto polymers. That is, Lewis acids would be physically enveloped by polymer thin films and, at the same time, immobilized by the interaction between π electrons of benzene rings of the polystyrene used as a polymer backbone and vacant orbitals of the Lewis acids [30]. $Sc(OTf)_3$ was first chosen as a Lewis acid to be immobilized [31]. Preparation of the microencapsulated $Sc(OTf)_3$ [MC $Sc(OTf)_3$] was performed as follows: Polystyrene (1.000 g) was dissolved in cyclohexane (20 ml) at 40 °C, and to this solution was added $Sc(OTf)_3$ (0.200 g) as a core. The mixture was stirred for 1 h at this temperature and then slowly cooled to 0 °C. Coacervates were found to envelop the core dispersed in the medium, and hexane (30 ml) was added to harden the capsule walls. The mixture was stirred at rt for 1 h, and the capsules were washed with acetonitrile several times and dried at 50 °C [32,33]. A scanning electron microscopy (SEM) micrograph and a scandium energy dispersive X-ray (EDX) map of MC $Sc(OTf)_3$ prepared are shown in Fig. 2. It was found that small capsules of MC $Sc(OTf)_3$ adhered to each other, probably due to the small size of the core. Judging from the scandium mapping,

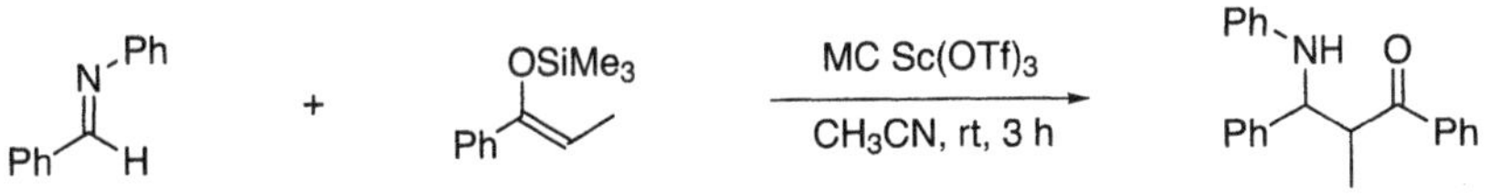

Use [a]	Yield/%
1	90
2	90
3	88
4	89
5	89
6	88
7	90

[a] Recovered catalyst was used successfully (Runs 2,3,4...)

Scheme 16. Imino aldol reaction (flow system)

$Sc(OTf)_3$ was located *all over the polymer surface*. The importance of the benzene rings of the polystyrene in immobilizing $Sc(OTf)_3$ was demonstrated by control experiments using polybutadiene or polyethylene instead of polystyrene. While 43% of $Sc(OTf)_3$ [100%=the amount of $Sc(OTf)_3$ immobilized by polystyrene] was bound using polybutadiene, no $Sc(OTf)_3$ was observed in the microcapsules prepared using polyethylene. These results demonstrate that the interaction between $Sc(OTf)_3$ and the benzene rings of polystyrene is a key to immobilizing $Sc(OTf)_3$. Steric factors (physical envelopment) would also contribute to the remarkable immobilizing ability of polystyrene compared to poor immobilization of polybutadiene and polyethylene.

MC $Sc(OTf)_3$ was used in several fundamental and important Lewis acid catalyzed carbon–carbon bond-forming reactions. All the reactions were carried out on a 0.5 mmol scale in acetonitrile [34] using MC $Sc(OTf)_3$ containing ca. 0.120 g $Sc(OTf)_3$. The reactions were carried out in both batch (using normal vessels) and flow systems (using circulating columns). It was found that MC $Sc(OTf)_3$ effectively activated aldimines. Imino aldol (Sheme 16), aza Diels–Alder (Scheme 17), cyanation (Scheme 18), and allylation (Scheme 19) reactions of aldimines proceeded smoothly using MC $Sc(OTf)_3$ to afford, respectively, the synthetically useful β-amino ester, tetrahydroquinoline, α-aminonitrile, and homoallylic amine derivatives in high yields.

Although it is well known that most Lewis acids are trapped and sometimes decomposed by basic aldimines, MC $Sc(OTf)_3$ worked well in all cases. One of the most remarkable and exciting points is that the activating ability of

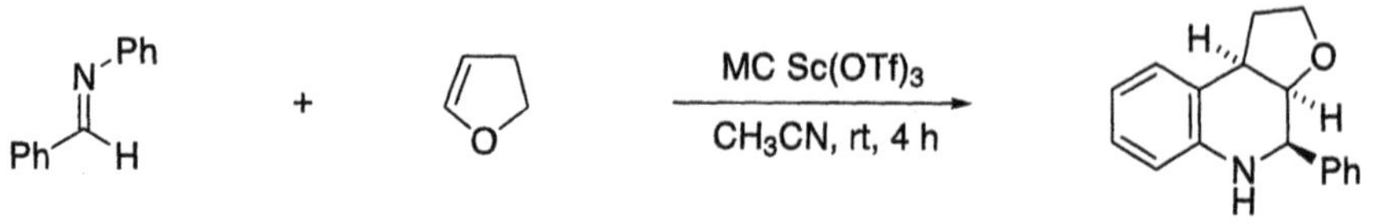

Use	Yield/%
1	80
2	78
3	78

Scheme 17. Aza-Diels–Alder reaction (flow system)

Use	Yield/%
1	77
2	77
3	76

Scheme 18. Cyanation reaction (flow system)

Use	Yield/%
1	85
2	87
3	83

Scheme 19. Allylation reaction of imine (flow system)

MC $Sc(OTf)_3$ for aldimines was revealed to be superior to monomeric $Sc(OTf)_3$ by preliminary kinetic studies (Fig. 3) [35].

The polymer catalyst was recovered quantitatively by simple filtration and could be reused. The activity of the recovered catalyst did not decrease even after seven uses (Scheme 16). MC $Sc(OTf)_3$ could also be successfully used in three-

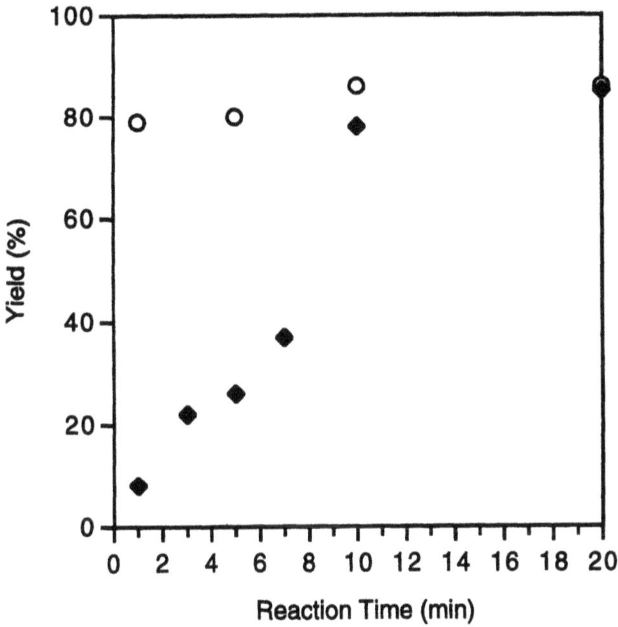

Fig. 3. Comparison of MC $Sc(OTf)_3$ (O) and $Sc(OTf)_3$ (◆) in the reaction of *N*-benzylideneaniline with (*Z*)-1-phenyl-1-trimethylsiloxypropene. The reaction was carried out using *N*-benzylideneaniline (0.5 mmol), (*Z*)-1-phenyl-1-trimethylsiloxypropene (0.60 mmol), and MC $Sc(OTf)_3$ [1.167 g, containing $Sc(OTf)_3$ 0.120 g] or $Sc(OTf)_3$ (0.120 g) in CH_3CN (5 ml) at rt

PhCHO + $PhNH_2$ + Ph(OSiMe$_3$)C=CHCH$_3$ —[MC $Sc(OTf)_3$; CH_3CN, rt, 3 h]→ PhCH(NHPh)CH(CH$_3$)C(=O)Ph

Use	Yield/%
1	90
2	96
3	93

Scheme 20. Mannich-type reaction (flow system)

component reactions such as Mannich-type (Scheme 20), Strecker (Scheme 21), and quinoline-forming reactions (Scheme 22). These reactions are known to be useful for the synthesis of biologically interesting compound libraries [36].

It was also found that MC $Sc(OTf)_3$ was effective for the activation of carbonyl compounds. The polymer scandium catalysts such as Nafion-Sc, polyallylscan-

PhCHO + $PhNH_2$ + Me_3SiCN $\xrightarrow[\text{CH}_3\text{CN, rt, 4 h}]{\text{MC Sc(OTf)}_3}$ PhCH(NHPh)CN

Use	Yield/%
1	70
2	71
3	75

Scheme 21. Strecker reaction (flow system)

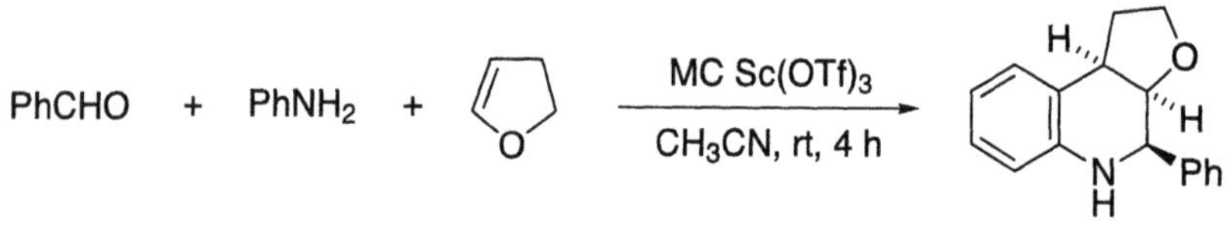

PhCHO + $PhNH_2$ + dihydrofuran $\xrightarrow[\text{CH}_3\text{CN, rt, 4 h}]{\text{MC Sc(OTf)}_3}$ tetrahydrofuroquinoline

Use	Yield/%
1	68
2	69
3	69

Scheme 22. Quinoline synthesis (flow system)

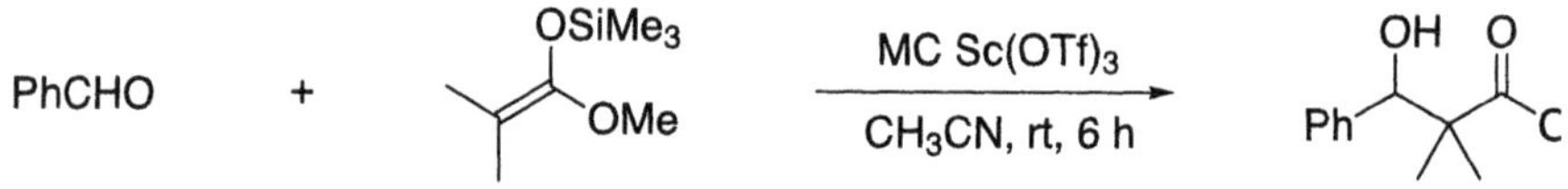

PhCHO + $Me_2C{=}C(OSiMe_3)OMe$ $\xrightarrow[\text{CH}_3\text{CN, rt, 6 h}]{\text{MC Sc(OTf)}_3}$ aldol product

Use	Yield/%
1	92
2	97
3	95

Scheme 23. Aldol reaction (batch system)

dium trifylamide ditriflate (PA-Sc-TAD) were found to have only a low ability to activate aldehydes, as mentioned previously. On the other hand, MC $Sc(OTf)_3$ efficiently activated aldehydes, and aldol (Scheme 23), cyanation (Scheme 24) and allylation (Scheme 25) reactions of aldehydes proceeded smoothly to give the corresponding aldol, cyanohydrin, and homoallylic alcohol derivatives in high yields.

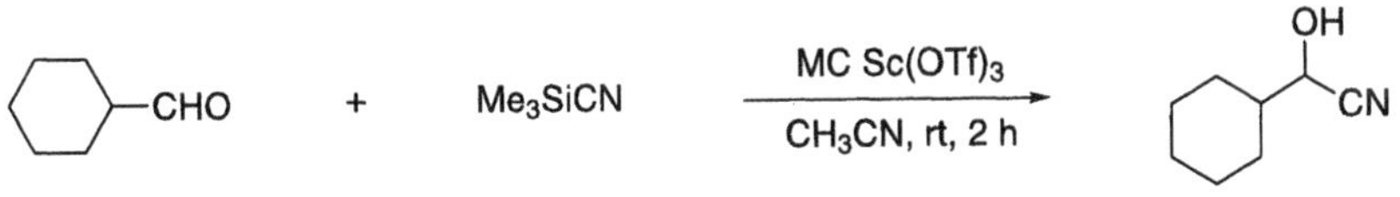

Cyclohexyl-CHO + Me_3SiCN → (MC $Sc(OTf)_3$; CH_3CN, rt, 2 h) → cyclohexyl-CH(OH)CN

Use	Yield/%
1	79
2	78
3	74

Scheme 24. Cyanation reaction of an aldehyde (batch system)

PhCHO + (allyl)$_4$Sn → (MC $Sc(OTf)_3$; CH_3CN, rt, 2 h) → PhCH(OH)CH$_2$CH=CH$_2$

Use [a]	Yield/%
1	92
2	91
3	90

Scheme 25. Allylation reaction of an aldehyde (batch system)

PhCOCH=CHPh + Me_2C=C(OSiMe$_3$)OMe → (MC $Sc(OTf)_3$; CH_3CN, rt, 6 h) → PhCOCH$_2$CH(Ph)C(Me)$_2$CO$_2$Me

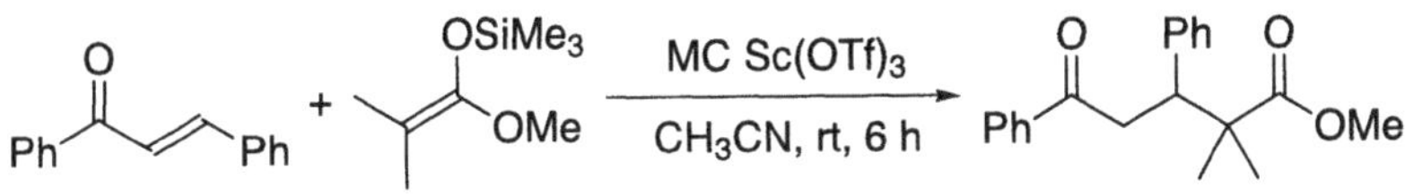

Use	Yield/%
1	92
2	97
3	95

Scheme 26. Michael reaction (batch system)

Michael reaction of an α,β-unsaturated ketone with a silyl enol ether (Scheme 26) and a Diels–Alder reaction of an oxazolidinone derivative with cyclopentadiene (Scheme 27) also worked well using MC $Sc(OTf)_3$. Moreover, a Friedel–Crafts acylation proceeded smoothly to produce an aromatic ketone in a good yield (Scheme 28). Friedel–Crafts alkylation and acylation reactions are fundamental and important processes in organic synthesis as well as in indus-

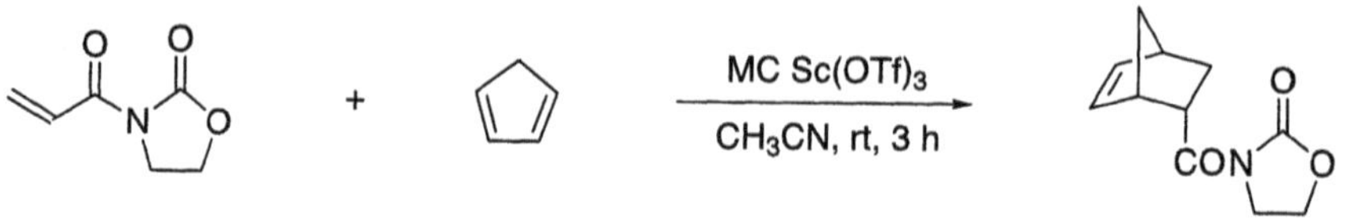

Use	Yield/%
1	77
2	79
3	80

Scheme 27. Diels–Alder reaction (batch system)

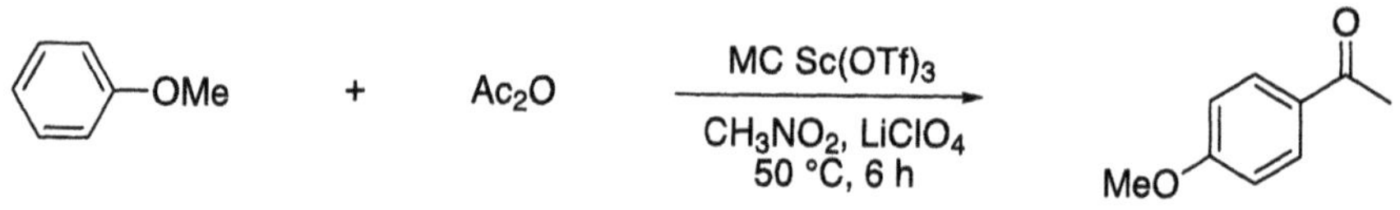

Use	Yield/%
1	76
2	76
3	81

Scheme 28. Friedel–Crafts acylation (batch system)

trial chemistry [37]. While the alkylation reaction proceeds in the presence of a catalytic amount of a Lewis acid, the acylation reaction generally requires more than a stoichiometric amount of a Lewis acid, usually aluminum chloride ($AlCl_3$), due to consumption of the Lewis acid by coordination to the aromatic ketone products. It should be noted that MC $Sc(OTf)_3$ has high activity in the Friedel–Crafts acylation and can be recovered easily by simple filtration and reused without loss of activity.

Similarly, microencapsulated ytterbium triflate (MC $Yb(OTf)_3$) has been prepared [38]. MC $Yb(OTf)_3$ was found to be effective in the aza-Diels–Alder reaction using ethyl vinyl ether as a dienophile (Scheme 29). In the presence of MC $Yb(OTf)_3$, *N*-benzylideneaniline reacted with ethyl vinyl ether to afford the corresponding tetrahydroquinoline derivative in a good yield. MC $Yb(OTf)_3$ could also be recovered quantitatively and could be reused. The yield of the 2nd and 3rd runs were comparable to that of the 1st run. In a similar reaction using $Sc(OTf)_3$ or MC $Sc(OTf)_3$ as a catalyst, ethyl vinyl ether polymerized rapidly and no desired adduct was obtained.

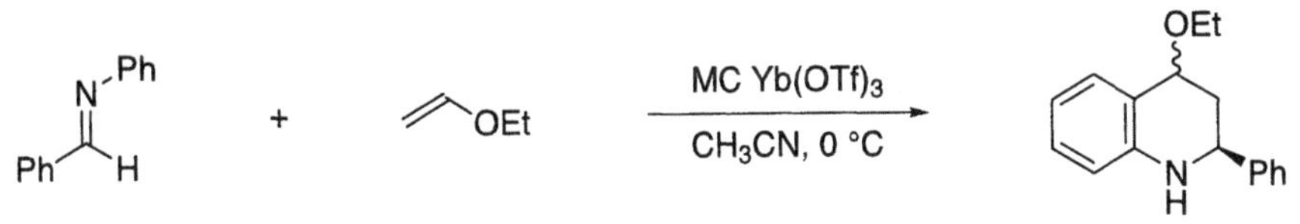

Use	Yield/%
1	78
2	70
3	77

Scheme 29. Aza-Diels–Alder reaction using $Yb(Otf)_3$ (batch system)

Thus, a microencapsulation technique has been shown to be quite effective for binding catalysts to polymers. Utilizing this technique, unprecedented polymer-supported, microencapsulated rare earth Lewis acids have been prepared. The catalysts thus prepared have been successfully used in many useful carbon–carbon bond-forming reactions. In all cases, the catalysts were recovered quantitatively by simple filtration and reused without loss of activity. This new technique for binding nonpolymer compounds to polymers will be applicable to the preparation of many other polymer-supported catalysts and reagents.

Acknowledgments. Our work in this area was partially supported by CREST, Japan Science and Technology Corporation (JST), and a Grant-in-Aid for Scientific Research from the Ministry of Education, Science, Sports and Culture, Japan. The author thanks and expresses his deep gratitude to his co-workers whose names appear in the references.

5 References and Notes

1. (a) Bailey DC, Langer SH (1981) Chem Rev 81:109. (b) Akelah A, Sherrington DC (1981) Chem Rev 81:557. (c) Frechet JM (1981) J Tetrahedron 37:663. (d) Smith K (ed) (1992) Solid supports and catalysts in organic synthesis. Ellis Horwood, Chichester. (e) Sherrington DC, Hodge P (eds) (1988) Synthesis and separations using functional polymers. John Wiley, New York. (f) Blazer HU, Pugin B (1995) In: Jannes G, Dubois J (eds) Chiral reactions in heterogeneous catalysis. Plenum Press, New York
2. Recently, Lewis acid mediated reactions have been especially focused upon because of their unique reactivities, selectivities, and mild conditions. Schinzer D (ed) (1989) Selectivities in Lewis acid promoted reactions. Kluwer, Dordrecht; Santelli M, Pons J-M (1995) Lewis acids and selectivity in organic synthesis. CRC Press, Boca Raton
3. Immobilization of $AlCl_3$ onto polymers or inorganic support materials has been reported. (a) Neckers DC, Kooistra DA, Green GW (1972) J Am Chem Soc 94:9284. (b) Krzywicki A, Marczewski M (1980) J Chem Soc Faraday Trans 1:1311. (c) Drago RS, Getty EE (1988) J Am Chem Soc 110:3311. (d) Clark JH, Martin K, Teasdale AJ, Barlow SJ (1995) J Chem Soc Chem Commun 2037. Montmorillonite-supported zinc chloride. (e) Clark JH, Cullen SR, Barlow SJ, Bastock TW (1994) J Chem Soc Perkin Trans 2 1117. Immobi-

lized trityl salts. (f) Mukaiyama T, Iwakiri H (1985) Chem Lett 1363. Polymer- or dendrimer-bound Ti-TADDOLates. (g) Seebach D, Marti RE, Hintermann T (1996) Helv Chim Acta 79:1710. (h) Rheiner PB, Sellner H, Seebach D (1997) Helv Chim Acta 80:2027. (i) Comina PJ, Beck AK, Seebach D (1998) Org Pro. Res Dev 2:18
4. (a) Kobayashi S (1991) Chem Lett 2187. (b) Kobayashi S, Hachiya I (1994) J Org Chem 59:3590. (c) Kobayashi S (1994) Synlett 689
5. Besides these examples, rare earth catalysts supported on ion-exchange resins have recently been reported. Yu L, Chen D, Li J, Wang PG (1997) J Org Chem 62:3575
6. Kobayashi S, Hachiya I, Araki M, Ishitani H (1993) Tetrahedron Lett 34:3755
7. (a) Kobayashi S, Hachiya I, Ishitani H, Araki M (1993) Synlett 472. (b) Hachiya I, Kobayashi S (1993) J Org Chem 58:6958. (c) Hachiya I, Kobayashi S (1994) Tetrahedron Lett 35:3319. (d) Kawada A, Mitamura S, Kobayashi S (1994) Synlett 545. (e) Kobayashi S, Araki M, Hachiya I (1994) J Org Chem 59:3758. (f) Kobayashi S, Moriwaki M, Hachiya I (1995) J Chem Soc Chem Commun 1527. (g) Kobayashi S, Moriwaki M, Hachiya I (1995) Synlett 1153. (h) Kobayashi M, Hachiya I, Yasuda M (1996) Tetrahedron Lett 37:5569. (i) Kobayashi S, Moriwaki M, Hachiya I (1997) Bull Chem Soc Jpn 70:267. (j) Kobayashi S, Wakabayashi T, Nagayama S, Oyamada H (1997) Tetrahedron Lett 38:4559. (k) Kobayashi S, Wakabayashi T, Oyamada H (1997) Chem Lett 831
8. (a) Kobayashi S, Araki M, Ishitani H, Nagayama S, Hachiya I (1995) Synlett 233. (b) Kobayashi S, Ishitani H, Nagayama S (1995) Chem Lett 423. (c) Kobayashi S, Ishitani H, Nagayama S (1995) Synthesis 1195. (d) Kobayashi S, Hachiya I, Suzuki S, Moriwaki M (1996) Tetrahedron Lett 37:2809. (e) Kobayashi S, Ishitani H, Komiyama S, Oniciu DC, Katritzky AR (1996) Tetrahedron Lett 37:3731. (f) Kobayashi S, Moriwaki M, Akiyama R, Suzuki S, Hachiya I (1996) Tetrahedron Lett 37:7783. (g) Kobayashi S, Ishitani H, Ueno M (1997) Synlett 115. (h) Kobayashi S, Moriwaki M (1997) Tetrahedron Lett 38:4251. (i) Kobayashi S, Akiyama R, Moriwaki M (1997) Tetrahedron Lett 38:4819. (j) Kobayashi S, Akiyama R, Kawamura M, Ishitani H (1997) Chem Lett 1039. (k) Kobayashi S, Busujima T, Nagayama S (1998) J Chem Soc Chem Commun 19. (l) Oyamada H, Kobayashi S (1998) Synlett 249
9. As for metal Nafions, Hg: (a) Olah GA, Meidar D (1978) Synthesis 671. (b) Saimoto H, Hiyama T, Nozaki H (1985) Bull Chem Soc Jpn 56:3078. Si: (c) Murata S, Noyori R (1980) Tetrahedron Lett 21:767. Cr, Ce: (d) Kanemoto S, Saimoto H, Oshima K, Nozaki H (1984) ibid. 25:3317. Al,: (e) Waller FJ (1984) Br Polym J 16:239. Ta: (f) Olah GA, Gupta B, Farina M, Felberg, JD, Ip WM, Husain A, Karpeles R, Lammertsma K, Melhotra AK, Trivedi NJ (1985) J Am Chem Soc 107:7097. See also: (g) Waller F (1986) J Catal Rev Sci Eng 28:1
10. $ScCl_3{\cdot}6H_2O$ (519 mg, 2.0 mmol) and Nafion (5 g, 1.2 meq/g) were combined in acetonitrile (10 ml) under reflux for 40 h. After cooling to rt, the polymer was filtered, washed with acetonitrile (20 ml×3), and then dried under reduced pressure for 24 h. A trial to prepare Nafion-Sc from Sc_2O_3 and Nafion failed (only 0.1% Sc was included in the polymer (ICP analysis))
11. Kobayashi S, Nagayama S (1996) J Org Chem 61:2256
12. Yamamoto Y, Asao N (1993) Chem Rev 93:2207
13. Nafion 50 itself has much lower activity in this reaction. For example, only 19% yield of the adduct was obtained in the reaction of acetophenone with tetraallyltin
14. (a) Schmid W, Whitesides GM (1991) J Am Chem Soc 113:6674. (b) Kim, E, Gordon DM, Schmid W, Whitesides GM (1993) J Org Chem 58:5500. (c) Li C-J, Chan T-H (1997) Organic reactions in aqueous media. Wiley, New York
15. Kobayashi S, Hachiya I, Yamanoi Y (1994) Bull Chem Soc Jpn 67:2342
16. Nafion-Sc (2 g) and sea sand (2 g, 15–20 mesh) were combined and the mixture was packed in a 6-mm glass tube. A carbonyl compound (0.5 mmol) and tetreallyltin (0.5 mmol) were dissolved in acetonitrile (4 ml) and passed through the column using a peristaltic pump (EYELA, MP-3)

17. (a) Takahashi H, Kawakita T, Ohno M, Yoshioka M, Kobayashi S (1992) Tetrahedron Lett 48:5691. (b) Corey EJ, Imwinkelried R, Pikul S, Xiang YB (1991) J Am Chem Soc 111:5493
18. PA-Sc-TAD (3) is fully characterized by elementary analysis. 3: C, 23.51; H, 3.62; N, 3.33; F, 12.5; S, 6.74; Sc, 2.0%
19. Kobayashi S, Nagayama S (1996) J Am Chem Soc 118:8977
20. Reviews: (a) Thompson LA, Ellman JA (1996) Chem Rev 96:555. (b) Früchtel JS, Jung G (1996) Angew Chem Int Ed Engl 35:17. (c) Terrett NK, Gardner M, Gordon DW, Kobylecki RJ, Steele J (1995) Tetrahedron 51:8135. (d) Lowe G (1995) Chem Soc Rev 37:309. (e) Gallop MA, Barrett RW, Dower WJ, Fodor SPA, Gordon EM (1994) J Med Chem 37:1233. (f) Gordon EM, Barrett RW, Dower WJ, Fodor SPA, Gallop MA (1994) J Med Chem 37:1385
21. Electron-deficient dienophiles will not work under the conditions, because the present reactions are based on inverse electron-demand aza-Diels–Alder reactions
22. Kobayashi S, Nagayama S, Busujima T (1996) Tetrahedron Lett 37:9221
23. For example, (a) Bailar JC Jr (1974) Catal Rev Sci Eng 10:17. (b) Pittman CU, Jacobson SE (1978) J Mol. Catal 3:293. (c) Dawans F, Morel D (1978) ibid. 3:403. (d) Carlini C, Braca G, Ciardelli F, Sbrana G (1977) ibid. 2:379. (e) Capka M, Svoboda P, Cerny M, Hetflejs J (1971) Tetrahedron Lett 3787. (f) Brown JM, Morinari H (1979) Tetrahedron Lett 2933. (g) Pinna F, Bonivento M, Strukul G, Graziani M, Cernia E, Palladino N (1976) J Mol Catal 1:309
24. Kobayashi S, Nagayama S (1997) Synlett 653
25. For example, Yamaguchi M (1991) In: Trost BM (ed) Comprehensive organic synthesis, vol 1. Pergamon, New York, chap 1.11
26. (a) Kobayashi S, Nagayama S (1997) J Org Chem 62:232. (b) Kobayashi S, Nagayama S (1997) J Am Chem Soc 119:10049
27. Ciardelli F, Tsuchida E, Wöhrle D (eds) (1996) Macromolecule-metal complexes. Springer, Berlin Heidelberg New York
28. Donbrow M (1992) Microcapsules and nanoparticles in medicine and pharmacy. CRC Press, Boca Raton
29. Marty JJ, Oppenheim RC, Speiser P (1978) Pharm Acta Helv 53:17; Kreuter J (1978) Pharm Acta Helv 53:33
30. (a) Greoffrey F, Cloke N (1995) In: Lappert MF (ed) Comprehensive organometallic chemistry II. Pergamon, Oxford, vol 4, p 1. (b) Edelmann FT (1995) In: Lappert MF (ed) Comprehensive organometallic chemistry II. Pergamon, Oxford, vol 4, p 11
31. Kobayashi S, Nagayama S (1998) J Am Chem Soc 120:2985
32. Judging from the recovered $Sc(OTf)_3$ (0.080 g), 0.120 g of $Sc(OTf)_3$ was microencapsulated according to this procedure. The weight of the capsules was 1.167 g which contained acetonitrile. MC $Sc(OTf)_3$ thus prepared can be stored at rt for more than a few months
33. IR (KBr) 3062, 3030 (νCH), 1946, 1873, 1805 (δCH), 1601, 1493 (benzene rings). 1255 ($\nu_{as}SO_2$), 1029 (ν_sSO_2), 756 (νC-S), 696 (νS-O) cm^{-1}. Cf. $Sc(OTf)_3$: 1259 ($\nu_{as}SO_2$), 1032 (ν_sSO_2), 769 (νC-S), 647 (νS-O) cm^{-1}; polystyrene: 3062, 3026 (νCH), 1944, 1873, 1803 (δCH), 1600, 1491 (benzene rings) cm^{-1}
34. Nitromethane was used in Friedel–Crafts acylation (Scheme 28)
35. Although the origin of the high activity of MC $Sc(OTf)_3$ is not clear at this stage, it was reported that aldimine–Lewis acid complexes were stabilized by using a polymer-supported Lewis acid. Cf. [24]
36. Cf. Ugi I, Domling A, Hörl W (1994) Endeavour 18:115; Armstrong RW, Combs AP, Tempest PA, Brown SD, Keating TA (1996) Acc Chem Res 29:123
37. For leading references on Friedel–Crafts acylation, Olah GA (1973) Friedel–Crafts chemistry. Wiley-Interscience, New York; Heaney H (1991) In: Trost BM (ed) Comprehensive organic synthesis. Pergamon, Oxford, vol 2, p 733; Olah GA, Krishnamurti R, Prakash GKS (1991) In: Trost BM (ed) Comprehensive organic synthesis. Pergamon, Oxford, vol 3, p 293
38. Kobayashi S, Nagayama S, Endo M, unpublished results

Author Index Volumes 1–2

Anwander R (1999) Principles in Organolanthanide Chemistry. *2*: 1–62

Fürstner A (1998) Ruthenium-Catalyzed Metathesis Reactions in Organic Synthesis. *1*: 37–72

Gibson SE (née Thomas), Keen SP (1998) Cross-Metathesis. *1*: 155–181

Hou Z, Wakatsuki Y (1999) Reactions of Ketones with Low-Valent Lanthanides: Isolation and Reactivity of Lanthanide Ketyl and Ketone Dianion Complexes. *2*: 233–253

Hoveyda AH (1998) Catalytic Ring-Closing Metathesis and the Development of Enantioselective Processes. *1*: 105–132

Kagan H, Namy JL (1999) Influence of Solvents or Additives on the Organic Chemistry Mediated by Diiodosamarium. *2*: 155–198

Kiessling LL, Strong LE (1998) Bioactive Polymers. *1*: 199–231

Kobayashi S (1999) Lanthanide Triflate-Catalyzed Carbon-Carbon Bond-Forming Reactions in Organic Synthesis. *2*: 63–118

Kobayashi S (1999) Polymer-Supported Rare Earth Catalysts Used in Organic Synthesis. *2*: 285–305

Molander G, Dowdy EC (1999) Lanthanide- and Group 3 Metallocene Catalysis in Small Molecule Synthesis. *2*: 119–154

Mori M (1998) Enyne Metathesis. *1*: 133–154

Nicolaou KC, King NP, He Y (1998) Ring-Closing Metathesis in the Synthesis of Epothilones and Polyether Natural Products. *1*: 73–104

Schrock RR (1998) Olefin Metathesis by Well-Defined Complexes of Molybdenum and Tungsten. *1*: 1–36

Shibasaki M, Gröger H (1999) Chiral Heterobimetallic Lanthanoid Complexes: Highly Efficient Multifunctional Catalysts for the Asymmetric Formation of C-C, C-O and C-P Bonds. *2*: 199–232

Tindall D, Pawlow JH, Wagener KB (1998) Recent Advances in ADMET Chemistry. *1*: 183–198

Yasuda H (1999) Organo Rare Earth Metal Catalysis for the Living Polymerizations of Polar and Nonpolar Monomers. *2*: 255–283

GPSR Compliance
The European Union's (EU) General Product Safety Regulation (GPSR) is a set of rules that requires consumer products to be safe and our obligations to ensure this.

If you have any concerns about our products, you can contact us on

ProductSafety@springernature.com

In case Publisher is established outside the EU, the EU authorized representative is:

Springer Nature Customer Service Center GmbH
Europaplatz 3
69115 Heidelberg, Germany

www.ingramcontent.com/pod-product-compliance
Ingram Content Group UK Ltd.
Pitfield, Milton Keynes, MK11 3LW, UK
UKHW021833190726
13853UKWH00003B/1294

* 9 7 8 3 6 6 2 1 4 2 3 0 1 *